Cognitive Networks

Applications and Deployments

OTHER COMMUNICATIONS BOOKS FROM AUERBACH

Advances in Biometrics for Secure Human Authentication and Recognition
Edited by Dakshina Ranjan Kisku, Phalguni Gupta, and Jamuna Kanta Sing
ISBN 978-1-4665-8242-2

Advances in Visual Data Compression and Communication: Meeting the Requirements of New Applications
Feng Wu
ISBN 978-1-4822-3413-8

Anonymous Communication Networks: Protecting Privacy on the Web
Kun Peng
ISBN 978-1-4398-8157-6

Case Studies in System of Systems, Enterprise Systems, and Complex Systems Engineering
Edited by Alex Gorod, Brian E. White, Vernon Ireland, S. Jimmy Gandhi, and Brian Sauser
ISBN 978-1-4665-0239-0

Cyber-Physical Systems: Integrated Computing and Engineering Design
Fei Hu
ISBN 978-1-4665-7700-8

Evolutionary Dynamics of Complex Communications Networks
Vasileios Karyotis, Eleni Stai, and Symeon Papavassiliou
ISBN 978-1-4665-1840-7

Fading and Interference Mitigation in Wireless Communications
Stefan Panic, Mihajlo Stefanovic, Jelena Anastasov, and Petar Spalevic
ISBN 978-1-4665-0841-5

Green Networking and Communications: ICT for Sustainability
Edited by Shafiullah Khan and Jaime Lloret Mauri
ISBN 978-1-4665-6874-7

Image Encryption: A Communication Perspective
Fathi E. Abd El-Samie, Hossam Eldin H. Ahmed, Ibrahim F. Elashry, Mai H. Shahieen, Osama S. Faragallah, El-Sayed M. El-Rabaie, and Saleh A. Alshebeili
ISBN 978-1-4665-7698-8

Intrusion Detection in Wireless Ad-Hoc Networks
Edited by Nabendu Chaki and Rituparna Chaki
ISBN 978-1-4665-1565-9

Machine-to-Machine Communications: Architectures, Technology, Standards, and Applications
Edited by Vojislav B. Misic and Jelena Misic
ISBN 978-1-4665-61236

MIMO Processing for 4G and Beyond: Fundamentals and Evolution
Edited by Mário Marques da Silva and Francisco A. Monteiro
ISBN 978-1-4665-9807-2

Network Innovation through OpenFlow and SDN: Principles and Design
Edited by Fei Hu
ISBN 978-1-4665-7209-6

Opportunistic Mobile Social Networks
Edited by Jie Wu and Yunsheng Wang
ISBN 978-1-4665-9494-4

Physical Layer Security in Wireless Communications
Edited by Xiangyun Zhou, Lingyang Song, and Yan Zhang
ISBN 978-1-4665-6700-9

SC-FDMA for Mobile Communications
Fathi E. Abd El-Samie, Faisal S. Al-kamali, Azzam Y. Al-nahari, and Moawad I. Dessouky
ISBN 978-1-4665-1071-5

Security and Privacy in Smart Grids
Edited by Yang Xiao
ISBN 978-1-4398-7783-8

Security for Multihop Wireless Networks
Edited by Shafiullah Khan and Jaime Lloret Mauri
ISBN 978-1-4665-7803-6

Self-Healing Systems and Wireless Networks Management
Junaid Ahsenali Chaudhry
ISBN 978-1-4665-5648-5

The State of the Art in Intrusion Prevention and Detection
Edited by Al-Sakib Khan Pathan
ISBN 978-1-4822-0351-6

Wireless Ad Hoc and Sensor Networks: Management, Performance, and Applications
Jing (Selina) He, Mr. Shouling Ji, Yingshu Li, and Yi Pan
ISBN 978-1-4665-5694-2

Wireless Sensor Networks: From Theory to Applications
Edited by Ibrahiem M. M. El Emary and S. Ramakrishnan
ISBN 978-1-4665-1810-0

ZigBee® Network Protocols and Applications
Edited by Chonggang Wang, Tao Jiang, and Qian Zhang
ISBN 978-1-4398-1601-1

AUERBACH PUBLICATIONS
www.auerbach-publications.com
To Order Call: 1-800-272-7737 • Fax: 1-800-374-3401 • E-mail: orders@crcpress.com

Cognitive Networks

Applications and Deployments

Edited by
Jaime Lloret Mauri • Kayhan Zrar Ghafoor
Danda B. Rawat • Javier Manuel Aguiar Perez

CRC Press
Taylor & Francis Group
Boca Raton   London   New York

CRC Press is an imprint of the
Taylor & Francis Group, an **Informa** business

CRC Press
Taylor & Francis Group
6000 Broken Sound Parkway NW, Suite 300
Boca Raton, FL 33487-2742

First issued in paperback 2016

© 2015 by Taylor & Francis Group, LLC
CRC Press is an imprint of Taylor & Francis Group, an Informa business

No claim to original U.S. Government works

Version Date: 20150507

ISBN 13: 978-1-138-03420-4 (pbk)
ISBN 13: 978-1-4822-3699-6 (hbk)

Library of Congress Cataloging-in-Publication Data

Cognitive networks (Lloret Mauri)
    Cognitive networks : applications and deployments / editors, Jaime Lloret Mauri, Kayhan Zrar Ghafoor, Danda B. Rawat, Javier Manuel Aguiar Perez.
        pages cm
    Includes bibliographical references and index.
    ISBN 978-1-4822-3699-6 (hardcover : alk. paper) 1. Cognitive radio networks. I. Lloret Mauri, Jaime. II. Title.

TK5103.4815.C457 2015
621.3841'5--dc23   2015002200

Visit the Taylor & Francis Web site at
http://www.taylorandfrancis.com

and the CRC Press Web site at
http://www.crcpress.com

Contents

PART V DYNAMIC RADIO SPECTRUM ACCESS

PART VI VEHICULAR COGNITIVE NETWORKS AND APPLICATIONS

Contributors

Sherine M. Abdel-Kader
Computers and Systems
   Department
Electronics Research Institute
Cairo, Egypt

Salah M. Abdel-Mageid
Computers and Systems
   Department
Al-Azhar University
Cairo, Egypt

Kamalrulnizam Abu Bakar
Faculty of Computing
Universiti Teknologi Malaysia
Johor, Malaysia

Javier Aguiar Pérez
Departamento de Teoria de
   la Señal y Comunicaciones
   e Ingenieria
Telemáatica Universidad de
   Valladolid
Valladolid, Spain

Athar Ali Khan
Department of Electrical
   Engineering
COMSATS Institute
   of Information
   Technology
Wah Cantt, Pakistan

Mariette Awad
Department of Electrical
   and Computer
   Engineering
American University
   of Beirut
Beirut, Lebanon

Leire Azpilicueta
Department of Electrical
   and Electronics
   Engineering
Universidad Pública de Navarra
Navarra, Spain

Ashish Bagwari
Department of Electronics
    and Communication
    Engineering
Uttarakhand Technical
    University
Dehradun, India

Chandra Bajracharya
Department of Electrical
    and Computer Engineering
Old Dominion University
Norfolk, Virginia

Diana Bri
Research Institute for Integrated
    Management of Coastal Areas
Universitat Politècnica de València
València, Spain

Stefano Busanelli
R&D Department
Guglielmo Srl
Reggio Emilia, Italy

Alessandro Colazzo
Azcom Technology Srl
Milan, Italy

Jean-Michel Dricot
OPERA Department
Université Libre de Bruxelles
Brussels, Belgium

Francisco Falcone
Department of Electrical and
    Electronics Engineering
Universidad Pública de Navarra
Navarra, Spain

Gianluigi Ferrari
Department of Information
    Engineering
University of Parma
Parma, Italy

Norshelia Binti Fisal
Faculty of Electrical
    Engineering
Universiti Teknologi Malaysia
Johor, Malaysia

Miguel Garcia
Department of Computer Science
Universitat Politècnica de València
València, Spain

T. R. Gopalakrishnan Nair
Saudi ARAMCO Endowed
    Chair of Technology
    and Information
    Management
Prince Mohammad Bin Fahd
    University
Al Khobar, Saudi Arabia

and

Research and Industry
    Incubation Center
Dayananda Sagar Institutions
Bangalore, India

Ling Hou
Department of Electronic
    Engineering
City University of Hong Kong
Kowloon, Hong Kong

Shengchun Huang
College of Electronic Science
and Engineering
National University of Defense
Technology
Changsha, People's Republic
of China

Mubashir Husain Rehmani
Department of Electrical
Engineering
COMSATS Institute
of Information Technology
Wah Cantt, Pakistan

Youssef A. Jaffal
Department of Electrical
and Computer
Engineering
American University of Beirut
Beirut, Lebanon

José Javier Astráin
Department of Mathematics
and Computer
Engineering
Universidad Pública de Navarra
Navarra, Spain

Jaime Lloret Mauri
Research Institute for Integrated
Management of Coastal
Areas
Universitat Politècnica de València
València, Spain

Salma Malik Tabassum
Department of Electrical
Engineering
COMSATS Institute
of Information Technology
Wah Cantt, Pakistan

Youssef Nasser
Department of Electrical and
Computer Engineering
American University of Beirut
Beirut, Lebanon

Francisco Ramos
València Nanophotonics
Technology Center
Universitat Politècnica de València
València, Spain

Danda B. Rawat
Department of Electrical
Engineering
Georgia Southern University
Statesboro, Georgia

Ali Safa Sadiq
Faculty of Computing
Universiti Teknologi Malaysia
Johor, Malaysia

Yasir Saleem
Sunway University
Selangor, Malaysia

Tarek M. Salem
Computers and Systems
Department
Electronics Research Institute
Cairo, Egypt

Geetam Singh Tomar
Machine Intelligence Research
    Labs
Gwalior, India

and

Department of Electrical
    Engineering
University of West Indies
St. Augustine, Trinidad and Tobago

Kavitha Sooda
Advanced Networking Research
    Group
Dayananda Sagar Institutions
Bangalore, India

and

Department of Computer
    Science and Engineering
Nitte Meenakshi Institute
    of Technology
Bangalore, India

Athanasios V. Vasilakos
Department of Computer
    and Telecommunications
    Engineering
University of Western Macedonia
Kozani, Greece

and

Department of Electrical and
    Computer Engineering
National Technical University
    of Athens
Athens, Greece

Jesús Villadangos
Department of Mathematics and
    Computer Science Engineering
Universidad Pública de Navarra
Navarra, Spain

Shan Wang
College of Electronic Science
    and Engineering
National University of Defense
    Technology
Changsha, People's Republic
    of China

Angus K. Y. Wong
School of Science and Technology
The Open University of
    Hong Kong
Hong Kong

Xianzhong Xie
Chongqing Key Lab on Mobile
    Communication
Institute of Personal
    Communications
Chongqing University of Posts
    and Telecommunications
Chongqing, People's Republic
    of China

Helin Yang
Chongqing Key Lab on Mobile
    Communication
Institute of Personal
    Communication
Chongqing University of Posts
    and Telecommunications
Chongqing, People's Republic
    of China

Ali Yassin
Department of Electrical and
    Computer Engineering
American University of Beirut
Beirut, Lebanon

K. H. Yeung
Department of Electronic
    Engineering
City University of Hong Kong
Kowloon, Hong Kong

Mohamed Zaki
Computers and Systems
    Department
Al-Azhar University
Cairo, Egypt

Shaojie Zhang
College of Electronic Science
    and Engineering
National University of Defense
    Technology
Changsha, People's Republic
    of China

Haitao Zhao
College of Electronic Science
    and Engineering
National University of Defense
    Technology
Changsha, People's Republic
    of China

Kayhan Zrar Ghafoor
Faculty of Engineering
Koya University
Koya, Iraq

Muhammad Zubair Farooqi
Department of Electrical
    Engineering
COMSATS Institute
    of Information Technology
Wah Cantt, Pakistan

PART I

INTRODUCTION

1

EFFICIENT SPECTRUM MANAGEMENT

Challenges and Solutions

TAREK M. SALEM, SHERINE M. ABD EL-KADER, SALAH M. ABDEL-MAGEID, AND MOHAMED ZAKI

Contents

1.1 Introduction

The usage of radio spectrum resources and the regulation of radio emissions are coordinated by national regulatory bodies such as the Federal Communications Commission (FCC). The FCC assigns spectrum to licensed users, also known as primary users (PUs), on a long-term basis for large geographical regions. However, a large portion of the assigned spectrum remains underutilized as illustrated in Figure 1.1. The inefficient usage of the limited spectrum necessitates the development of dynamic spectrum access techniques [1], where users who have no spectrum licenses, also known as secondary users, are allowed to use the temporarily unused licensed spectrum. In recent years, the FCC has been considering more flexible and comprehensive uses of the available spectrum through the use of cognitive radio (CR) technology [2].

The limitations in spectrum access due to the static spectrum licensing scheme can be summarized as follows (Figure 1.1):

> *Fixed type of spectrum usage*: In the current spectrum licensing scheme, the type of spectrum use cannot be changed. For example, a TV band in Egypt cannot be used by digital TV

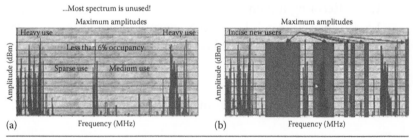

Figure 1.1   Spectrum is wasted. Opportunistic spectrum access can provide improvements in spectrum utilization. (a) Spectrum usage by traditional spectrum management, (b) spectrum usage by utilizing spectrum holes.

broadcast or broadband wireless access technologies. However, this TV band could remain largely unused due to cable TV systems.

Licensed for a large region: When a spectrum is licensed, it is usually allocated to a particular user or wireless service provider in a large region (e.g., an entire city or state). However, the wireless service provider may use the spectrum only in areas with a good number of subscribers to gain the highest return on investment. Consequently, the allocated frequency spectrum remains unused in other areas, and other users or service providers are prohibited from accessing this spectrum.

Large chunk of licensed spectrum: A wireless service provider is generally licensed with a large chunk of radio spectrum (e.g., 50 MHz). For a service provider, it may not be possible to obtain license for a small spectrum band to use in a certain area for a short period of time to meet a temporary peak traffic load. For example, a CDMA2000 cellular service provider may require a spectrum with a bandwidth of 1.25 or 3.75 MHz to provide temporary wireless access service in a hotspot area.

Prohibit spectrum access by unlicensed users: In the current spectrum licensing scheme, only a licensed user can access the corresponding radio spectrum and unlicensed users are prohibited from accessing the spectrum, even though it is unoccupied by the licensed users. For example, in a cellular system, there could be areas in a cell without any users. In such a case, unlicensed users with short-range wireless communications would not be able to access the spectrum, even though their transmission would not interfere with cellular users.

The term *cognitive radio* was defined in [3] as follows: "Cognitive radio is an intelligent wireless communication system that is aware of its ambient environment. This cognitive radio will learn from the environment and adapt its internal states to statistical variations in the existing RF environment by adjusting the transmission parameters (e.g. frequency band, modulation mode, and transmit power) in real-time." A CR network enables us to establish communications among CR nodes or users. The communication parameters can be adjusted according to the change in the environment, topology, operating conditions, or user requirements. From this definition, two main characteristics of the CR can be defined as follows:

- *Cognitive capability*: It refers to the ability of the radio technology to capture or sense the information from its radio environment. This capability cannot simply be realized by monitoring the power in some frequency bands of interest, but more sophisticated techniques, such as autonomous learning and action decision, are required in order to capture the temporal and spatial variations in the radio environment and avoid interference with other users.
- *Reconfigurability*: The cognitive capability provides spectrum awareness, whereas reconfigurability enables the radio to be dynamically programmed according to the radio environment [36]. More specifically, the CR can be programmed to transmit and receive signals at various frequencies and to use different transmission access technologies supported by its hardware design.

The ultimate objective of the CR is to obtain the best available spectrum through cognitive capability and reconfigurability as described earlier. Since most of the spectrum is already assigned, the most important challenge is to share the licensed spectrum without interfering with the transmission of other licensed users as illustrated in Figure 1.2. The CR enables the usage of a temporarily unused spectrum, which is referred to as a spectrum hole or a white space [3]. If this band is further utilized by a licensed user, the CR moves to another

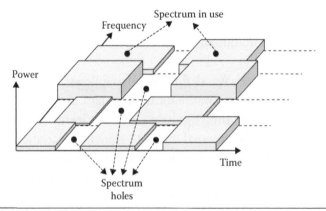

Figure 1.2   Spectrum holes concept.

spectrum hole or stays in the same band, altering its transmission power level or modulation scheme to avoid interference.

According to the network architecture, CR networks can be classified as infrastructure-based CR networks and CR ad hoc networks (CRAHNs) [3]. An infrastructure-based CR network has a central network entity such as a base station in cellular networks or an access point in wireless local area networks (LANs), whereas a CRAHN does not have any infrastructure backbone. Thus, a CR user can communicate with other CR users through ad hoc connection on both licensed and unlicensed spectrum bands.

In infrastructure-based CR networks, the observations and analysis performed by each CR user feed the central CR base station, so that it can make decisions on how to avoid interfering with primary networks. According to this decision, each CR user reconfigures its communication parameters, as shown in Figure 1.3a. On the contrary, in CRAHNs, each user needs to have all CR capabilities and is responsible for determining its actions based on the local observation, as shown in Figure 1.3b. Because the CR user cannot predict the influence of its actions on the entire network with its local observation, cooperation schemes are essential, where the observed information can be exchanged among devices to broaden the knowledge on the network.

In this chapter, an up-to-date survey of the key researches on spectrum management in CRAHNs is provided. We also identify and discuss some of the key open research challenges related to each aspect of spectrum management. The remainder of this chapter is arranged as follows: The differences between CRAHNs and classical ad hoc networks are introduced in Section 1.2. A brief overview of the spectrum management framework for CRAHNs is provided in Section 1.3. The challenges associated with spectrum sensing are given and enabling spectrum sensing methods are explained in Section 1.4. An overview of the spectrum decision for CR networks with open research issues is presented in Section 1.5. Spectrum sharing for CRAHNs is introduced in Section 1.6. Spectrum mobility and proposed tool to solve spectrum management research challenges for CRAHNs are explained in Sections 1.7 and 1.8, respectively. Common control channels (CCCs) are declared in Section 1.8. Finally, Section 1.9 concludes the chapter.

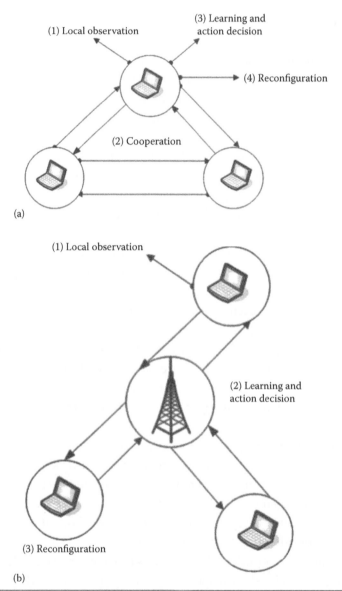

Figure 1.3 Comparison of CR capabilities between infrastructure-based CR networks (a) and CRAHNs (b).

1.2 Classical Ad Hoc Networks versus CRAHNs

The changing spectrum environment and the importance of protecting the transmission of the licensed users of the spectrum mainly differentiate classical ad hoc networks from CRAHNs. We describe these unique features of CRAHNs compared to classical ad hoc networks as follows:

- *Choice of transmission spectrum*: In CRAHNs, the available spectrum bands are distributed over a wide frequency range, which vary over time and space. Thus, each user shows different spectrum availability according to the PU activity. As opposed to this, classical ad hoc networks generally operate on a pre-decided channel that remains unchanged with time. For the ad hoc networks with multichannel support, all the channels are continuously available for transmission, although nodes may select few of the latter from this set based on self-interference constraints. A key distinguishing factor is the primary consideration of protecting the PU transmission, which is entirely missing in classical ad hoc networks.

- *Topology control*: Ad hoc networks lack centralized support, and hence must rely on local coordination to gather topology information. In classical ad hoc networks, this is easily accomplished by periodic beacon messages on the channel. However, in CRAHNs, as the licensed spectrum opportunity exists over a large range of frequencies, sending beacons over all the possible channels is not feasible. Thus, CRAHNs are highly probable to have incomplete topology information, which leads to an increase in collisions among CR users as well as interference with the PUs.

- *Multihop/multispectrum transmission*: The end-to-end route in CRAHNs consists of multiple hops having different channels according to the spectrum availability. Thus, CRAHNs require collaboration between routing and spectrum allocation in establishing these routes. Moreover, the spectrum switches on the links are frequent based on PU arrivals. As opposed to classical ad hoc networks, maintaining an end-to-end quality of service (QoS) involves not only the traffic load but also the number of different channels and possibly spectrum bands that are used in the path, the number of PU-induced spectrum change events, and the consideration of periodic spectrum sensing functions, among others.

- *Distinguishing mobility from PU activity*: In classical ad hoc networks, routes formed over multiple hops may periodically experience disconnections caused by node mobility. These cases may be detected when the next hop node in the path does not reply to messages and the retry limit is exceeded at the link layer. However, in CRAHNs, a node may not be able to transmit immediately if it detects the presence of a PU on the spectrum, even in the absence of mobility. Thus, correctly inferring mobility conditions and initiating an appropriate recovery mechanism in CRAHNs necessitate a different approach from the classical ad hoc networks.

1.3 Spectrum Management Framework for CRAHN

The components of the CRAHN architecture, as shown in Figure 1.4a, can be classified into two groups: the primary network and the CR network components. The primary network is referred to as an existing network, where the PUs have a license to operate in a certain spectrum band. If primary networks have an infrastructure support, the operations of the PUs are controlled through primary base stations. Due to their priority in spectrum access, the PUs should not be affected by unlicensed users. The CR network (or secondary network) does not have a license to operate in a desired band. Hence, additional functionality is required for CR users (or secondary users) to share the licensed spectrum band. Also, CR users are mobile and can communicate with each other in a multihop manner on both licensed and unlicensed spectrum bands. Usually, CR networks are assumed to function as stand-alone networks, which do not have direct communication channels with the primary networks. Thus, every action in CR networks depends on their local observations.

In order to adapt to a dynamic spectrum environment, the CRAHN necessitates the spectrum-aware operations, which form a cognitive cycle [4]. As shown in Figure 1.4b, the steps of the cognitive cycle consist of four spectrum management categories: *spectrum sensing*, *spectrum decision*, *spectrum sharing*, and *spectrum mobility*. To implement CR networks, each function needs to be incorporated into the classical layering protocols, as shown in Figure 1.5. The main features of spectrum management functions are as follows [3]:

Spectrum sensing: A CR user can be allocated to only an unused portion of the spectrum. Therefore, a CR user should monitor the available spectrum bands and then detect the spectrum holes.

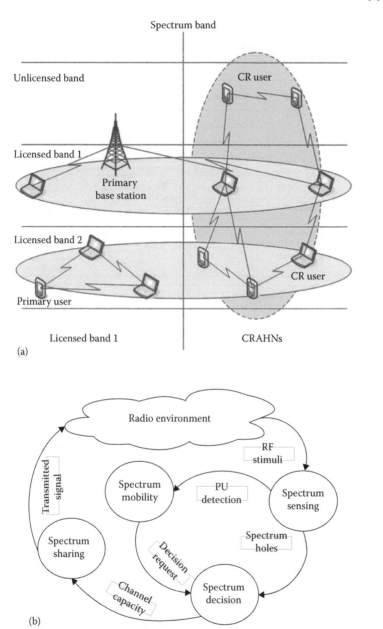

Figure 1.4   The CRAHN architecture (a) and the CR cycle (b).

Spectrum sensing is a basic functionality in CR networks, and hence it is closely related to other spectrum management functions as well as layering protocols to provide information on spectrum availability.

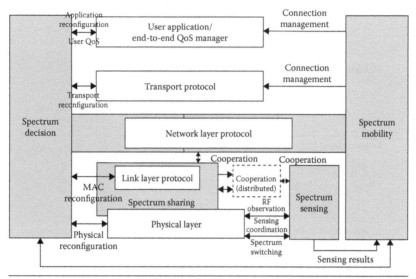

Figure 1.5   Spectrum management framework for CRNs.

Spectrum decision: Once the available spectrums are identified, it is essential that the CR users select the most appropriate band according to their QoS requirements. It is important to characterize the spectrum band in terms of both the radio environment and the statistical behaviors of the PUs. In order to design a decision algorithm that incorporates dynamic spectrum characteristics, we need to obtain a priori information regarding the PU activity. Furthermore, in CRAHNs, spectrum decision involves jointly undertaking spectrum selection and route formation.

Spectrum sharing: Since there may be multiple CR users trying to access the spectrum, their transmissions should be coordinated to prevent collisions in overlapping portions of the spectrum. Spectrum sharing provides the capability to share the spectrum resource opportunistically with multiple CR users, which includes resource allocation to avoid interference caused to the primary network. For this reason, game theoretical approaches have also been used to analyze the behavior of selfish CR users. Furthermore, this function necessitates a CR medium access control (MAC) protocol, which facilitates the sensing control to distribute the sensing task among the coordinating nodes as well as spectrum access to determine the timing for transmission.

Spectrum mobility: If a PU is detected in the specific portion of the spectrum in use, CR users should vacate the spectrum immediately and continue their communications in another vacant portion of the spectrum. For this reason, either a new spectrum must be chosen or the affected links may be circumvented entirely. Thus, spectrum mobility necessitates a spectrum handoff scheme to detect the link failure and to switch the current transmission to a new route or a new spectrum band with minimum quality degradation. This requires collaborating with spectrum sensing, neighbor discovery in a link layer, and routing protocols. Furthermore, this functionality needs a connection management scheme to sustain the performance of upper layer protocols by mitigating the influence of spectrum switching.

To overcome the drawback caused by the limited knowledge of the network, all of spectrum management categories are based on cooperative operations where CR users determine their actions based on the observed information exchanged with their neighbors. In the following Sections 1.4 through 1.7, spectrum management categories for CRAHNs are introduced. Then, we investigate how these spectrum management functions are integrated into the existing layering functionalities in ad hoc networks and address their challenges. Also, open research issues for this spectrum management are declared.

1.4 Spectrum Sensing for CR Networks

A CR is designed to be aware of and sensitive to the changes in its surrounding, which makes spectrum sensing an important requirement for the realization of CR networks. Spectrum sensing enables CR users to exploit the unused spectrum portion adaptively to the radio environment. This capability is required in the following cases: (1) CR users find available spectrum holes over a wide frequency range for their transmission (out-of-band sensing) and (2) CR users monitor the spectrum band during transmission and detect the presence of primary networks so as to avoid interference (in-band sensing).

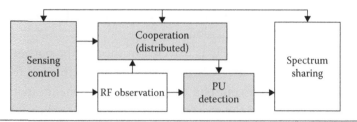

Figure 1.6   Spectrum sensing structure for CRAHNs.

As shown in Figure 1.6, the CRN necessitates the following functionalities for spectrum sensing:

- *PU detection*: The CR user observes and analyzes its local radio environment. Based on these location observations of itself and its neighbors, CR users determine the presence of PU transmissions, and accordingly identify the current spectrum availability.
- *Sensing control*: This function enables each CR user to perform its sensing operations adaptively to the dynamic radio environment.
- *Cooperation*: The observed information in each CR user is exchanged with its neighbors so as to improve sensing accuracy.

In order to achieve high spectrum utilization while avoiding interference, spectrum sensing needs to provide high detection accuracy. However, due to the lack of a central network entity, CR ad hoc users perform sensing operations independently of each other, leading to an adverse influence on sensing performance. We investigate these basic functionalities required for spectrum sensing to address this challenge in CRAHNs. In Sections 1.4.1 through 1.4.4, more details about functionalities for spectrum sensing are provided.

1.4.1 PU Detection

Since CR users are generally assumed not to have any real-time interaction with the PU transmitters and receivers, they do not know the exact information of the ongoing transmissions within the primary networks. Thus, PU detection depends only on the local radio observations of CR users. Generally, PU detection techniques for

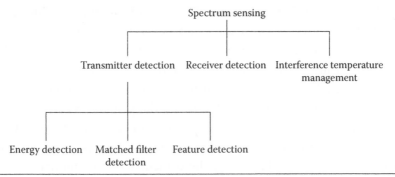

Figure 1.7 Classification of spectrum sensing.

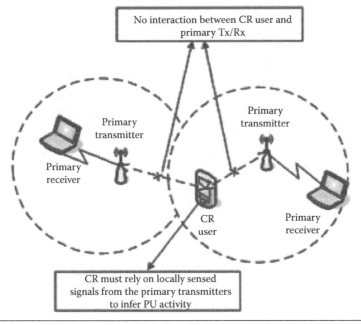

Figure 1.8 Spectrum sensing techniques. Tx, Transmitter, Rx, Receiver.

CRAHNs can be classified into three groups [3,5]: *primary transmitter detection*, *primary receiver detection*, and *interference temperature management* as declared in Figure 1.7. As shown in Figure 1.8, the primary transmitter detection is based on the detection of the weak signal from a primary transmitter through the local observations of CR users. The primary receiver detection aims at finding the PUs that receive data within the communication range of a CR user. Also, the local oscillator leakage power emitted by the radio-frequency (RF) front end of the primary receiver is usually exploited, which is typically weak. Thus, although it provides the most effective way to find spectrum holes, this method (i.e., primary receiver detection) is only feasible in the detection of the TV receivers. Interference temperature management accounts for the cumulative RF energy from multiple transmissions and sets a maximum cap on their aggregate level that the primary receiver could tolerate, called *an interference temperature limit* [6]. As long as CR users do not exceed this limit by their transmissions, they can use this spectrum band. However, the difficulty of this model lies in accurately measuring the interference temperature since CR users cannot distinguish between actual signals from the PU and noise or interference. For these reasons, most of the current research on spectrum sensing in CRAHNs has mainly focused on the primary transmitter detection.

Waleed et al. [7] presented a two-stage local spectrum sensing approach. In the first stage, each CR performs the existing spectrum sensing techniques, that is, energy detection, matched filter detection, and feature detection. In the second stage, the output from each technique is combined using fuzzy logic in order to deduce the presence or absence of a primary transmitter. Simulation results verify that the sensing approach technique outperforms the existing local spectrum sensing techniques. The sensing approach shows a significant improvement in sensing accuracy by exhibiting a higher probability of detection and low false alarms.

Ghasemi and Sousa [8] presented a scheme for cooperative spectrum sensing on distributed CR networks. A fuzzy logic rule-based inference system is used to estimate the presence possibility of the licensed user's signal based on the observed energy at each CR terminal. The estimated results are aggregated to make the final sensing decision at the fusion center.

1.4.2 Sensing Control

The main objective of spectrum sensing is to find more spectrum access opportunities without interfering with primary networks. To this end, the sensing operations of CR users are controlled and coordinated by a sensing controller, which considers two main issues:

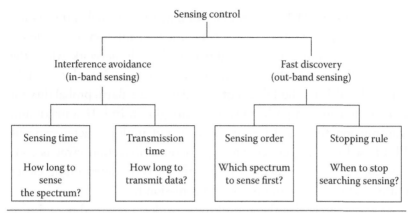

Figure 1.9 Configuration parameters coordinated by sensing control.

(1) how long and how frequently CR users should sense the spectrum to achieve sufficient sensing accuracy in in-band sensing and (2) how quickly CR users can find the available spectrum band in out-of-band sensing, which are summarized in Figure 1.9.

1.4.2.1 In-Band Sensing Control The first issue is related to the maximum spectrum opportunities as well as interference avoidance. The in-band sensing generally adopts the sensing period structure where CR users are allowed to access the spectrum only during the transmission period followed by sensing (observation) period. In the periodic sensing, longer sensing time leads to higher sensing accuracy, and hence less interference. But as the sensing time becomes longer, the transmission time of CR users will be decreased. Conversely, while longer transmission time increases the access opportunities, it causes higher interference due to the lack of sensing information. Thus, how to select the proper sensing and transmission times is an important issue in spectrum sensing.

Sensing time optimization is investigated in [9,10]. The sensing time is determined to maximize the channel efficiency while maintaining the required detection probability, which does not consider the influence of a false alarm probability. In [3], the sensing time is optimized for a multiple spectrum environment so as to maximize the throughput of CR users.

The focus of [11,12] is on determining the optimal transmission time. In [12], for a given sensing time, the transmission time is determined to maximize the throughput of the CR network while the packet collision probability for the primary network is under a certain threshold. This method does not consider a false alarm probability for estimating collision probability and throughput. In [11], a maximum transmission time is determined to protect multiple heterogeneous PUs based on the perfect sensing where no detection error is considered. All efforts stated above mainly focus on determining either optimal sensing time or optimal transmission time.

However, in [13], a theoretical framework is presented to optimize both sensing and transmission times simultaneously in such a way as to maximize the transmission efficiency subject to interference avoidance constraints where both parameters are determined adaptively depending on the time-varying cooperative gain.

In [14], a notification protocol based on in-band signaling is presented to disseminate the evacuation information among all CR users and thus evacuate the licensed spectrum reliably. This protocol uses the spreading code for its transmission, leading to tolerance in interference from both primary and other CR transmissions. Furthermore, due to its flooding-based routing scheme, it requires little prior information on the network topology and density, which does not consider the influence of a false alarm probability.

1.4.2.2 Out-of-Band Sensing Control When a CR user needs to find a new available spectrum band (out-of-band sensing), a spectrum discovery time is another crucial factor to determine the performance of CRAHNs. Thus, this spectrum sensing should have a coordination scheme not only to discover as many spectrum opportunities as possible but also to minimize the delay in finding them. This is also an important issue in spectrum mobility to reduce the switching time.

First, the proper selection of spectrum sensing order can help to reduce the spectrum discovery time in out-of-band sensing. In [15], an n-step serial search scheme is presented to mainly focus on correlated occupancy channel models, where the spectrum availability from all spectrum bands is assumed to be dependent on that of its adjacent spectrum bands. In [16,17], both transmission time and spectrum searching sequence are optimized by minimizing searching delay as well as maximizing spectrum opportunities.

Moreover, if the CR user senses more spectrum bands, it is highly probable to detect a better spectrum band, which results in longer spectrum searching time. To exploit this trade-off efficiently, a well-defined stopping rule of spectrum searching is essential in out-of-band sensing. In [18], an optimal stopping time is determined to maximize the expected capacity of CR users subject to the maximum number of spectrum bands a CR user can use simultaneously.

1.4.3 Cooperative Sensing

In CRAHNs, each CR user needs to determine the spectrum availability by itself depending only on its local observations. However, the observation range of the CR user is small and typically less than its transmission range. Thus, even though CR users find the unused spectrum portion, their transmission may cause interference at the primary receivers inside their transmission range, the so-called receiver uncertainty problem [2]. Furthermore, if the CR user receives a weak signal with a low signal-to-noise ratio (SNR) due to multipath fading, or it is located in a shadowing area, it cannot detect the signal of the PUs. Thus, in CRAHNs, spectrum sensing necessitates an efficient cooperation scheme in order to prevent interference with PUs outside the observation range of each CR user [2,19].

A common cooperative scheme is forming clusters to share the sensing information locally. Such a scheme for wireless mesh networks is presented in [20], where the mesh router and the mesh clients supported by it form a cluster. Here, the mesh clients send their individual sensing results to the mesh router, which are then combined to get the final sensing result. Since CRAHNs do not have the central network entity, this cooperation should be implemented in a distributed manner.

For cooperation, when a CR user detects the PU activities, it should notify its observations promptly to its neighbors to evacuate the busy spectrum. To this end, a reliable control channel is needed for discovering neighbors of a CR user as well as exchanging sensing information. In addition to this, asynchronous sensing and transmission schedules make it difficult to exchange sensing information between neighbors.

Thus, robust neighbor discovery and reliable information exchange are critical issues in implementing cooperative sensing in CRAHNs. This cooperation issue will also be leveraged by other spectrum management functions: spectrum decision, spectrum sharing, and spectrum mobility.

In [21], an optimal cooperative sensing strategy is presented, where the final decision is based on a linear combination of the local test statistics from individual CR users. The combining weight of each user's signal indicates its contribution to the cooperative decision making. For example, if a CR user receives a higher SNR signal and frequently makes its local decision consistent with the real hypothesis, then its test statistic has a larger weighting coefficient. In case of CR users in a deep fading channel, smaller weights are used to reduce their negative influence on the final decision. In Section 1.4.4, some of the key open research issues related to spectrum sensing are introduced.

1.4.4 Open Research Issues in Spectrum Sensing

- *Optimizing the period of spectrum sensing*: In spectrum sensing, the longer the observation period, the more accurate will be the spectrum sensing result. However, during sensing, a single-radio wireless transceiver cannot transmit signals in the same frequency band. Consequently, a longer observation period will result in lower system throughput. This performance trade-off can be optimized to achieve an optimal spectrum sensing solution. Classical optimization techniques (e.g., convex optimization) can be applied to obtain the optimal solution.
- *Spectrum sensing in multichannel networks*: Multichannel transmission [e.g., orthogonal frequency division multiplexing (OFDM)-based transmission] would be typical in a CR network. However, the number of available channels would be larger than the number of available interfaces at the radio transceiver. Therefore, only a fraction of the available channels can be sensed simultaneously. Selection of the channels (among all available channels) to be sensed will affect the performance of the system. Therefore, in a multichannel environment, selection of the channels should be optimized for spectrum sensing to achieve optimal system performance.

1.5 Spectrum Decision for CR Networks

CRNs require capabilities to decide on the best spectrum band among the available bands according to the QoS requirements of the applications. This notion is called spectrum decision and it is closely related to the channel characteristics and the operations of PUs. Spectrum decision usually consists of two steps: First, each spectrum band is characterized based on not only local observations of CR users but also statistical information of primary networks. Second, based on this characterization, the most appropriate spectrum band can be chosen.

Generally, CRAHNs have unique characteristics in spectrum decision due to the nature of multihop communication. Spectrum decision needs to consider the end-to-end route consisting of multiple hops. Furthermore, available spectrum bands in CR networks differ from one hop to the other. As a result, the connectivity is spectrum dependent, which makes it challenging to determine the best combination of the routing path and spectrum. Thus, spectrum decision in ad hoc networks should interact with routing protocols. The main functionalities required for spectrum decision are as follows:

- *Spectrum characterization*: Based on the observation, the CR users determine not only the characteristics of each available spectrum but also its PU activity model.
- *Spectrum selection*: The CR user finds the best spectrum band for each hop on the determined end-to-end route so as to satisfy its end-to-end QoS requirements.
- *Reconfiguration*: The CR users reconfigure communication protocol as well as communication hardware and RF front end according to the radio environment and user QoS requirements.

CR ad hoc users require spectrum decision in the beginning of the transmission. As depicted in Figure 1.10, through RF observation, CR users characterize the available spectrum bands by considering the received signal strength, the interference, and the number of users currently residing in the spectrum, which are also used for resource allocation in classical ad hoc networks. However, unlike classical ad hoc networks, each CR user observes heterogeneous spectrum availability that varies over time and space due to the PU activities.

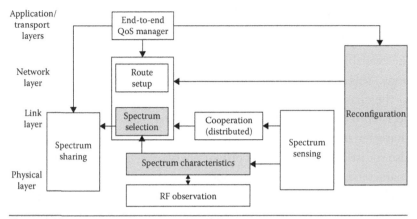

Figure 1.10 Spectrum decision structure for CRAHNs.

This changing nature of the spectrum usage is considered in the spectrum characterization. Based on this characterization, CR users determine the best available spectrum band to satisfy their QoS requirements. Furthermore, quality degradation of the current transmission can also initiate spectrum decision to maintain the quality of a current session. In Sections 1.5.1 through 1.5.4, more details about functionalities required for spectrum decision are provided.

1.5.1 Spectrum Characterization

In CRNs, multiple spectrum bands with different channel characteristics may be found to be available over a wide frequency range [22]. It is critical to first identify the characteristics of each available spectrum band. In Section 1.5.1.1, a spectrum characteristic in terms of radio environment and PU activity models is discussed.

1.5.1.1 Radio Environment Since the available spectrum holes show different characteristics, which vary over time, each spectrum hole should be characterized by considering both the time-varying radio environment and the spectrum parameters such as operating frequency and bandwidth. Hence, it is essential to define parameters that can represent a particular spectrum band as follows:

Interference: From the amount of the interference at the primary receiver, the permissible power of a CR user can be derived, which is used for the estimation of the channel capacity.

Path loss: The path loss is closely related to the distance and frequency. As the operating frequency increases, the path loss increases, which results in a decrease in the transmission range. If transmission power is increased to compensate for the increased path loss, interference at other users may increase.

Wireless link errors: Depending on the modulation scheme and the interference level of the spectrum band, the error rate of the channel changes.

Link layer delay: To address different path loss, wireless link error, and interference, different types of link layer protocols are required at different spectrum bands. This results in different link layer delays.

1.5.1.2 PU Activity In order to describe the dynamic nature of CR networks, a new metric is needed to capture the statistical behavior of primary networks, called PU activities. Since there is no guarantee that a spectrum band will be available during the entire communication of a CR user, the estimation of the PU activity is a very crucial issue in spectrum decision.

Most of CR research assumes that the PU activity is modeled by exponentially distributed interarrivals [23]. In this model, the PU traffic can be modeled as a two-state birth and death process. An ON (busy) state represents the period used by PUs and an OFF (idle) state represents the unused period [6,13,24]. Since each user arrival is independent, each transition follows the Poisson arrival process. Thus, the length of ON and OFF periods is exponentially distributed.

There are some efforts to model the PU activity in specific spectrum bands based on field experiments. In [25], the characteristics of primary usage in cellular networks are presented based on the call records collected by network systems, instead of real measurement. This analysis shows that an exponential call arrival model is adequate to capture the PU activity while the duration of wireless voice calls does not follow an exponential distribution. Furthermore, it is shown that a simpler random walk can be used to describe the PU activity under high traffic load conditions.

In [26], a statistical traffic model of wireless LANs based on a semi-Markov model is presented to describe the temporal behavior of wireless LANs. However, the complexity of this distribution hinders its practical implementation in CR functions.

1.5.2 Spectrum Selection

Once the available spectrum bands are characterized, the most appropriate spectrum band should be selected. Based on the user QoS requirements and the spectrum characteristics, the data rate, acceptable error rate, the delay bound, the transmission mode, and the bandwidth of the transmission can be determined. Then, according to a spectrum selection rule, the set of appropriate spectrum bands can be chosen. However, as stated earlier, since the entire communication session consists of multiple hops with heterogeneous spectrum availability, the spectrum selection rule is closely coupled with routing protocols in CRAHNs. Since there exist numerous combinations of route and spectrum between the source and the destination, it is infeasible to consider all possible links for spectrum decision. In order to determine the best route and spectrum more efficiently, spectrum decision necessitates the dynamic decision framework to adapt to the QoS requirements of the user and channel conditions. Furthermore, in recent research, the route selection is performed independent of the spectrum decision. Although this method is quite simple, it cannot provide an optimal route because spectrum availability on each hop is not considered during route establishment. Thus, the joint spectrum and routing decision method is essential for CRAHNs.

Furthermore, because of the operation of primary networks, CR users cannot obtain a reliable communication channel for long durations. Moreover, CR users may not detect any single spectrum band to meet the user's requirements. Therefore, they can adopt the multi-radio transmissions where each transceiver (radio interface) tunes to different noncontiguous spectrum bands for different users and transmits data simultaneously. This method can create a signal that is not only capable of high data throughput but also immune to the interference and the PU activity. Even if a PU appears in one of the current spectrum bands, or one of the next hop neighbors disappears, the rest of the connections continue their transmissions without any loss of connectivity [27,28]. In addition, transmission in multiple spectrum bands allows lower power to be used in each spectrum band. As a result, less interference with PUs is achieved, compared to the transmission on single spectrum band. For these reasons, spectrum decision should support multiple spectrum selection capabilities. For example, how to determine the number of spectrum bands and how to select the set of appropriate bands are still open research issues in CR networks.

1.5.3 Reconfiguration

Besides spectrum and route selection, spectrum decision involves reconfiguration in CRAHNs. The protocols for different layers of the network stack must adapt to the channel parameters of the operating frequency. Once the spectrum is decided, CR users need to select the proper communication modules such as physical layer technology and upper layer protocols adaptively dependent on application requirements as well as spectrum characteristics, and then reconfigure their communication system accordingly. In [29], the adaptive protocols are presented to determine the transmission power as well as the best combination of modulation and error correction code for a new spectrum band by considering changes in the propagation loss. In Section 1.5.4, some of the key open research issues related to spectrum decision are introduced.

1.5.4 Open Research Issues in Spectrum Decision

- Data dissemination in CR ad hoc networks, guaranteeing reliability of data dissemination in wireless networks, is a challenging task. Indeed, the characteristics and problems intrinsic to the wireless links add several issues in the shape of message losses, collisions, and broadcast storm problems, just to name a few. Channel selection strategy is required to solve this problem.
- Channel selection strategies are greatly influenced by the primary radio nodes activity. It is required to study the impact of the primary radio nodes activity on channel selection strategies.

- A decision model is required for spectrum access; stochastic optimization methods (e.g., the Markov decision process) will be an attractive tool to model and solve the spectrum access decision problem in CRNs.

1.6 Spectrum Sharing for CRAHNs

The shared nature of the wireless channel necessitates the coordination of transmission attempts between CR users. In this respect, spectrum sharing provides the capability to maintain the QoS of CR users without causing interference to the PUs by coordinating the multiple accesses of CR users as well as allocating communication resources adaptively to the changes of radio environment. Thus, spectrum sharing is performed in the middle of a communication session and within the spectrum band, and includes much functionality of a MAC protocol and resource allocation in classical ad hoc networks.

However, the unique characteristics of CRs such as the coexistence of CR users with PUs and the wide range of available spectrum incur substantially different challenges for spectrum sharing in CRAHNs. Spectrum sharing techniques are generally focused on two types of solutions, that is, spectrum sharing inside a CR network (intra-network spectrum sharing), and among multiple coexisting CR networks (inter-network spectrum sharing) [30]. However, since the CRAHNs do not have any infrastructure to coordinate inter-network operations, they are required to consider only the intra-network spectrum sharing functionality. Figure 1.11 depicts the functional blocks for spectrum sharing in CRAHNs. The unique features of spectrum sharing especially focus on resource allocation and spectrum access in CRAHNs.

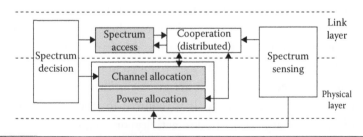

Figure 1.11   Spectrum sharing structure for CRNs.

In Sections 1.6.1 and 1.6.2, more details about functionalities for spectrum sharing are explained.

1.6.1 Resource Allocation

Based on the QoS monitoring results, CR users select the proper channels (channel allocation) and adjust their transmission power (power control) so as to achieve QoS requirements as well as resource fairness. Especially, in power control, sensing results need to be considered so as not to violate the interference constraints. In general, game theoretic approaches are exploited to determine the communication resources of each user in CRAHNs [31,32]. Each CR user has a common interest to use the spectrum resources as much as possible. However, CR users have competing interests to maximize their own share of the spectrum resources, that is, the activity of one CR user can impact the activities of the others. Furthermore, the rational decisions of a CR user must be undertaken while anticipating the responses of its rivals.

Game theory provides an efficient distributed spectrum sharing scheme by describing the conflict and cooperation among CR users, and hence allowing each CR user to rationally decide on its best action.

In [31], spectrum sharing for unlicensed band is presented based on the one-shot normal form and repeated games. Furthermore, it is shown that orthogonal power allocation, that is, assigning the channel to only one transmission to avoid co-channel interference with other neighbors, is optimal for maximizing the entire network capacity.

In [33], both single-channel and multichannel asynchronous distributed pricing (SC/MC-ADP) schemes are presented, where each CR user announces its interference price to other nodes. Using this information from its neighbors, the CR user can first allocate a channel and, in case there exist users in that channel, then determine its transmitting power. While there exist users using distinct channels, multiple users can share the same channel by adjusting their transmit power. Furthermore, the SC-ADP algorithm provides higher rates to users compared to selfish algorithms where users select the best channel without any knowledge about their neighbors' interference levels. While this method considers the channel and power allocation at the same time, it does not address the heterogeneous spectrum availability over time and space, which is a unique characteristic of CRAHNs.

1.6.2 *Spectrum Access*

It enables multiple CR users to share the spectrum resource by determining who will access the channel or when a user may access the channel. This is (most probably) a random access method due to the difficulty in synchronization. Spectrum sharing includes MAC functionality as well. However, unlike classical MAC protocols in ad hoc networks, CR MAC protocols are closely coupled with spectrum sensing, especially in sensing control described in Section 1.4.2.

In CRAHNs, the sensing schedules are determined and controlled by each user, and not being controlled and synchronized by the central network entity. Thus, instead of determined sensing schedules for all CRs, CR ad hoc users may adopt the a periodic or on-demand sensing triggered by only spectrum sharing operations can trigger the spectrum sensing, that is, when CR users want to transmit or are requested their spectrum availability by neighbor users. Furthermore, sensing and transmission intervals, determined by the sensing control in spectrum sensing, influence the performance of spectrum access [37–39].

Classification of MAC protocols based on the nature of channel access, that is, random access, time slotted, and a hybrid protocol that is a combination of the two, is shown in Figure 1.12.

In [18], MAC layer packet transmission in the hardware-constrained MAC (HC-MAC) protocol is presented. Typically, the radio can only sense a finite portion of the spectrum at a given time, and for single transceiver devices, sensing results in decreasing the data transmission rate. HC-MAC derives the optimal duration for sensing based on the reward obtained for correct results, as against the need aggressively scanning the spectrum at the cost of transmission time. A key difference of this protocol as against the previous work is that the sensing at either ends of the

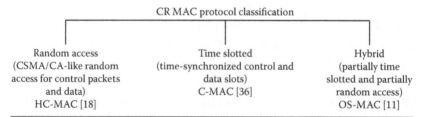

Figure 1.12 Classification of CR MAC protocols. HC-MAC, Hardware constraint-medium access control; C-MAC, Cognitive-medium access control; OS-MAC, Opportunistic spectrum-medium access control.

link is initiated after the channel contention on the dedicated CCC. The feasible channels at the two CR users on the link are then determined. However, the control messages used for channel negotiation may not be received by the neighboring nodes, and their transmission may influence the sensing results of the CR users that win the contention.

The presence of interferers that may cause jamming in the CR user frequencies is considered in the single-radio adaptive channel (SRAC) MAC protocol [34]. However, this work does not completely address the means to detect the presence of a jammer and how the ongoing data transmission is switched immediately to one of the possible backup channels when the user is suddenly interrupted. In Section 1.6.3, some of the key open research issues related to spectrum sharing are introduced.

1.6.3 Open Research Issues in Spectrum Sharing

Since spectrum sharing and sensing share some of the functionalities, most of the issues are similar to those of spectrum sensing, which are explained as follows:

Distributed power allocation: The CRAHN user determines the transmission power in a distributed manner without support of the central entity, which may cause interference due to the limitation of sensing area even if it does not detect any transmission in its observation range. Thus, spectrum sharing necessitates sophisticated power control methods for adapting to the time-varying radio environment so as to maximize the capacity with the protection of the transmissions of PUs.

Topology discovery: The use of nonuniform channels by different CR users makes topology discovery difficult. From Figure 1.13, we see that the CR users A and B experience different PU activities in their respective coverage areas and thus may only be allowed to transmit on mutually exclusive channels. The allowed channels for CR A (1,2) being different from those used by CR B (3) make it difficult to send out periodic beacons informing the nodes within the transmission range of their own ID and other location coordinates needed for networking.

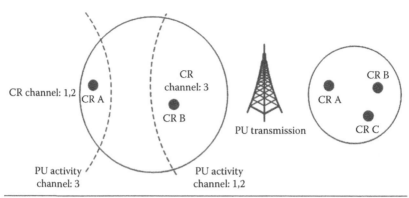

Figure 1.13   Spectrum sharing challenges in CRAHNs.

Evolution and learning: The occupancy history of the spectrum bands by the PUs may vary with the time of the day and location. It is desired that the MAC protocol learns the characteristic PU activity and accordingly alters its spectrum selection and data transmission strategy. Although the partially observable Markov decision process (POMDP) MAC protocol proposed in [43] takes the initial steps in this direction, more detailed and elaborate learning models are needed. How long should the learning duration be and its effect during the network operation are the issues that need to be investigated. Moreover, the problem of constructing a detailed channel occupancy needs further research, so that the different times of the day and the different locations traversed by the mobile CR user can be incorporated. The probabilistic spectrum selection algorithm that uses this history may be designed to guarantee performance bounds during long-term operation. For this reason, open challenges include how the theoretical research and network operation are combined, so that the gains arising from the choice of the spectrum at the link layer are appropriately weighted in each decision round, and the computational time for considering the past history is minimized.

Spectrum access and coordination: In classical ad hoc networks, the request-to-send (RTS) and clear-to-send (CTS) mechanisms are used to signal the control of the channel and reduce simultaneous transmissions to some extent. In CR networks, however, the available spectrum is dynamic and users may switch the channel after a given communicating pair of nodes has exchanged the channel access signal. Thus, a fresh set of RTS–CTS exchange may need to be undertaken in the new channel to enforce a silence zone among the neighboring CR users in the new spectrum. Moreover, the CR users monitoring the earlier channel are oblivious to the spectrum change on the link.

1.7 Spectrum Mobility for CRAHNs

CR users are generally regarded as *visitors* to the spectrum. Hence, if the specific portion of the spectrum in use is required by a PU, the communication needs to be continued in another vacant portion of the spectrum. This notion is called spectrum mobility. Spectrum mobility gives rise to a new type of handoff in CR networks, the so-called spectrum handoff, in which the users transfer their connections to an unused spectrum band. In CRAHNs, a spectrum handoff occurs when: (1) a PU is detected, (2) the CR user loses its connection due to the mobility of users involved in an ongoing communication, or (3) a current spectrum band cannot provide the QoS requirements. In the spectrum handoff, temporary communication break is inevitable due to the process for discovering a new available spectrum band. Since the available spectra are discontiguous and distributed over a wide frequency range, CR users may require the reconfiguration of operation frequency in their RF front end, which leads to significantly longer switching time. Figure 1.14 illustrates the functional blocks for spectrum mobility in CRAHNs.

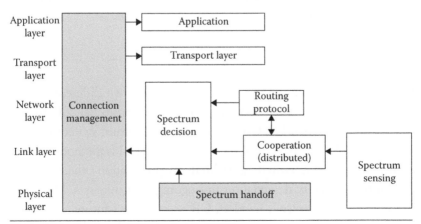

Figure 1.14 Spectrum mobility structure for CRAHNs.

The purpose of the spectrum mobility management in CRAHNs is to ensure smooth and fast transition leading to minimum performance degradation during a spectrum handoff. Furthermore, in spectrum mobility, the protocols for different layers of the network stack should be transparent to the spectrum handoff and the associated latency and adapt to the channel parameters of the operating frequency. Another intrinsic characteristic of spectrum mobility in CR networks is the interdependency with the routing protocols. Similar to the spectrum decision, the spectrum mobility needs to involve the recovery of link failure on the end-to-end route. Thus, it needs to interact with routing protocols to detect the link failure due to either user mobility or PU appearance. In the following subsections 1.7.1 through 1.7.3, the main functionalities required for spectrum mobility in the CRAHN are described.

As stated earlier, the spectrum mobility events can be detected as a link failure caused by user mobility as well as PU detection. Furthermore, the quality degradation of the current transmission also initiates spectrum mobility. When these spectrum mobility events are detected through spectrum sensing, neighbor discovery, and routing protocol, they trigger the spectrum mobility procedures. Figure 1.14 illustrates the functional blocks for spectrum mobility in CRAHNs. By collaborating with spectrum decision, a CR user determines a new spectrum band on the determined route, and switches its current session to the new spectrum (spectrum handoff). During the spectrum handoff, the CR user needs to maintain the current transmission that is not to be interfered by the switching latency. In Sections 1.7.1 and 1.7.2, we investigate the two main functionalities for spectrum mobility: spectrum handoff and connection management.

1.7.1 Spectrum Handoff

Spectrum handoff can be implemented based on two different strategies: reactive spectrum handoff and proactive spectrum handoff. In the reactive spectrum handoff, CR users perform spectrum switching after detecting the link failure due to spectrum mobility. This method requires immediate spectrum switching without any preparation time, resulting in significant quality degradation in ongoing transmissions. However, in the proactive spectrum handoff, CR users predict future activity in the current link and determine a new spectrum while maintaining the current transmission, and then perform spectrum switching before the link failure happens. Since the proactive spectrum handoff can maintain current transmissions while searching a new spectrum band, spectrum switching is faster but requires more complex algorithms for these concurrent operations. Depending on the events that trigger the spectrum mobility, different handoff strategies are needed.

While the reactive spectrum handoff is generally used in the event of a PU appearance, the proactive spectrum handoff is suitable for the events of user mobility or spectrum quality degradation. These events do not require immediate spectrum switching and can be easily predicted. Even in the PU appearance event, the proactive spectrum handoff may be used instead of the reactive scheme, but requires an accurate model for the PU activity to avoid an adverse influence on communication performance [35,40].

In addition, for seamless communication in dynamic radio environments, this spectrum handoff should support intelligent connection release and reestablishment procedures during spectrum switching. When a CR user moves, it needs to determine whether it should stay connected to its next-hop forwarder through power control or immediately switching to a new neighbor. This has to be undertaken ensuring that the network stays connected throughout the handoff procedure.

Spectrum handoff delay is the most crucial factor in determining the performance of spectrum mobility. This delay is dependent on the following operations in CR networks: First, the different layers of the protocol stack must adapt to the channel parameters of the operating frequency. Thus, each time a CR user changes its frequency, the network protocols may require modifications on the operation parameters, which may cause protocol reconfiguration delay.

Also we need to consider the spectrum and route recovery time and the actual switching time determined by the RF front-end reconfiguration. Furthermore, to find the new spectrum and route, CR users need to perform out-of-band sensing and neighbor discovery. Recent research has explored the minimization of the delay in out-of-band sensing through the search-sequence optimization, which is explained in Section 1.4.2.2. Furthermore, for more efficient spectrum discovery in out-of-band sensing, IEEE 802.22 adopts the backup channel lists that are selected and maintained so as to provide the highest probability of finding an available spectrum band within the shortest time [4].

In [23], an algorithm for updating the backup channel lists is proposed to support fast and reliable opportunity discovery with the cooperation of neighbor users. To mitigate the delay effect on the ongoing transmission, connection management needs to coordinate the spectrum switching by collaborating with upper layer protocols.

1.7.2 Connection Management

When the current operational frequency becomes busy in the middle of a communication by a CR user, then applications running in this node have to be transferred to another available frequency band. However, the selection of new operational frequency may take time. An important requirement of connection management protocols is the information about the duration of a spectrum handoff. Once the latency information is available, the CR user can predict the influence of the temporary disconnection on each protocol layer, and accordingly preserve the ongoing communications with only minimum performance degradation through the reconfiguration of each protocol layer and an error control scheme [41,42]. Consequently, multilayer mobility management protocols are required to accomplish the spectrum mobility functionalities. These protocols support mobility management adaptive to different types of applications. For example, a transmission control protocol (TCP) connection can be put to a wait state until the spectrum handoff is over. Moreover, since the TCP parameters will change after a spectrum handoff, it is essential to learn the new parameters and ensure that the transitions from the old parameters to the new parameters are carried out rapidly. In Section 1.7.3, some of the key open search issues are declared.

1.7.3 Open Research Issues in Spectrum Mobility

To the best of our knowledge, there exists no research effort to address the problems of spectrum mobility in CRAHNs to date. Although the routing mechanisms that have been investigated in the classical ad hoc networks may lay the groundwork in this area, there still exist many open research topics:

> *Switching delay management*: The spectrum switching delay is closely related not only to hardware, such as an RF front end, but also to algorithm development for spectrums sensing,

spectrum decision, link layer, and routing. Thus, it is desirable to design the spectrum mobility in a cross-layer approach to reduce the operational overhead among the functionalities and to achieve a faster switching time. Furthermore, the estimation of accurate latency in spectrum handoff is essential for reliable connection management.

Flexible spectrum handoff framework: As stated earlier, there are two different spectrum handoff strategies: reactive and proactive spectrum handoffs, which show different influence on the communication performance. Furthermore, according to the mobility event, a spectrum switching time will change. For example, since a PU activity region is typically larger than the transmission range of CR users, multiple hops may be influenced by spectrum mobility events at the same time, which makes the recovery time much longer. Furthermore, spectrum handoff should be performed while adapting to the type of applications and network environment. In case of a delay-sensitive application, CR users can use a proactive switching, instead of a reactive switching. In this method, through the prediction of PU activities, CR users switch the spectrum before PUs appear, which helps to reduce the spectrum switching time significantly. However, energy-constrained devices such as sensors need reactive spectrum switching. Thus, we need to develop a flexible spectrum handoff framework to exploit different switching strategies.

The different CR functionalities of spectrum sensing, decision, sharing, and mobility need to be implemented within the protocol stack of a wireless device. Specifically, in the following Section 1.8, a control channel in the link layer unique to CR networks will be introduced.

1.8 Common Control Channel

The CCC is used for supporting the *transmission coordination* and *spectrum*-related information exchange between the CR users. It facilitates neighbor discovery and helps in spectrum sensing coordination, control signaling, and exchange of local measurements between the CR users. The classification for CCC is declared in Figure 1.15.

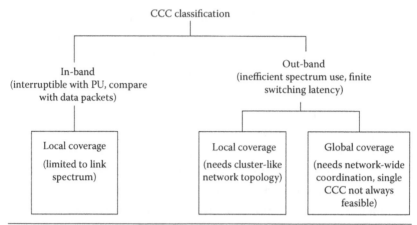

Figure 1.15 Classification of CCC.

The operation of the CCC is different from the data transmission over the licensed band in the following aspects:

CR users may optimize their channel use over a number of constraints, such as channel quality, access time, observed PU activity, and network load, among others during CR data transmission. However, these parameters are not known to the CR users in advance at the start of the network operation, and, thus, it is a challenge to choose the CCC with a minimum or no exchange of network information.

Spectrum bands that are currently used for data transfer may suddenly become unavailable when a PU appears. While the data communication is interrupted, the affected CR users need to coordinate with a new spectrum that does not interfere with the PUs on either end of the link. This control information used in the new spectrum selection must be sent reliably, and, thus, an always-on CCC is needed.

1.9 Conclusion

CR technology has been proposed in recent years as a revolutionary solution toward more efficient utilization of the scarce spectrum resources in an adaptive and intelligent way. By tuning the frequency to the temporarily unused licensed band and adapting the operating parameters to environment variations, CR technology provides future wireless devices with additional bandwidth, reliable broadband communications, and versatility for rapidly growing data applications. To realize the goal of spectrum-aware communication, the CR devices need to incorporate the spectrum sensing, decision, sharing, and mobility functionalities. The main challenge in CRAHNs is to integrate these functions in the layers of the protocol stack, so that the CR users can communicate reliably in a distributed manner over a multihop/multispectrum environment without an infrastructure support. The discussions provided in this book strongly support cooperative spectrum-aware communication protocols that consider the spectrum management functionalities. The proposed tool and simulator in this book gives insight in choosing the best suitable tool that fits for different categories of spectrum management.

References

1. T. M. Salem, S. A. El-kader, S. M. Ramadan, and M. Zaki, "Opportunistic spectrum access in cognitive radio ad hoc networks," *International Journal of Computer Science Issues (IJCSI)*, vol. 11, no. 1, 2014.
2. I. F. Akyildiz, W.-Y. Lee, M. C. Vuran, and S. Mohanty, "NeXt generation/ dynamic spectrum access/cognitive radio wireless networks: A survey," *Elsevier Computer Networks*, vol. 50, no. 13, pp. 2127–2159, 2006.
3. S. Haykin, "Cognitive radio: Brain-empowered wireless communications," *IEEE Journal on Selected Areas in Communications,* vol. 23, no. 2, pp. 201–220, 2005.
4. M. A. McHency, P. A. Tenhula, D. McCloskey, D. A. Reberson, and C. S. Hood, "Chicago spectrum occupancy measurements and analysis and a long-term studies proposal protocol," *Proceedings of the 1st International Workshop on Technology and Policy for Accessing Spectrum*, Boston, MA, August 2006.
5. M. Nekovee, "A survey of cognitive radio access on TV white spaces," *Hindawi International Journal of Digital Multimedia Broadcasting*, Article ID: 236568, pp. 1–11, 2010.
6. G. Yuan, R. Grammenos, Y. Yang, and W. Wang, "Performance analysis of selective opportunistic spectrum access with traffic prediction," *IEEE Transactions on Vehicular Technology*, vol. 59, no. 4, pp. 1949–1959, 2011.
7. E. Waleed et al., "Improved local spectrum sensing for cognitive radio networks," *EURASIP Journal on Advances in Signal Processing*, vol. 2, no. 1, pp. 1–12, 2012.
8. A. Ghasemi and E. S. Sousa, "Spectrum sensing in cognitive radio networks: Requirements, challenges and design trade-offs," *IEEE Communications Magazine*, vol. 46, no. 4, pp. 32–39, 2008.

9. A. Ghasemi and E. S. Sousa, "Optimization of spectrum sensing for opportunistic spectrum access in cognitive radio networks," *Proceedings of the IEEE Consumer Communications and Networking Conference (CCNC)*, pp. 1022–1026, Las Vegas, NV, January 2009.

10. P. Wang, L. Xiao, S. Zhou, and J. Wang, "Optimization of detection time for channel efficiency in cognitive radio systems," *Proceedings of the IEEE Wireless Communications and Networking Conference (WCNC)*, pp. 111–115, Hong Kong, People's Republic of China, March 2010.

11. E. Jung and X. Liu, "Opportunistic spectrum access in heterogeneous user environments," *Proceedings of the IEEE DySPAN*, Chicago, IL, October 2010.

12. Y. Pei, A. T. Hoang, and Y.-C. Liang, "Sensing throughput tradeoff in cognitive radio networks: How frequently should spectrum sensing be carried out?" *Proceedings of the IEEE International Symposium on Personal, Indoor and Mobile Radio Communications (PIMRC)*, Athens, Greece, July 2010.

13. W.-Y. Lee and I. F. Akyildiz, "Optimal spectrum sensing framework for cognitive radio networks," *IEEE Transactions on Wireless Communications*, vol. 7, no. 10, pp. 3845–3857, 2010.

14. X. Liu and Z. Ding, "ESCAPE: A channel evacuation protocol for spectrum agile networks," *Proceedings of the IEEE DySPAN*, Dublin, Ireland, April 2010.

15. L. Luo and S. Roy, "Analysis of search schemes in cognitive radio," *Proceedings of the IEEE SECON*, pp. 647–654, San Diego, CA, June 2010.

16. H. Kim and K. G. Shin, "Efficient discovery of spectrum opportunities with MAC-layer sensing in cognitive radio networks," *IEEE Transactions on Mobile Computing*, vol. 7, no. 5, pp. 533–545, 2010.

17. H. Kim and K. G. Shin, "Fast discovery of spectrum opportunities in cognitive radio networks," *Proceedings of the IEEE DySPAN*, Chicago, IL, October 2010.

18. J. Jia, Q. Zhang, and X. Shen, "HC-MAC: A hardware-constrained cognitive MAC for efficient spectrum management," *IEEE Journal on Selected Areas in Communications*, vol. 26, no. 1, pp. 106–117, 2010.

19. C. Cordeiro and K. Challapali, "C-MAC: A cognitive MAC protocol for multi-channel wireless networks," *Proceedings of the IEEE DySPAN*, pp. 147–157, Dublin, Ireland, April 2011.

20. F. Digham, M. Alouini, and M. Simon, "On the energy detection of unknown signals over fading channels," *Proceedings of the IEEE ICC*, vol. 5, pp. 3575–3579, IEEE, Anchorage, AK, May 2010.

21. Z. Quan, S. Cui, and A. H. Sayed, "Optimal linear cooperation for spectrum sensing in cognitive radio networks," *IEEE Journal of Selected in Signal Processing*, vol. 2, no. 1, pp. 28–40, 2012.

22. M. Gandetto and C. Regazzoni, "Spectrum sensing: A distributed approach for cognitive terminals," *IEEE Journal on Selected Areas in Communications*, vol. 25, no. 3, pp. 546–557, 2007.

23. C. Chou, S. Shankar, H. Kim, and K. G. Shin, "What and how much to gain by spectrum agility?" *IEEE Journal on Selected Areas in Communications,* vol. 25, no. 3, pp. 576–588, 2010.
24. A. W. Min and K. G. Shin, "Exploiting multi-channel diversity in spectrum-agile networks," *Proceedings of the INFOCOM,* pp. 1921–1929, Phoenix, AZ, April 13–18, 2008.
25. D. Willkomm, S. Machiraju, J. Bolot, and A. Wolisz, "Primary users in cellular networks: A large-scale measurement study," *Proceedings of the IEEE DySPAN,* Chicago, IL, October 2010.
26. S. Geirhofer, L. Tong, and B. M. Sadler, "Dynamic spectrum access in the time domain: Modeling and exploiting white space," *IEEE Communications Magazine,* vol. 45, no. 5, pp. 66–72, 2007.
27. R. W. Brodersen et al., "Corvus: A cognitive radio approach for usage of virtual unlicensed spectrum," *Berkeley Wireless Research Center (BWRC) White paper,* 2008. Available: http://www.bwrc.eecs.berkeley.edu/MCMA.
28. W.-Y. Lee et al., "A spectrum decision framework for cognitive radio networks," *IEEE Transaction on Mobile Computing,* vol. 10, no. 2, pp. 161–174, 2011.
29. M. B. Pursley and T. C. Royster IV, "Low-complexity adaptive transmission for cognitive radios in dynamic spectrum access networks," *IEEE Journal on Selected Areas in Communications,* vol. 26, no. 1, pp. 83–94, 2010.
30. G. D. Nguyen and S. Kompella, "Channel sharing in cognitive radio networks," *Military Communications Conference,* pp. 2268–2273, San Jose, CA, October 31–November 3, 2010.
31. R. Etkin, A. Parekh, and D. Tse, "Spectrum sharing for unlicensed bands," *IEEE Journal of Selected Areas in Communications,* vol. 25, no. 3, pp. 517–528, 2010.
32. Z. Ji and K. J. R. Liu, "Dynamic spectrum sharing: A dynamic spectrum sharing: A game theoretical overview," *IEEE Communications Magazine,* vol. 4, no. 5, pp. 88–94, Baltimore, MD, November 8–11, 2009.
33. J. Huang, R. A. Berry, and M. L. Honig, "Spectrum sharing with distributed interference compensation," *Proceedings of the IEEE DySPAN,* pp. 88–93, November 2005.
34. L. Ma, C.-C. Shen, and B. Ryu, "Single-radio adaptive channel algorithm for spectrum agile wireless ad hoc networks," *Proceedings of the IEEE DySPAN,* pp. 547–558, Dublin, Ireland, April 2009.
35. L. Yang, L. Cao, and H. Zheng, "Proactive channel access in dynamic spectrum network," *Proceedings of the Cognitive Radio Oriented Wireless Networks and Communications (CROWNCOM),* pp. 482–486, Orlando, FL, August 2009.
36. F. K. Jondral, "Software-defined radio: Basic and evolution to cognitive radio," *EURASIP Journal on Wireless Communication and Networking,* vol. 2, no. 1, pp. 275–283, 2008.
37. K. Zhang, Y. Mao, and S. Leng, "Analysis of cognitive radio spectrum access with constraining interference," *International Conference on Computational Problem-Solving (ICCP),* pp. 226–231, Chengdu, People's Republic of China, 2011.

38. N. Baldo and M. Zorzi, "Cognitive network access using fuzzy decision making," *IEEE Transactions on Wireless Communications*, vol. 8, no. 7, pp. 3523–3535, 2010.

39. Y. Shi et al., "Distributed cross-layer optimization for cognitive radio networks," *IEEE Journal on Selected Areas in Communications*, vol. 59, no. 8, pp. 4058–4069, 2010.

40. S. Alrabaee, M. Khasawneh, A. Agarwal, N. Goel, and M. Zaman, "A game theory approach: Dynamic behaviours for spectrum management in cognitive radio network," IEEE *GLOBECOM Workshops (GC Wkshps)*, pp. 919–924, Anaheim, CA, December 3–7, 2012.

41. C. F. Shih, W. Liao, and H. Chao, "Joint routing and spectrum allocation for multi-hop cognitive radio networks with route robustness consideration," *IEEE Transactions on Wireless Communications*, vol. 10, no. 9, pp. 2940–2949, 2011.

42. S. Ju and J. B. Evans, "Scalable cognitive routing protocol for mobile ad-hoc networks," *Proceedings of the IEEE GLOBECOM*, Miami, FL, December 6–10, 2010.

43. Q. Zhao, L. Tong et al., "Decentrallized cognitive MAC opportunistic spectrum access in ad hoc networks: A POMDP framework," *IEEE Journal on Selected Areas in Communications*, vol. 25, no. 3, pp. 589–600, 2007.

Biographical Sketches

Tarek M. Salem (tareksalem@eri.sci.eg) completed his BS in systems and computer engineering, Faculty of Engineering, Al-Azhar University, in May 2013. During 2005–2012, he joined the Arabic Organization for Industrialization. In 2013, he was a research assistant at the Electronics Research Institute (ERI) in Egypt. He is an assistant researcher in computers and systems department at the ERI in Egypt. Now he is pursuing his PhD.

Sherine M. Abdel-Kader (sherine@eri.sci.eg) obtained her MSc and PhD degrees from the electronics and communications department and computers department, Faculty of Engineering, Cairo University, Giza, Egypt in 1998 and 2003, respectively. Dr. El-Kader is an associate professor of computers and systems department at the Electronics Research Institute (ERI) in Egypt. She is currently supervising 3 PhD students and 10 MSc students. She has published more than 25 papers and 4 book chapters on computer networking area. She is working on many computer networking hot topics such as WiMAX, WiFi, internet protocol (IP) mobility, quality of service (QoS), wireless sensor networks, ad hoc networking, real-time traffics, Bluetooth, and IPv6. She was an associate professor at the Faculty of Engineering, Akhbar El Yom Academy, Cairo, Egypt from 2007 to 2009. Also she is a technical reviewer for many international journals. She is heading the Internet and Networking unit at the ERI from 2003 until now.

Salah M. Abdel-Mageid (smageid@azhar.edu.eg) received his MS and PhD in systems and computers engineering from Al-Azhar University, Cairo, Egypt, in 2002 and 2005, respectively. Since 2012, he has been an associate professor in systems and computers engineering at Al-Azhar University. He performed his postdoctoral research in 2007 and 2008 in the computer science and engineering department, School of Engineering, Southern Methodist University at Dallas, Texas. His research interests include mobile computing, cellular networks, sensor networks, cognitive radio networks, and Internet services and applications.

Mohamed Zaki (azhar@mailer.scu.eun.eg) received his BSc and MSc degrees in electrical engineering from Cairo University, Giza, Egypt in 1968 and 1973, respectively. He received his PhD degree in computer engineering from Warsaw Technical University, Poland, in 1977. He is a professor of software engineering, computer and system engineering department, Faculty of Engineering, Al-Azhar University, Cairo. His fields of interest include artificial intelligence, soft computing, and distributed system.

2

A SURVEY ON JOINT ROUTING AND DYNAMIC SPECTRUM ACCESS IN COGNITIVE RADIO NETWORKS

XIANZHONG XIE, HELIN YANG, AND ATHANASIOS V. VASILAKOS

Contents

2.1 Introduction

The utilization of wireless spectrum has been growing rapidly with the dramatic development of the wireless telecommunication industry in the past few decades. Recently, the Federal Communications Commission (FCC) has recognized that traditional fixed spectrum allocation can be very inefficient with the bandwidth demands varying highly along the time or space dimension. Therefore, the FCC has considered more flexible and comprehensive uses of the available spectrum [1] through the use of cognitive radio (CR) technology [2].

CR technologies have the potential to provide wireless devices with various capabilities, such as frequency agility, adaptive modulation, transmit power control, and localization, which make spectrum access (SA) more efficient and intensive. With the development of CR technologies, dynamic SA (DSA) becomes a promising approach to increase the efficiency of spectrum usage [2,3]. In DSA, a piece of spectrum can be allocated to one or more users such as primary users (PUs) that have higher priority in using it; however, the use of that spectrum is not exclusively granted to these users. Other users such as secondary users (SUs) can also access the spectrum as long as the PUs are not temporally using it or the PUs can properly be protected. By the way, the radio spectrum can be reused in an opportunistic manner or shared all the time; thus, the spectrum utilization efficiency can significantly be improved.

In traditional wireless networks, all network nodes will be provided with a certain fixed spectrum band. However, there may be no such preallocated spectrum that can be used by every node at any time in DSA networks. This new feature of DSA network imposes even greater challenges on wireless networking, especially on routing.

Therefore, most of the routing schemes in CRNs focus on the joint dynamic spectrum allocation and path establishment by monitoring the PU activity pattern [4]. Taking the PU avoidance proposed in [5] as the first consideration, SEARCH routing protocol leverages the geographic information for greedy forwarding. Based on the interference model, a joint routing, opportunistic spectrum scheduling, and time-sharing scheduling algorithm to minimize the aggregate interference from SUs to a PU with the aberration was proposed in [6]. A routing and dynamic spectrum allocation (ROSA) algorithm was proposed in [7], which is a joint spectrum allocation scheduling and transmission power control scheme to maximize the network throughput and limit the physical interference to the users.

In addition, the dynamic nature of the radio spectrum calls for the development of novel spectrum-aware routing algorithms, with multihop communication requirements in CRNs. In fact, spectrum occupancy is location dependent; therefore, the available spectrum bands may be different at each relay node in a multihop path. Hence, controlling the interaction between the routing and the spectrum management is of fundamental importance.

A stability-aware routing protocol (STRAP) was developed for multihop DSA networks to utilize the unused frequency bands without compromising the stability of the network [8]. Another proposal is the spectrum-aware routing protocol (SPEAR), which can establish the robust paths even in the diverse spectrum environment under rather stringent latency conditions [9]. These protocols address similar problems as they operate in a multichannel context and face the multiple channel hidden terminal problems [10].

With respect to these previous studies, our main objective is to provide readers with a systematic overview of joint routing and SA for CRNs. The application of SUs to DSA has been actively studied over the past several years, covering diverse scenarios. However, because the spectrum availability may change from time to time and hop by hop, minimizing the aggregate interference from SUs to PUs with the aberration is a challenge. Therefore, the routing and spectrum access algorithm enabling the throughput maximization, taking care of the interference minimization, and maximizing the weighted sum of differential backlogs the system stay stable. Also, media access control (MAC) has an important role in several CR functions: spectrum mobility, routing, and SA.

In this chapter, we address this problem and thus help map the design space of joint routing and DSA in CRNs. We can summarize the contributions of this work as follows:

First, we present the key enabling technologies of CRNs and dissect mainly the proposed several issues about SA through the description of CRN architecture.

We also identify the most common techniques that are used for solving the SA problem in CRNs. Further, we summarize their characteristics, advantages, and disadvantages of the existing SA schemes: Monrovian model, graph theory, game theory, and artificial intelligence algorithms.

We then proceed to map the opportunistic routing design space by drawing the three general classical routing solutions (multihop routing, ad hoc network routing, and adaptive routing) for CRNs. Moreover, we identify that the ad hoc network routing is further classified into three subclasses: optimization modeling, probabilistic modeling, and graph modeling.

Based on the presentation of the SA and routing, we break through the existing routing classification criteria such as multipath routing algorithm, quality-of-service (QoS) routing algorithm, routing algorithm based on geographical information, and power-aware routing algorithm, and summarize the latest research of routing technology research for DSA.

Finally, we discuss the open research issues and challenges of joint routing and SA in order to provide the basic direction for future research works.

The rest of the chapter is structured as follows: Sections 2.2 and 2.3 provides a brief overview of the CR technology, CRN architecture and its functionalities. Section 2.4 presents a comprehensive survey of SA algorithms and routing schemes for CRNs in detail. It also discusses the advantages and limitations of the existing works. Section 2.5 reviews the existing works in joint routing and SA, and presents a global CR routing protocol classification for SA. Section 2.6 presents some challenges that must be addressed to enable the performance of CRNs with some open issues for future research in this area. Finally, Section 2.7 concludes the chapter.

2.2 CR Technology

The key enabling technologies of CRNs are the CR techniques that provide the capability to share the spectrum in an opportunistic manner. Formally, a CR can change its transmitter parameters based on the interaction with its environment [1]. A typical duty cycle of CR, as illustrated in Figure 2.1, includes detecting spectrum white space, selecting the best frequency bands, coordinating SA with other users, and vacating the frequency when a PU appears.

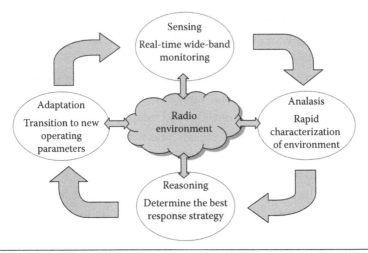

Figure 2.1   Cognitive cycle.

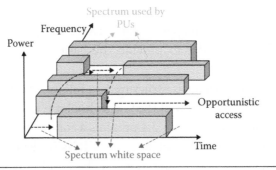

Figure 2.2   The spectrum hole concept.

From this definition, two main characteristics of CR can be defined as follows [11]:

- *Cognitive capability*: Through spectrum sensing and analysis, on the one hand, SUs can detect the spectrum white space (see Figure 2.2) when a portion of frequency band is not being used by the PUs. On the other hand, when PUs start using the licensed spectrum again, SUs can detect their activity. Therefore, no harmful interference is generated due to SU's transmission. After recognizing the spectrum white space by sensing, spectrum management and handoff function of CR enable SUs to choose the best frequency band and hop to meet various QoS requirements [12]. Consequently, the best spectrum can be selected, shared with other users, and exploited without interference to the PUs. For instance, when a PU reuses its frequency band, the SUs using the licensed band can direct their transmission to other available frequencies, according to the channel capacity determined by the noise and interference levels, path loss, channel error rate, and holding time.

- *Reconfigurability*: A CR can be programmed to transmit and receive signals at various frequencies using different access technologies supported by its hardware design [13]. Through this capability, the best spectrum band and the most appropriate operating parameters can be selected and reconfigured.

2.3 CRN Architecture

2.3.1 Network Components

A CRN architecture basically consists of primary networks (PNs) and secondary networks (SNs) [12,14], as shown in Figure 2.3.

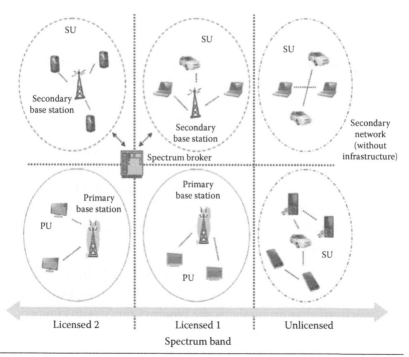

Figure 2.3 Network architecture of dynamic spectrum sharing.

A PN is composed of a set of PUs and one or more primary base stations, where PUs are authorized to use certain licensed spectrum bands under the coordination of primary base stations. An SN is a cognitive network whose components do not have license to access any frequency bands. SNs can be split into infrastructure and ad hoc networks that are operated by network operators or stand-alone users, respectively. In addition, their transmission should not be interfered by SNs. PUs and primary base stations are in general not equipped with CR functions. Therefore, if an SN shares a licensed spectrum band with a PN, besides detecting the spectrum white space and utilizing the best spectrum band, it immediately detects the presence of a PU and directs the secondary transmission to another available band.

In infrastructure mode, secondary base stations provide one-hop communication to SUs, have the ability to discover spectrum holes, and operate in the most suitable available band in order to avoid interfering with the PNs. An example of infrastructure CRN architecture is the IEEE 802.22 network [15].

2.3.2 Spectrum Heterogeneity

SUs are capable of accessing both the licensed portions of the spectrum used by PUs and the unlicensed portions of the spectrum. Consequently, the operation types of CRNs can be classified into licensed band operation and unlicensed band operation.

- *Licensed band operation*: The licensed band is primarily used by the PN. Hence, CRNs are mainly used for SUs for the detection of PUs in this case. That means, if PUs appear in the spectrum band occupied by SUs, SUs should vacate that spectrum band and move to an available spectrum immediately.

- *Unlicensed band operation*: In the absence of PUs, SUs have the same right to access the spectrum. Hence, sophisticated spectrum sharing methods are required for SUs to compete for the unlicensed band.

2.3.3 Network Heterogeneity

The SUs have the opportunity to perform the following three different access types:

- *CRN access*: SUs can access their own CR base station on both licensed and unlicensed spectrum bands. Because all interactions occur inside the CRNs, their spectrum sharing policy can be independent of that of the PNs.
- *CR ad hoc access*: SUs can communicate with other SUs through an ad hoc connection on both licensed and unlicensed spectrum bands.
- *PN access*: SUs can also access the primary base station through the licensed band. Unlike for other access types, they require an adaptive MAC protocol, which enables roaming over multiple PNs with different access technologies.

2.3.4 Spectrum Assignment Framework

2.3.4.1 Spectrum Sharing Architecture Both centralized and distributed spectrum sharing techniques are enabled in CRNs. They are described as follows:

- *Centralized spectrum sharing*: A centralized entity controls the spectrum allocation and access procedures, with a spectrum allocation map being constructed based on the measurements from its controlled entities.
- *Distributed spectrum sharing*: In the cases where constructing an infrastructure is not preferable, distributed spectrum sharing will be performed, with each SU being responsible for spectrum allocation or access based on its local measurements.
- *External sensing*: An external agent performs the sensing and broadcasts the channel occupancy information to the CR.

2.3.4.2 Spectrum Allocation The available spectrum can be allocated to a CR user in either a cooperative manner or a noncooperative manner. Once a spectrum band is allocated, the receiver of this communication must be informed about the spectrum allocation, that is, a transmitter–receiver handshake protocol is essentially required.

- *Cooperative spectrum allocation*: In this allocation, local measurements (e.g., interference measurement) of each SU are shared with the other users to improve the spectrum utilization.
- *Noncooperative spectrum allocation*: Also known as a selfish spectrum allocation scheme, it performs spectrum sharing by considering only the user at hand.

2.3.4.3 Spectrum Access   In the presence of multiple SUs trying to access the same spectrum simultaneously, SA should be coordinated in order to prevent multiple users colliding in overlapping portions of the spectrum. Both overlay and underlay SA techniques are enabled in CRNs.

- *Overlay SA*: An SU accesses the network using spectrum bands that are not used by PUs.
- *Underlay SA*: Sophisticated spread spectrum techniques are exploited to improve the spectrum utilization, with the signal of SUs being regarded as noise by the PUs.

2.3.4.4 Spectrum Sharing   This technology is generally focused on two types of solutions: intranetwork (spectrum sharing inside a CRN) and internetwork (spectrum sharing among multiple coexisting CRNs).

- *Intranetwork spectrum sharing*: It focuses on spectrum allocation among the users of a CRN.
- *Internetwork spectrum sharing*: It enables multiple systems to be developed in overlapping locations and spectrum.

In a more general sense, a CRN can be perceived as an intelligent overlay network that contains multiple coexisting networks. Thus, building a fully functioning CRN can be a very challenging task, due to the difficulties in designing multiple system components, including, but not limited to, physical (PHY) layer signal processing, MAC layer spectrum management, and network layer routing and statistical control. Furthermore, these system components often interact in complex ways, which may require cross-layer design and control frameworks.

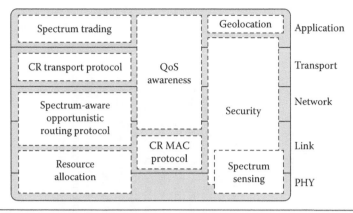

Figure 2.4   The architecture of CR in the layered model.

Figure 2.4 illustrates the key functions of the PHY, MAC, and network layers in a CRN. In the PHY layer, spectrum sensing is the essential component that enables SUs to identify spectrum holes. Cognitive SA is carried out through transceiver optimization and reconfiguration. The specific tasks that the MAC layer of an SU must perform include sensing scheduling and spectrum-aware access control. The spectrum sensing scheduler controls the sensing operations, whereas the spectrum-aware access control governs the SA to the identified spectrum holes. The sensing–access coordinator controls the operations of these two functions on a time basis by taking care of the trade-off between the sensing requirement and the SA opportunity that the SUs may achieve.

Finally, the spectrum manager links the three layers and supports the access of available spectrum in a dynamic and efficient manner. Although the aforementioned architecture is by no means the only architecture that can be designed for CRNs, instead, it serves as the functional architecture for this chapter. For other architectures, see [12] and [16].

2.4 SA for CRNs

In this section, the most common techniques used for solving the SA problem in CRNs are presented. Table 2.1 presents a summary of these techniques, their characteristics, advantages, and disadvantages.

Table 2.1　Summary of Main SA Techniques

TECHNIQUE	CHARACTERISTICS	STRENGTHS	LIMITATIONS
Markovian model	Based on the previous knowledge, the Markovian model determines the observation duration for the specific application and makes the SA decision	Can model complicated statistical process; good for decision strategy; easily scalable; can predict based on experience	Requires good training sequence; computationally complex
Graph theory	CRN can be modeled as a graph, using dynamic interference graph model to capture the SA situation. Graph coloring is used for spectrum decision making	Characterizes the impact of interference among users easily	Simplified assumptions; cannot guarantee all necessary parameters in CRN, such as adjacent-channel interference and QoS requirement; high computational complex
Game theory	It analyzes the strategic behavior of PUs and SUs. It sets the utility functions that account for fairness and efficiency among the players. Solution is found through Nash Equilibrium (NE)	Reflects the behaviors of decision makers easily; considers the fairness among users; can be used for both cooperative and noncooperative decisions between SUs	NE point is not always guaranteed; the utility function and the game formulation are not always structured actually due to other factors, such as cheating users
Artificial intelligence algorithms	These are stochastic search methods that mimic natural evolution and social behavior. They find the optimal parameters of SA model	Excellent for parameter optimization and learning invoicing relationship between parameter values; conceptually easy to scale; can handle the arbitrary kinds of constraints and objectives	Training may be slow depending on the network size; a possible risk includes finding local minimas

2.4.1 Monrovian Model

An approach using reinforcement learning for the detection of spectral resources in a multiband CR scenario is investigated in [17], which introduces the optimal detection strategy by solving a Markov decision process (MDP). A medium access control layer SA algorithm based on knowledge reasoning is considered, where the optimal range of channels is determined through proactive accurate access and channel quality information.

2.4.1.1 Decentralized Partially Observable MDP Framework   An SU can obtain a better estimate about the primary channel occupancy based on the observations of the channels not only on the current but also on the past observations of the channels, if the primary traffic exhibits some temporal correlation. In particular, if a channel is characterized at each instant to be either idle (state 0) or busy (state 1), the state transitions may be modeled as a Markov chain, where optimal sensing policy can be obtained by modeling the system as a partially observable MDP (POMDP) [18], which can optimize SUs' performance, accommodate spectrum sensing error, and protect PUs from harmful interference. Considering the complexity of the POMDP scheme [19], a modified scheme based on POMDP framework reduces the complexity of the POMDP formulation by decoupling the design of sensing strategy from the design of the access strategy. In addition, the authors showed that a separation principle decouples the design of sensing policy from that of SA policy. A somewhat similar approach can also be found in [20], which considers an unspotted PN. An extension of the scheme in [18] is presented in [21], which incorporates the SUs' residual energy and buffer state in the POMDP formulation for spectrum sensing and access.

Without loss of generally, most of the POMDP schemes are in the decentralized cognitive MAC. Zhao et al. proposed a channel sensing/access policy that considers the partial knowledge of licensed channel state at SUs and handles spectrum sensing errors limiting interference to PN. Hence, the proposed decentralized cognitive (DC)-MAC operations are represented in Figure 2.5 [18], which are resumed as follows:

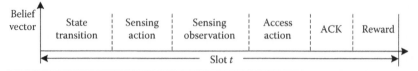

Figure 2.5   The sequence of operations in a slot.

1. At the beginning of each slot, the transmitter and the receiver select one channel to sense according to the belief vector. The two users exploit the same belief vector to ensure that they tune to the same channel.

2. If the sensed channel is available, the transmitter generates a random back-off time and the sender continues to monitor the channel during this period. If the channel remains idle, the transmitter starts a request-to-send/clear-to-send (RTS/CTS) handshake to verify whether the sensed channel is also available at the receiver side.

3. The transmitter sends data over channel. If the data are successfully received, the receiver transmits an acknowledgment message. Finally, both the sender and the receiver update their belief vector.

DC-MAC is one of the few opportunistic MAC protocols that include sensing errors in its design; however, its implementation is limited by the assumption that the transition probability in the Markov channel model is known. In practice, this may not be available for high collision frequency and high drop rate among SUs.

However, these existing results in those studies do not take into account the fading of wireless channels, which may affect the spectrum sensing polices. Also, the time consumption of spectrum sensing and the computational complexity have not been considered. Therefore, Wang et al. [22] proposed a heuristic policy to find the optimal sensing policy based on the theory of POMDP with comparable performance and low complexity. Moreover, strong coupling between channel fading and PU occupancy exists and impacts the trade-offs under the new setting. The scheduling process, in the presence of such coupling, can become even more complicated when temporal correlation in both sets of system states are included, particularly when such memory bears a long-term correlation structure. Wang and

Zhang [23] formulated an optimal POMDP to examine the intricate trade-offs in the optimal scheduling process, when incorporating the temporal correlation in both the channel fading and PU occupancy states.

2.4.1.2 Centralized POMDP Framework   The POMDP framework for the DC-MAC requires global synchronization, which gives a very low throughput when collision probability is high and the number of channels to be sensed per time slot is not optimized. In order to solve the problem, Prabhjot et al. [24] considered probabilistic channel availability in case of licensed channel detection for single-channel allocation, whereas variable data rates are considered using channel aggregation technique in the multiple channel access models. These models are designed for a centralized architecture to enable dynamic spectrum allocation and are compared based on access latency and service duration. Figure 2.6 shows the Markov queuing models.

Prabhjot et al. modeled the functionalities of media access at Base Station (BS) using distributed queuing where the first queue is equivalent of sorting overlapped detections received from all SUs in the network, updating its database, contention, and conflict management. We call this queue as SA queue (SAQ). This queue receives the channel requests from the SU at an arrival rate of λ_1 and from SU at an arrival rate of λ_2. The requests once sorted will be considered for channel allocation, which is modeled as channel allocation queue (CAQ). Thus, at the output of CAQ there is a random splitting with probability p if the serving request is a feedback to SAQ and with probability q if it is a feedback to CAQ, where, for both p and q, $0 < p, q < 1$.

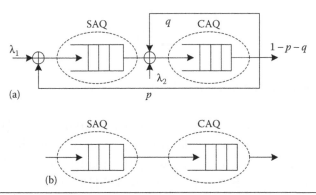

Figure 2.6   Queuing models for SCDQMAC (a) and MCDQMAC (b).

Due to the consideration of allocating only one channel to an SU at any given time, it is referred to as single-channel distributed queuing-based MAC (SCDQMAC) model. This model is shown in Figure 2.6a.

Due to this feature of the model to allocate multiple channels to an SU and the use of distributed queuing, it is referred to as multichannel distributed queuing-based MAC (MCDQMAC) model. The queuing model for MCDQMAC is shown in Figure 2.6b.

Also, Kaur and Khosla [25] proposed a Monrovian queuing model for dynamic spectrum allocation. In a centralized architecture to overcome the hidden terminal problem and to obtain complete information of unused frequency (spectrum hole), sensing is considered to be decentralized (Figure 2.7).

These queues are special cases of stochastic processes, characterized by an arrival process of service requests and a waiting list of requests to be processed. The queue stacking all entries of SUs is referred to as SAQ, and all the requests entering this queue are served on first-come first-serve (FCFS) basis. At any time when bandwidth needs be allocated to the SU, the head considers both the requests from the SU and the PU who need its licensed channel. Thus, while distributing a number of frequencies to the PU and SU, the arrival rates of both the users are summed to access the frequencies with the Head. The queue so formed is referred to as CAQ. They use Markov process to analyze the queuing models. The blocking probability P_i for the bandwidth request of the channel state i made by the CR that finds all the channels with the Head as occupied is given by Erlang-B formula as given below:

$$P_i = \frac{\rho_2^i}{i! \sum_{i=0}^{S} (\rho_2^i / i!)}$$

where:

ρ_2 is the traffic intensity for CAQ

2.4.1.3 Hidden Markov Model The hidden Markov model (HMM) [26] can then be used to identify the sequences of observations with the same pattern by choosing the model that would most likely produce the observed sequences. HMMs have been applied to CR spectrum occupancy prediction research [27,28]. Akbar and

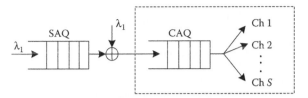

Figure 2.7 The equivalent model of network of queues. (P. Kaur and A. Khosla, *IACSIT International Journal of Engineering and Technology,* 3(1), 1–4, 2011.)

Tranter [27] developed an HMM-based DSA algorithm where the HMM models and predicts the spectrum occupancy of the licensed radio bands for CRNs. In addition, a channel status predictor using HMM-based pattern recognition was proposed in [28]. However, none of the previous work takes the time delay incurred by hardware platforms into consideration for prediction. Moreover, the ideas of using HMM for prediction in the previous work are all based on pattern recognition. In fact, HMM can be exploited beyond pattern recognition. Therefore, a modified HMM-based single SU prediction was proposed and examined in [29]. The scheme considered the time delays that undermine the accuracy of spectrum sensing. Also, the HMM has some limitations: It requires good training sequence and it is computationally complex. Therefore, Li et al. [30] proposed an HMM scheme based on myopic channel sensing policy to estimate the parameters of primary channel Markov models with a linear complexity (low complexity).

2.4.2 Game Theory

Different approaches that have been used to model the strategic interactions for spectrum sharing are shown in Figure 2.8. Game theory serves as a powerful tool in order to model the strategic behavior of PUs and SUs for their coexistence. The economic models include principles such as setting the price of the spectrum available in order to maximize the revenue of the PUs, choosing the best seller for the spectrum in order to maximize the satisfaction from the usage of the spectrum for SUs, modeling the market competition, and so on. Therefore, many of the existing literature on modeling the economic interactions in wireless networks use a combination of game theory, price theory, and market theory.

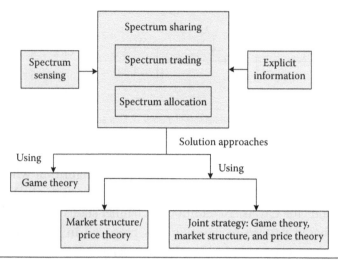

Figure 2.8 Solution for spectrum sharing in CRNs.

Game theory is a well-developed mathematical framework that has been widely applied to the intelligent behaviors of rational decision makers in strategic interactions, where the action of one actor or player impacts (and perhaps conflicts with) that of other components. In a CRN, wireless services are provided to SUs that seek to achieve the maximum performance if they do not belong to the same network entity. Therefore, game theoretical approaches can be used to solve the DSA problem [31,32]. There are two types of games—cooperative and noncooperative—based on whether the users exchange information regarding their decisions or not, respectively. In the literature, most related SA algorithms formulate a game and try to find the optimal solution through the Nash equilibrium (NE) [33].

A no-regret learning game approach [34] was proposed to analyze the behavior of cognitive users in distributed adaptive channel allocation, and both cooperative and noncooperative scenarios are considered.

2.4.2.1 Repeated Game The simplest game theoretic model that captures these concepts is that of a repeated game. A repeated game is one in which each stage of the game is repeated, usually with an innate time horizon.

R. Etkin et al. proposed a repeated game approach [35–37] for spectrum allocations. The spectrum sharing strategy could be enforced using the NE of dynamic games. In this model, SUs will have to vacate their current channel whenever a PU becomes active, and R. Etkin et al. considered a cost associated with switching channels.

Channel selection in opportunistic SA has also often been modeled as a repeated game. Mitola [2] modeled the sharing of open spectrum as a repeated game. They considered a punishment scheme and showed that a more efficient equilibrium can be reached when autonomous radios interact repeatedly, as opposed to when they interact in a single-stage game (in general, a well-known result in game theory, an example of which is the repeated prisoner's dilemma). They continued further and also considered the incentives for CRs to truthfully report their operating conditions in negotiating access to spectrum: relying on mechanism design, Mitola [2] designed cheat-proof strategies for dynamic spectrum sharing. The selection of the best spectrum opportunities by SUs of some spectrum band is modeled as a repeated game in [3]. In this model [3], SUs will have to vacate their current channel whenever a PU becomes active, and Akbar and Tranter [27] consider a cost associated with switching channels. A subgame perfect equilibrium, an NE that is also equilibrium for every proper subgame of the original game, is one way to characterize the likely outcome of such a game.

In [4], C.-T. Chou et al. use repeated games to model the evolution of reputation among SUs, when one of them is chosen to manage the spectrum made available by the PU. In several of the applications mentioned earlier, repeated interactions among a set of CRs allow for the design of incentive mechanisms that lead to a more efficient equilibrium. A different question is whether there are simple ways for a radio, by observing others' actions and the utility resulting from its own actions, to converge to an NE. We treat that question next.

Relatively recent developments in games of imperfect public and private information have the potential to yield new insight into what we have termed the price of ignorance [38]. Therefore, spectrum pricing problem with analysis of the market equilibrium is studied in [39] and [40]. The utility functions that account for fairness and efficiency among the players were presented in [41] and [42] to prove that cooperative game can achieve better spectrum sharing performance than noncooperative game.

2.4.2.2 Potential Games   The class of games called potential games is of particular interest in the context of learning. If a dynamic adaptation problem can be modeled as a potential game, then if radios follow a simple adaptation algorithm (which we will discuss in more detail shortly), they are guaranteed to reach a solution that is stable from the point of view of the entire network.

Even when players have utility functions that reflect their own selfish interests, rather than those of the network, in a number of cases of interest to DSA and cognitive network games the model results in a potential game. Thomas et al. [43], for example, modeled the topology control problem for an ad hoc network where nodes can select a channel to operate on from a finite set of available channels. This topology control mechanism consists of two phases: in the first phase, radios select a transmit power level with energy efficiency and network connectivity in mind, and in the second phase, they select channels with interference minimization objectives. Thomas et al. [43] are able to show that both problems (power control and channel selection) can be formulated as ordinal potential games, and best-response dynamics are guaranteed to converge to equilibrium.

However, the computational complexity for DSA and the need of centralized controllers can often be overwhelming or even prohibitive. To address these problems, a recent work [44] has proposed a two-tier market model based on the decentralized bargain theory, where the spectrum is traded from a PU to multiple SUs on a larger timescale, and then redistributed among SUs on a smaller timescale. A mechanism design was proposed in [45] and [46] to suppress the cheating behavior of SUs in open SA by introducing the function to user's utility.

In general, game theory has been widely used in cognitive SA algorithms, which can be used for both cooperative and noncooperative decisions between SUs. The main disadvantage of this approach is that the utility function and the game formulation must be very carefully structured to achieve equilibrium. Furthermore, the performance in the equilibrium is also affected by the game formulation and the utility function.

Gale–Shapley theorem: This theorem for SA in CRNs has also been proposed with a game theory technique called *stable matching* in [47]. The stable matching theory was proposed for studying the stability of marriage and the one-to-one function. This approach does not consider the dynamic nature of CRNs, which results from the mobility of the users, the dynamic service requirements of the users, and the existence of primary transmissions.

Auction mechanism: This mechanism for spectrum sharing has also been proposed in [48]. It is an applied branch of economics studying the way people behave in auction markets. A real-time spectrum auction framework was proposed in [49] to assign spectrum packages to proper wireless users under interference constraints. In [50] and [51], a belief-assisted double auction mechanism was proposed to achieve efficient dynamic spectrum allocation, with collusion-resistant strategies that combat possible user collusive behavior using optimal reserve prices. In [52], the auction framework that allows SUs share the available spectrum of PUs acts as a resource provider announcing a price and a reserve bid to allocate the received power as a function of the bids. Effective mechanisms to suppress dishonest or collusive behaviors are also considered in [53], in case SUs distort their valuations about spectrum resources and interference relationships.

2.4.3 Graph Theory

Graph theory algorithms: A cognitive network can be modeled as a graph $G = (U, E_C, L_B)$, where U is the set of users sharing the spectrum, L_B represents the channel availability list at each vertex, and E_C is the set of edges modeling the interference constraints. A list coloring spectrum allocation algorithm based on graph coloring theory was proposed in [54].

How to effectively manage and use these spectrums based on graph theory becomes the key problem in CR system. A more scalable scheme called dynamic interference graph allocation (DIGA) can be used [42,55,56], which is based on a dynamic interference graph that captures the aggregated interference effects when multiple transmissions simultaneously happen on a channel. Figure 2.9 shows an example of how to construct an interference graph. Assume that there are three PUs, each occupying one of the three channels. An SU within the coverage area of a PU cannot use the channel occupied by that PU. For example, channels available to SU *A* are (1,2). Furthermore, neighboring SUs (indicated by a line connecting two

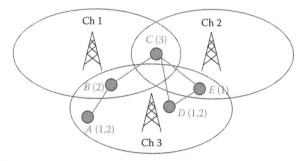

Figure 2.9 Spatial spectrum opportunity sharing among SUs formulated as a graph coloring problem.

SUs in the figure) interfere with each other if they access the same channel. The problem is how to allocate available channels to SUs to optimize certain network utility such as sum capacity under fairness constraints. Zheng and Peng [57] proposed different centralized and distributed strategies to optimize system throughput and fairness while minimizing interference. For instance, given $u,v \in U$, if u and v interfere when using simultaneously a channel m, an edge between u and v is labeled m, where m is an element of E_C. Figure 2.10 illustrates an example of a graph [57] according to this representation. In the proposed schemes, channel assignment follows the order of the nodes to maximize the system utility.

Interference graphs are commonly used in centralized approaches where the spectrum server constructs the graph and assigns the channels. In distributed approaches, the SUs themselves form the sets of available channels and negotiate with their neighbors which spectrum bands to select in order to avoid interference between the links and

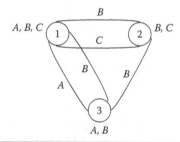

Figure 2.10 A Network color graph (NCG) representation of channel availability and interference constraints. (H. Zheng and C. Peng, "Collaboration and fairness in opportunistic spectrum access," *Proceedings of the IEEE International Conference on Communications*, vol. 5, pp. 3132–3136, May 2005.)

maximize their performance. Double hopping [58] is a DIGA scheme that proposes a distributed algorithm to generate a hopping pattern that minimizes the number of used frequencies. Maximizing the number of used channels for a transmission would reduce the interference to PUs. Although there are many algorithms proposed so far in the literature using interference graphs, to our knowledge, they all consider a network with SUs only. Future works should be extended to also include PUs in the graphs to limit the sets of available channels in links in the neighborhood of operating PUs.

Graph coloring is a commonly used technique for decision making [59–61], where the cognitive network is mapped to a graph, which is either unidirectional or bidirectional according to the characteristics of the algorithms. The algorithm considers the color list and the color rewards (bandwidth and throughput). In each stage, the algorithm labels all the nodes with a nonempty color list according to a labeling rule. Each label is associated with a color. The algorithm selects the node with the highest label and assigns the color (e.g., color). It then deletes the color from the node's color list, and also from the colonists of its color-constrained neighbors. When a node allocates colors, the change of the color list of its neighbors should be registered. The labels of the colored node and its neighbor nodes are modified according to the new graph. The algorithm enters the next stage until every node's color list becomes empty. In the following, we introduce two labeling rules. Although this is generally accepted, it does not reflect the adjacent-channel Interference (ACI) that causes severe performance degradation when links are close to each other. Future work in graph coloring-based cognitive SA should incorporate another layer in the color graph corresponding to the ACI and prevent the connected vertices from using either the same or adjacent spectrum bands.

Many previous works have illustrated the problem of spatial spectrum opportunity sharing, which is equivalent to graph coloring shown in [57] and [62]. Specifically, SUs form vertices in a graph and an edge between two vertices indicates two interfering users. Treating each channel as a color leads to a graph coloring problem: Color each vertex using a number of colors from its color list under the constraint that two vertices linked by an edge cannot share the same color to obtain a color assignment that maximizes a given utility function. Obtaining the optimal coloring is known to be NP-hard.

In addition, in order to improve the performance, many researchers also used evolutionary algorithms to enhance the graph coloring performance. Game theory provides another approach to spatial opportunity allocation [63]. An interesting connection between the resultant colored graph and the NE of the corresponding game is noted in [63]. Nguyen and Lee [61] proposed a suboptimal heuristic greedy algorithm with a much lower complexity and reasonable performance, which is based on the coloring interference graph constructed for unnerved SUs under the negative influence of all serviced SUs and active PUs. Also, the paper [61] proposed the greedy joint scheduling and power control algorithm, which is completely based on the adaptive construction of the coloring interference graph for all channels. This algorithm tries to fast update the coloring interference graph after each SU is served.

Recently, a color-sensitive graph coloring spectrum allocation algorithm has been proposed in [56] and [64], which takes the combination of benefit and fairness into account. But the running time was too long in allocating channels. In [57], an optimal algorithm to compute maximum throughput solutions was proposed based on a multichannel contention graph to characterize the impact of interference. The algorithm proposed in [65] regarded that the different users would have different reward values after obtaining different channels. However, it had a poor adaptability in consideration of the dynamic changes of the open spectrum management system. Therefore, the algorithm based on graph coloring model could achieve a good trade-off between throughput and fairness while ensuring interference-free transmission.

2.4.4 Artificial Intelligence

Artificial intelligence algorithms are stochastic search methods that mimic the natural evolution and the social behavior of species, such as artificial neural network (ANN) algorithm, genetic algorithm (GA), ant colony optimization (ACO) algorithm, and artificial bee colony (ABC) algorithm. The algorithms are widely used to optimize the radio parameters for DSA and solve the SA problem.

The idea of ANN was then applied to computational models, which is nothing more than a set of nonlinear functions with adjustable parameters to give a desired output. Because of their ability to dynamically adapt and be trained at any time, ANNs are able to *learn* patterns, features, and attributes of the system they describe. They have also been used for radio parameter adaptation in CR [66,67] and SA behavior learning [68]. In [66,67], the ANN determines the radio parameters for given channel states with three optimization goals, including meeting the BER, maximizing the throughput, and minimizing the transmit power. In [68], a practical spectrum allocation behavior learning method based on multilayer perception ANN is introduced, by which the state of different channels in future timeslots (either idle or busy) can be forecasted through supervised learning such that CR nodes can create a handover channel list in advance without interruption to its ongoing transmission.

GAs are random search techniques used for finding optimal solutions to problems such as cognitive SA. Park et al. [69] validated the applicability of GA-based radio parameter adaptation for the CDMA2000 forward link in a realistic scenario with Rician fading. Thilakawardana and Moessner [70] investigated a GA-based cell-by-cell dynamic spectrum allocation scheme to achieve a better spectral efficiency than the fixed spectrum allocation scheme. Kim et al. [71] implemented a software for CR with the spectrum-sensing capability and a GA to optimize the radio parameters for DSA. They are based on the principles of evolution and genetics that are different from other optimization techniques. This means that the *fitter* individual has higher probability to survive. To solve optimization problems, GA uses fitness functions and requires the parameters to be coded as chromosomes or finite-length strings over a finite alphabet, which are collected in groups called *populations*. The advantage of using GAs to solve the optimization problem of spectrum assignment in CR is that they can handle arbitrary kinds of constraints and objectives. However, one of the major disadvantages associated with GA is that the process for finding the optimal solution is quite slow and there is always the risk of finding not the globally optimal solution.

Moreover, Changchang and Yunxiao [72] presented a niche adaptive GA for CR spectrum allocation. The adaptive niche GA has the following features: (1) adaptively and dynamically adjusting the crossover probability and mutation probability and (2) dividing the population into several niches, each niche evolving parallelly.

Considering the fact that the cognitive ability of SUs in CRN is similar to the intelligent behaviors of ant colonies in a biological environment. In [73–75], ant colony algorithm is used to solve the spectrum allocation problem in CRNs. ACO is a well-known metaheuristic in which a colony of ants cooperates in exploring good solutions to a combinatorial optimization problem. In addition, the proposed algorithm is distributed with each node selecting its channel based on the pheromones received by its neighbors. The target objective is to maximize the total probability of successful transmissions, but something that does not guarantee the QoS of the transmissions.

ABC algorithm is a swarm intelligent optimization algorithm inspired by honey bee foraging, which introduces the concepts of employed bees, onlookers, and scouts, and enhances the performance of a search method by using a memory structure. Where, employed bees are equal to the number of food sources. Onlookers share the information of the food sources and explore. Scouts are employed bees that search new food sources, abandoning their own. In [75], a general model and utility functions for optimizing efficiency and fairness in spectrum allocation of CR by ABC algorithm are defined. In order to decrease the search space, a mapping process is proposed between the channel assignment matrix and the position of the bees of ABC based on the characteristics of the channel availability and the interference constraints.

Fuzzy algorithm is a commonly used technique for decision making and optimization algorithms in SA [76–78]. Fuzzy inference theory is also employed for two goals: One is the introduction of preference information on diverse assessment variables. The preference given in the form of a weight vector indicates an unequal importance between variables. The other goal is to overcome the limitation of nonuniform measurement standards due to multiple variables, and then calculate the integrated utility on SA decision.

Figure 2.11 shows the structure of a fuzzy logic system (FLS) [76]. When an input is applied to an FLS, the inference engine computes the output set corresponding to each rule. The input to the FLS can be the arrival rate of the PUs or SUs, the channel availability, the distance between users (either PUs or SUs), the velocity (if SUs are moving), the conflict graph (or any other model that captures the interference relationship between users), and so on. These parameters

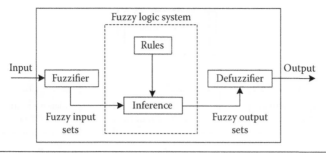

Figure 2.11 The structure of an FLS.

are the input to the fuzzy controller, which takes decisions based on the set of predefined rules as to which SU will select which spectrum band. The rules use membership functions as weighting factors to determine their influence on the final fuzzy output sets regarding the spectrum utilization. Fuzzy logic is used mostly in cases where the configuration of the CRN is known a priori. Only in these cases, for a given set of values, the spectrum assignment can be performed automatically. However, a fuzzy system is not scalable because a large number of rules is normally required for performing SA and considering all the different parameters that can affect the SA decisions; these rules are mainly subjective. All these parameters that should be used as input can make the formulation of rules very difficult. Finally, the membership functions can affect the results dramatically if not structured properly.

2.5 Routing Schemes of CRNs

While CRNs have attracted huge attention from the research community, most of the research work focuses on the PHY and MAC layers [79]. However, routing algorithm design, as a key aspect of networking technologies, plays an important role in improving network performance in multihop CRNs and should be carefully considered. A unique challenge for routing design in CRNs is the dynamics of spectrum availability. From a routing perspective, it is expected that data packets are routed via a stable and reliable path to avoid frequent rerouting problem, since frequent rerouting may include broadcast storm to the network, waste the scarce radio resource, and degrade the end-to-end (e2e) network performance such as throughput and

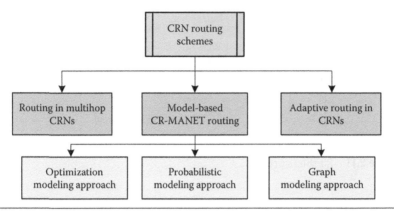

Figure 2.12   CRNs routing schemes.

fairness. In this chapter, we present a brief overview of routing schemes for CRNs. As illustrated in Figure 2.12, there are three general classical routing solutions for CRNs. MANET, mobile ad hoc networks.

2.5.1 Routing in Multihop CRNs

It is clear from previous studies [80] that cognitive nodes can rely on each other to form a heterogeneous network spreading across different primary radio networks' cells. Multihop CRNs can be constructed whereby CRs relay information between a CR sender and a CR receiver. In multihop CRNs, the farthest neighbor routing algorithm using the underlay access [81] to find a multihop route from the cognitive source (CS) to the cognitive destination (CD) was proposed in [82]. If the primary source (PS) transmits the data to the primary destination (PD), the maximum hop transmission distance subject to the QoS requirement of the primary transmission is calculated. Otherwise, the maximum hop transmission distance equals to the maximum transmission distance subject to the QoS requirement of the cognitive transmission.

The multihop cognitive network scenario is shown in Figure 2.13. It consists of a PN and a CR ad hoc network (CRAHN), where there are the PS and the PD. In the CRAHN, there are the CS, the CD, and many other cognitive nodes. The CS and the CD are assumed to be located on a line that is parallel to the x-axis. Each node is assumed to have a single antenna because of the size and the power constraint, where nodes A_i and B_j represent the cognitive receiver in the ith hop

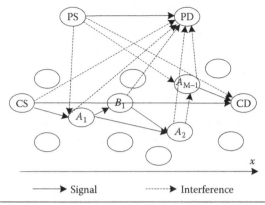

Figure 2.13 The primary transmission coexists with the multihop cognitive transmission.

and the jth cognitive relay, respectively. For the route found between the CS and the CD, let M and N denote the number of cognitive receivers and cognitive relays, respectively.

This task becomes even more challenging because the CR domain still lacks many defining rules and principles. Clearly, routing is the first issue to deal with in order to construct CR-aware multihop networks. Recently, several routing schemes have been proposed for multihop CRNs. For example, a spectrum tree-based on-demand routing protocol (STOD-RP) was proposed in [83], in which all nodes are assumed to be stationary or move very slowly, and the statistics of PU activities and available spectrum band information are assumed to be always available. A spectrum-aware on-demand routing protocol (SORP) is proposed in [84], which considers the inter-flow interference and channel switching delay, and is a cumulative delay-based routing protocol. However, there are several challenges existing in the CRN routing protocol design, three of which are as follows:

1. The *first* challenge is the integration of route discovery with the spectrum decision [85]. Due to the time-varying and intermittent spectrum availability, the spectrum channel information needs to be known when selecting the route.
2. The *second* challenge is the lack of a stable common control channel (CCC). Since a CR node has to vacate the channel as soon as a PU appears on that channel, however, the implementation of a fixed CCC may be infeasible.

3. The *third* challenge is the failure of spectrum-adaptive route recovery. In addition to node mobility, link failure in multihop CRNs may happen when PU activities are detected around CRN users [85].

Therefore, Huang et al. [83] designed a routing scheme for a multihop CRN that overcomes the aforementioned three challenges by considering the path stability and node capacity. The scheme makes the four following contributions:

1. A realistic mobility model is proposed to describe the movement of highly mobile nodes and estimate the link stability performance based on node movement patterns.
2. A CRN topology management scheme is proposed based on a clustering model that considers the radio link availability, the average number of hops, and channel switching from the member nodes to the cluster head (CH).
3. Two new CCC selection schemes are proposed based on the node contraction concept and the discrete particle swarm optimization algorithm. The intercluster control channels and gateways are selected from the CHs based on the node degree level.
4. A novel routing scheme is proposed that tightly integrates with the channel assignment scheme based on the node capacity.

From the above discussion, it is true that CRNs present some resemblance to multiradio multichannel networks, but multihop CRN technologies add the new challenges of having to deal with transmissions over parallel channels and handling routing protocol, PR-to-CR interference. A relay path is often composed of multiple user terminals. Due to multihop transmission, the e2e (PR to CR is one situation) throughput could be degraded. One obvious solution is to allow concurrency among multihop transmissions, which would, nevertheless, increase the interference and potentially degenerate the per-hop capacity. As such, it is uncertain whether employing concurrency in multihop relaying is beneficial for the e2e throughput. For instance, in [86], multihop systems are proved to provide substantial throughput improvement by operating the concurrent relaying transmission in association with the nonconcurrent transmission. In addition, in [82], the use of relays in the CRN with interference control is addressed. In particular, based on the transparency and reliability conditions, Xie et al. derived the maximum reachable distance of a CU in one-hop, proposed two routing schemes to perform relaying, investigated the influence of concurrent transmission among multiple relays, and analyzed the e2e channel utilization, energy efficiency, and delay advantages of CRNs with and without relaying.

Moreover, SUs possess physical capabilities that, if efficiently exploited, allow them to sense, switch, and transmit over many bands of the spectrum, thus removing some physical constraints considered in previous wireless networks. Nevertheless, if the primary spectrum band, once available, remains usable for an unlimited duration (e.g., counted in hours or days), the obtained network model does not differ in essence from any wireless environment considered today. In fact, the routing problem becomes very similar to the one defined and resolved in a multihop multichannel mesh network. Besides, if the environment imposed by the primary nodes' behavior gets more dynamic, new cognitive-specific approaches need to be proposed. Such dynamic routing approaches shall be used until the sporadic availability of primary bands becomes on average smaller than (short) communication duration. Practically, this last dynamic environment requires per-packet routing solutions since a path cannot be considered for whole flow duration. Therefore, in such cases an opportunistic forwarding approach based on the instantaneously available primary bands is a potential candidate to replace e2e outing approaches. Indeed, the overhead and time duration required to establish a path for a short period of use make traditional routing an unthinkable solution. Therefore, in [87], Abbagnale and Cuomo designed a routing scheme that captures the connectivity characteristics of paths and suitably selects the best route in uncertain and high variable scenarios, which make these paths *shorter* with respect to the paths where nodes have less spectrum availability. Therefore, the routing scheme solved the overhead and time duration requirement problem.

In multihop CRNs, the increase of the number of cognitive receivers results in a lower e2e reliability and a lower e2e throughput. To combat these disadvantages, the number of cognitive receivers should be reduced. Recently, the research of the cooperative routing has been done in [88] and [89]. However, none of them deals with the reduction of the number of cognitive receivers with the consideration of the coexistence of the primary and cognitive transmissions. Therefore, Lin and Sasase [90] proposed the primary traffic-based multihop routing with the cooperative transmission that reduces the average number of cognitive receivers on the route from the CS to the CD by using the cooperative transmission and has better performance in terms of the e2e reliability, the e2e throughput, and the average required transmission power of transmitting the data from the CS to the CD.

In general, multihop CRN is one of the most advanced areas in wireless communication situation. For this reason, we focus on multihop CRNs and the routing in this chapter. If SUs are exploited over well-defined primary spectrum when considering the activity and holding time, the simple routing approach can be used well. However, if CR implementation will play more flexibly when the available spectrum bands and their usage happen, the perfect routing scheme should be judiciously chosen first for every situation, depending on the traffic happening and the availability of the primary bands, even the history in the considered wireless environment.

In addition, there is another question: the presence of multiple channels for parallel transmissions, or the channel selection and spectrum management, should be considered in dynamic routing in CRNs at the MAC layer. In practice, the interaction between routing and MAC layers should be considered when the channel decision happens at the MAC layer. The selection of different channels may lead to different neighbors if the channel in the cognitive MAC implies the selection of the next hop. This should be optimized based on the MAC-related information to select all next hops to the destination. But the obtained paths may not be optimal for all flows in the CRN, because the selection situation only considers local information as well as lacks a global vision. Therefore, a good approach is to assist the routing from lower layers that remains for a short period of time. This metric proposed in [80] also reflected the spectrum availability and its quality.

2.5.2 Routing in CRAHNs

Several classification methods have been proposed to categorize the existing routing schemes for CRAHNs. In this section, we present a new classification of CRAHN routing schemes based on the approaches used for establishing the e2e routes. The CRAHN routing schemes are further classified into three subclasses: optimization modeling, probabilistic modeling, and graph modeling (see Figure 2.2).

2.5.2.1 Optimization Modeling Approach A representative routing scheme was proposed in [91], which briefly discusses other CRAHN routing schemes based on optimization modeling, such as in [7,92,93].

Caleffi et al. [92] presented two main existing routing metrics in CRAHNs: (1) They are often based on heuristics without considering optimization and (2) they are unable to measure the actual cost of a route because they do not consider the route diversity effects. Therefore, an optimal routing metric called OPERA is designed to achieve two features: (1) optimality—using both Dijkstra- and Bellman–Ford-based routing protocols to optimize the OPERA— and (2) accuracy—OPERA exploits the route diversity from the intermediate nodes to guarantee the actual e2e delay. The performance evaluation can improve the performance of the routing metric for CRAHNs.

Hou [91] assumed that all SUs have a set of spectrum bands available for transmission. Every spectrum band can be divided into several sub-bands of different bandwidths. The spectrum selection scheme also considers a signal interference model; this method allows at most one SU to use a sub-band in the area caused by the interference ranges of the transmitter and the receiver (see Figure 2.14). Nagaraju et al.

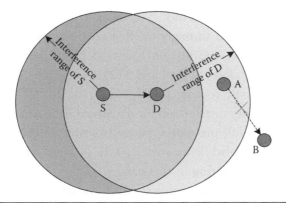

Figure 2.14 Signal interference constraint.

[93] proposed an optimized CRAHN routing scheme by setting the objective of maximizing the network throughput.

In addition, recently, researchers have defined a new routing metric: an efficient shortest path algorithm by discovering the lowest weight route between any pair of nodes in a CRN. Sampath et al. [94] presented a scheme that combines the switching delays and the link stability. However, this metric is not optimal, which was proved in [95]. The paper [96,97] proposed a routing metric and a routing protocol to achieve high throughput efficiency by considering the route stability. Wang et al. [98] proposed an asymptotic study of the capacity and delay for CRNs by comparing shortest path routing, network coding techniques, and multi-path routing. However, the quality of a route was ignored in their work. Therefore, Lin and Chen [99] presented an opportunistic path metric that accounts for the route diversity effects. However, both the closed form expression and the optimality are not taken into account in their schemes. Abbagnale and Cuomo [100] and Chowdhury and Akyildiz [101] proposed a novel CR metric by minimizing the interference considering path stability and availability over time. Caleffi et al. [92] proposed a novel CR routing metric with the objective to overcome the issues: non-optimality and in-accuracy caused by the interference.

2.5.2.2 Probabilistic Modeling Approach   Recently, researchers have proposed some probabilistic approaches designed for multichannel multihop networks [102–104]. The probability distribution odds-on mean (OOM) is used widely to improve the selection probabilities. Further, there are two shortcomings when the schemes used in centralized routing of network traffic: (1) delay requirement due to information back and forth throughout effect in the network and (2) the complexity and the information overhead grow quickly with the network size. Therefore, Soltani and Mutka [102] presented a decentralized probabilistic routing to solve the problem by utilizing the ArgMax probability distribution for dynamic CRN environments.

Khalife et al. [80] presented a novel routing protocol for CRNs that employs a probabilistic metric. CRNs have specific properties; consequently, routing in this particular environment should be designed in a way to ensure route stability and availability. However, the proposed routing metric is able to capture effectively through probabilistic calculation, the specific constraints imposed on a cognitive node by the activity, and the density of the primary nodes. The protocol algorithm favors the use of multiple frequencies between two communicating CR nodes (PUs and SUs), which is an important feature of CR environments. Cagatay Talay and Altilar [105] augured that, if a channel was reliable before its presentment, it is more likely to be reliable in the future. Based on this argument, the weight of a channel is defined as a function of the history of temporal usage of the channel, and the probabilistic routing schemes are proposed. In the future work, the researchers may aim to study real PUs' data in order to extract measurement information about primary nodes' locations and behaviors. It is also important to study the interaction between the dynamics of the arrivals or departures of the connections and the dynamics of the cognitive networks. Also, we believe that this work could open a new line of future research on a different variety of network protocols designed for a dynamic environment that incorporate probability distribution in their decision making.

2.5.2.3 Graph Modeling Approach Several recent works have analyzed and proposed routing algorithms based on the graph modeling approach for CRAHNs [100,106–108]. We aim at presenting the schemes reported in those studies in this section.

In [108], a topology formation algorithm was proposed to construct a layered graph model (LGM) that the number of layers corresponds to the maximum number of channels in order to be sensed by a node. Jun et al. [107] used the color theory for distinct channels to form a colored multigraph model by applying a novel shortest path algorithm to the model. The scheme in [109] used a novel topology formation algorithm to enable nodes with both adjacent hop interference and the hop count being minimized.

Two routing schemes based on path connectivity for CRAHNs were proposed in [100,106]. Abbagnale and Cuomo [100] presented the foundation to form a new routing scheme to model and evaluate the connectivity of routing paths based on path connectivity in CRAHN. Furthermore, considering that the Gymkhana routing on a number of topologies with different PUs' activities plays an important role in CRAHNs, they [100] used the Gymkhana model to derive a utility function for comparing routing paths and provided a distributed protocol [106] for efficiently routing data in a CRAHN. The proposed utility function measures the connectivity of different paths, by taking into account the PU's behavior, the cost of the channel switching, and the hop count. The proposed model had a practical value in CRN scenarios with the benefits of reduced complexity and attractive performance behaviors.

The path selection scheme in [87] is shown in Figure 2.15. The metric classifies the paths on the basis of their connectivity and selects one of them for routing by using the information contained in the received route requests (RREQs).

However, almost all the studies on CRAHN routing using the graph modeling approach only address the networks with static links and do not explicitly consider the time-varying nature of link availability. Another interesting aspect of CRAHN routing design is the exploitation of the channel diversity in CRAHNs to improve routing efficiency, since two SUs in the CRAHNs are typically connected by multiple paths through different intermediate SUs and channels. In future, researchers will apply novel routing schemes to solve the problems.

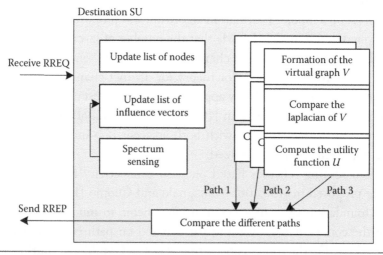

Figure 2.15   Gymkhana operations performed at the destination. RREP, route reply.

2.5.3 *Adaptive Routing in Cognitive Radio Networks*

Adaptive routing [110], as a critical functionality of wireless self-organized networks, has already been studied considerably in recent decades. In consideration of the unique characteristics of the radio environment in CRNs, there are several new researches that may be solved by adaptive routing in designing cognitive routing algorithms.

1. First, the fixed CCC used in traditional routing protocols is infeasible in CRNs, with some specific routing functionalities, such as neighbor discovery, route discovery, and route establishment, difficult to facilitate.

2. Second, intermittent connectivity between neighboring nodes may happen frequently not only because of the node mobility and dynamic characteristics of wireless signals but also because of the dynamic changes of available spectrum in CRNs. Besides, the QoS of PUs (e.g., e2e latency, packet loss probability) may also be degraded because of the issues of spectrum handoff and interference. Therefore, various important factors, such as the time-variant channel attenuation, dynamic changes of spectrum, and the frequency–space domain characteristics of the radio environment should be considered jointly in constituting multihop CR routes.

3. Last but not least, spectrum-aware rerouting algorithms should also be developed to adapt to spectrum fluctuations and enable optimization or maintenance of CR routes.

In order to tackle the problem of the above researches, there are some papers proving their adaptive models.

Huang et al. [83] proposed a novel CRN adaptive routing scheme to solve the high mobility that considers the path stability and node capacity by reducing the complexity.

An adaptive traffic route control scheme was presented in [111] in order to offer a stable and flexible wireless communication, which provides high QoS for CR technology and examines the performance of the proposed scheme. Figure 2.16 shows the frame of the proposed adaptive traffic route control scheme.

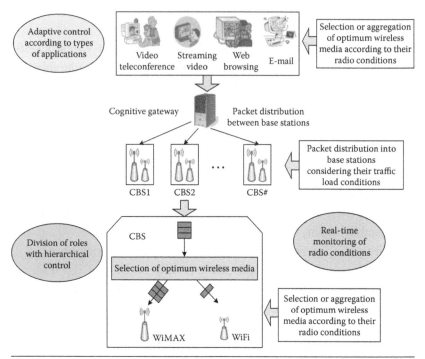

Figure 2.16   The frame of adaptive route control.

2.6 Routing Technology Research for DSA CRNs

The quality and distribution of the spectrum resources must be considered in routing technology research for DSA networks. In this chapter, we break through the traditional routing classification criteria such as multipath routing algorithm, QoS routing algorithm, routing algorithm based on geographical information, and power-aware routing algorithm; summarize the latest research of routing technology research for DAS networks by using a new classification criterion that reflects the routing design affected by spectrum sensing and spectrum management technologies in CRN environment; and analyze and compare some representative of the routing algorithms.

2.6.1 Spectrum-Aware Opportunistic Routing Protocols

For wireless networks, new routing algorithms are necessary due to the dynamic nature of the physical layer (in both time and frequency domains) in CRNs [12]. However, routing for DSA is still a less-explored area. There are two issues that should be taken into account in jointing routing and SA:

1. In a changing environment, it is important to be aware of the dynamically available spectrum holes and adapt their spectrum allocation operations to routing algorithms and protocols.
2. Routing algorithms should consider the routing paths through which the generated interference is minimized if it interacts with dynamic spectrum allocation routines.

For the first issue, researchers also considered the SA problem in the context of dynamic industrial, scientific, and medical (ISM) band operations [112,113]. However, Murthy and Manoj [114] proved that the packet routing algorithms, particularly algorithms, may be unreliable in the domain of ad hoc networks. There is also a natural issue on trust and cooperation was proposed when joint coding or multi-hop routing [63,115].

In addition, several routing protocols have been applied for SA in CRNs. For example, Yang et al. [9] presented a SPEAR metric that establishes robust paths in the diverse spectrum environment. Another proposal called STRAP was proposed in [8]; this scheme can apply unused frequency bands without compromising the stability of the CRN. Moreover, the ROSA algorithm is one of the advanced proposals to maximize the throughput by minimizing the interference and maximizing the weighted sum of differential backlogs in [97].

For the second issue, it includes the multiple layer constraints and multiple paths for load balancing, which are important for routing in the presence of DSA [5,99,116–118]. In [5], a geographic forwarding based routing protocol called SEARCH was proposed for ad hoc networks. Furthermore, the scheme uses Kalman filters to handle node mobility for selecting shorter paths. In addition, Pefkianakis et al. proposed a metric called spectrum-aware mesh routing (SAMER) [116] based on a mesh network framework, so that the routing algorithm passes traffic across paths with higher spectrum allocation availability and opportunistic performance. Moreover, S.-C. Lin et al. toke the frequency and transmission control into the optimization parameters in order to minimize the interference in the network [99,118].

2.6.2 Considering Control Channel for Routing Algorithms

DSA networks are divided into two major categories: unlicensed CR in unlicensed bands and unlicensed CR in licensed bands.

These two types of CR are required to avoid interference to the PU network especially in the second category. Cognitive nodes need to switch working band or adjust transmitter power or modulation timely, when the PU begins to occupy the current band using by cognitive nodes. This special requirement raises a challenge to the communication coordination among cognitive nodes such as the broadcast message transfer and the route message transfer.

In [84,94,119], it is assumed that the global shared control channel that is used by an interface of stationary cognitive nodes to complete communication coordination, exchange of spectrum information, neighbor discovery, and some other operations is configured in DSA networks; however, how to select the global shared control channel has not been explained in these references. Because the available spectrum resources change dynamically as the time and positions change, the selection of control channel will need to be carefully designed whether it is global or local so that broadcast messages can cover the network entirely.

The distribution of the available spectrum of cognitive nodes is asymmetrical and discrete; therefore, cognitive nodes transmit neighbor discovery messages in all available channels and get a collection of the available spectra of neighbor nodes and current working channel, thus discovering the network topology in [85,120–125].

This approach can lead the DSA networks to adapt changes in the environment quickly, but the cost of maintaining the neighbor lists and conveying broadcast messages is relatively large.

2.6.3 Joint Routing and Spectrum Selection Algorithm

Wang and Zheng [121] analyzed and compared the hierarchical design of spectrum and routing selection; as a result, routing selection is independent of the spectrum selection that can simplify the protocol design in a hierarchical scheme, but the performance of its e2e is not as good as joint design in the DSA dynamic spectrum change networks. Therefore, the cross-layer routing design technique has better adaptability.

In [84,94,119,120,122], the design ideas of joint ROSA were adopted. Cognitive nodes transfer the latest available spectrum information of the nodes chosen among the path in the route discovery, thus making a full use of the spectrum information for routing selection and spectrum allocation when routing has been established.

Cheng et al. [119] consider the channel switching delay related to the relative distance of spectrum so that a route with the minimum path delay can be selected by calculating the time delay from the different spectrum options. Then they analyzed the model of cross data flow and proposed a strategy for spectrum switching among cross nodes (On-demand routing protocol, DORP) [119]. Ma et al. [120] proposed a multihop single-radio cognitive routing protocol, which considers that delay can be reduced by avoiding switch, but the use of different channels can improve network throughput. Therefore, the selected spectrum allocation strategy should be the trade-off between switch delay and throughput.

Joint routing and spectrum selection algorithm is beneficial to optimizing e2e transmission performance in DSA networks. Spectrum allocation occurs in the route response stages by using both multihop single-transceiver CR routing protocol (MSCRP) algorithms [120] and DORP algorithms [119], which depend on spectrum selection of the destination nodes intensively.

According to the allocated spectrum information on the path, cognitive nodes calculate the delay and other indicators; however, the large computation of the nodes increases the delay of route establishing, which does not apply to high real-time requirements of the network.

2.6.4 Considering Channel Allocation for Routing Algorithms

The performance of routing algorithms is affected by channel allocation directly due to joint design of channel allocation and routing algorithms in DSA networks.

According to the allocation granularity, multichannel distribution strategy can be divided into three kinds of packet-based groups in the existing wireless network based on the package, the link, and the data flow. Packet-based strategy cannot be applied into the actual networks due to switching channel frequent switching of channels. Channel allocation strategy based on link is more flexibly used by the most majority of DSA networks. However, all the cognitive nodes from the source to the destination use the same channel that is required in the channel strategy based on data flow.

Fuji et al. [124] proposed a multiband routing scheme based on data flow that sends the route request messages on various bands based on on-demand routing. Neighbor nodes also forward route request messages based on on-demand routing. Destination nodes create route on multiple bands after receiving route request messages according to the principle of minimum hops.

Channel allocation strategy based on the link that causes larger time delay due to frequent switching was adopted in [94]. The advantages of both channel allocation based on comprehensive data streams and link were synthesized in [122]; therefore, according to the distribution characteristics of spectrum, segmentation-based channel allocation utilizing the flexibility of link and small delay based on data flow to allocate channels was proposed.

2.6.5 PU Interference Avoidance for Routing Algorithm

Interference caused to cognitive nodes by PU is changed dynamically as the time and positions change in unlicensed CR in licensed band DSA networks. Traditional wireless networks cannot adapt to this dynamic spectrum environment. Cognitive nodes can take the following measures to deal with when PU appears in the bands occupied by cognitive nodes: (1) bypass the interference regions and create a new route, (2) switch the band and find idle bands for communication, and (3) change the transmit power or modulation to avoid interference to the communication of PU.

Fujii and Suzuki [125] proposed a distributed automatic retransmission request technology using space–time block code (STBC)-distributed Automatic Repeat-reQuest (AQR), which identifies the interference proactively and selects the route adaptively by bypassing the interference region, but cannot be applied into energy-constrained networks due to the large energy consumption of the nodes caused by omnidirectional broadcast of the package.

Route created in multibands was proposed in r [124]; therefore, if the current detected bands are unavailable, alternate bands can be enabled. Bian and Jung-Min [122] proposed an allocation scheme based on segment splitting and merging under the segmentation-based channel allocation strategy, according to switching bands and rebuilding route in accordance with varying conditions of spectrum. SPEAR algorithm designed by Sampath et al. [94] was used by the reservation channels on the path for each link, while local repair strategy was used to reselect new channels for broken link under each node keeping e2e routing indicators such as throughput and delay to meet the situation; if the local repair fails, a new route will be created by source nodes.

2.6.6 Joint Dynamic Routing and Cross-Layer Opportunistic Spectrum Access

A new metric called the ROSA was proposed to maximize throughput through in [7]. ROSA uses queuing and spectrum dynamics opportunistically to calculate the next hop based on the SA utility function. Then, each packet will potentially follow a different path according to the queuing and spectrum dynamics. Furthermore, the paper [7] proposed a new scheme for the ROSA's MAC, called collaborative virtual sensing (CVS), that aims at providing nodes with accurate spectrum information based on a combination of physical sensing and local exchange of information. However, the ROSA's MAC may cause collision without considering the SA management. Therefore, we propose a new scheme, called distributed collaborative scheme in cognitive MAC (DCCMAC) protocol, which provides nodes with accurate spectrum information based on a physical sensing and local exchange of information. DCCMAC is achieved by combining scanning results and information from control packets exchanged on CCC that contain information about SA used on different minibands.

For simplicity, we assume that each node is equipped with two transceivers: one is employed on the CCC and the other one is a reconfigurable transceiver. Handshakes on the CCC are conducted in parallel with data transmission on the DC. Also, each node considers the cooperative transmission with relay link selections.

DCCMAC is illustrated in Figure 2.17 three-way RTS, CTS, and relay preparation-to-relay (RTR) handshakes, backlogged nodes contend for SA on the CCC. In addition, RTS is needed for the transmitter (the user) to collect and announce the accessed spectrum information to its neighbors. Therefore, the RTS/CTS/RTR packets

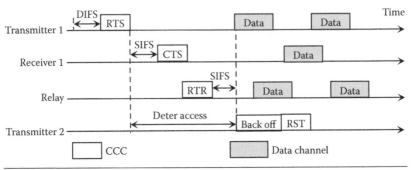

Figure 2.17 The distributed collaborative design in cognitive MAC protocols.

include channel access information to allow the nodes to make adaptive decisions.

From Figure 2.3, we can see that the transmitter uses an RTS packet to announce the receiver and the spectrum slot of the relay link selection. Once the RTS packet is received, the receiver responds to the transmitter by a CTS packet in short interframe space (SIFS) and modifies the specified transmission data from the transceiver to the RTS in a packet. The selected relay will send a new RTR packet after receiving the RTS and CTS packets, and the new RTR packet will announce the accessed spectrum information to neighbors and the statue of the selected relay to the receiver.

Once RTS/CTS/RTR is successfully exchanged, the sender and the receiver tune their transceivers to the selected spectrum portion. It is necessary to sense the selected spectrum before transmitting, and, if it is idle, the sender begins data transmission without further delay. Due to the presence of PUs or by conflicting reservations caused by losses of control packets, it may make the sender or the receiver find the selected spectrum busy just before data transmission. In this case, the SU gives up the selected spectrum and goes back to the control channel for further negotiation. During the period of the RTS/CTS/RTR exchange, if the sender-selected spectrum cannot be entirely used, as well as the receiver senses the presence of PU or SU, the receiver will not send the CTS packet. When the waiting state for CTS timer expires and exceeds the RTS retransmission limit, the sender will go back to the control channel for further negotiation.

In particular, this scheme is affected by both the presence of PUs and the multiple channel hidden terminal problems. As we know, the signaling and data transmissions are managed efficiently by the two radios of the CCC scheme; the sensing and signaling exchange process may be interrupted when only one radio decreases the device costs and energy consumption data transmissions to perform. Moreover, a single transceiver in the MAC protocol addresses the multiple channel hidden terminal problems that may lead to collisions between packets transmitted by different SUs. An SU with a single transceiver cannot handle transmission and listening at the same time. Consider, for instance, Figure 2.18 that shows the scenario: node *A* will start the RTS/CTS handshake on the control channel (channel 1) when it wants to send a packet to node *B*. After negotiation, node *A* starts communication when channel 2 is selected. However, node *C* does not hear the RTS/CTS messages due to its listening to channel 3, so it decides to initiate a transmission on channel 2, which may cause a collision.

Therefore, it is necessary to present a novel MAC approach to solve the problem in multihop CRAHNs. We first describe the concept of the novel MAC approach. A node pair first executes a handshake scheme (RTS/CTS) and a channel announcement scheme (DCI/ access network control [ANC]) before a message communications (Data/ACK). DCI informs a node on DC that this DC is idle and can be used. Note that senders are also supposed to receive ACK in a multihop CRAHN, so the DC selected must be idle for both the sender and the receiver. Handshake scheme is used to negotiate a list of expected idle data channels (EIDCs) by this node pair, while channel announcement is used to select an actually idle DC in the final EIDC channel list and to help all their idle neighbors update

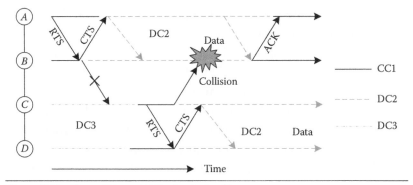

Figure 2.18   The multichannel hidden terminal problem.

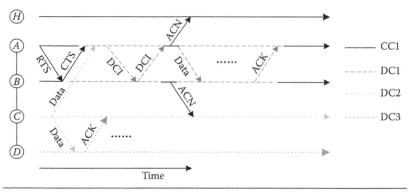

Figure 2.19   Multihop MAC protocol.

their channel usage information. Figure 2.19 describes an execution example of multihop MAC protocol. There is one CC and three DCs. Three node pairs, that is, AB, BC, and CD, are communicating on DC3, DC2, and DC1, respectively. Node H (the central coordinator described in Section 2.6.4) is responsible for controlling the common channel. So the scheme in MAC protocols can be followed as

Handshake phase: According to its channel usage information (CUI), a computes an EIDC channel list (q) recording that DC1 and DC3 are idle, and then node A sends a text to speech (TTS) with an EIDC channel list to B. When B receives this RTS, it computes its EIDC channel list, then computes the final EIDC channel list and the EIDC channel list of A and B, and finally sends a CTS with a final EIDC channel list to A.

Channel announcement period: Assume that DC1 is the first DC in the final EIDC channel list and then both A and B switch to DC1 and listen for time slots t and $2t$, where t is selected according to the maximum data packet size. Because DC1 is occupied by CD, both A and B may receive a packet from C or D, which indicates that DC1 is busy. Therefore, both A and B continue to switch to DC2 without sending information to them, since they both are aware of the fact that C or D is their common neighbor. After monitoring DC3, A and B exchange DCI to make sure that DC3 is idle for both nodes. Then, A and B switch to the CC and sequentially send the same ANC about this channel election, which helps their idle neighbors (e.g., H) to update their CUI.

Data communication phase: *A* and *B* switch back to DC3 and communicate with Data/ACK exchanging. When this exchanging is finished, *A* and *B* switch to the CC again and update their CUI by overhearing the ACN sent on the CC by their communicating neighbors.

2.6.7 Considering Network Routing Model for Routing Algorithms

Traditional network routing model under a static spectrum environment cannot be used in dynamic spectrum network directly as the same available spectrum resource of its nodes. Xin et al. [108] proposed a new LGM of the network [108]. Each channel of the network maps the different layers of the model. Cognitive nodes in the layers are expressed by child nodes, while cognitive nodes in the same layer were connected with horizontal edge. Channel allocation is identified by the vertical edge; however, the connection of vertical edges is restricted by the number of idle interfaces of nodes. The stream-oriented interface assignment scheme used in the dynamic layered graph (DLG) model was proposed in [126]. Compared to static allocation scheme, dynamic allocation scheme allows the interface to switch the channel for different data streams. Both LGM and DLG models calculate the route through assigning different weights to horizontal and vertical edges and updating models in the routing process.

Khalife et al. [80] used probabilistic models to design the route metric. The interference caused by PU, the received signal strength, and the data transmission rate can be achieved are calculated according to probability models as to be the weight of the link, then optimal path is selected by the conventional distance vector algorithm.

Zhu et al. [85] constructed a spectral tree model to calculate the route. Spectral tree is built by cognitive nodes according to its available channel. The routing tables of other cognitive nodes are maintained by the root node, whereas that of the destination node can be obtained through accessing the root node. However, the spectral tree model cannot be applied to the complex spectrum network and the tree model optimization is also a need to be explored sequentially.

2.7 Challenges of Jointing Routing and DSA in CRNs

The research of CR technology utilizes spectrum resources efficiently. A compatible coexistence between heterogeneous networks and interactive applications shows great prospects. To make cognitive devices transmit information efficiently in the most optimal way in a limited signal space, cross-layer design of routing algorithms for DSA networks must be adopted, which allocates spectrum and selects the route through spectrum sensing information provided by the physical layer and spectrum scheduling information provided by the link layer synthesized by the network layer.

The close collaboration between the physical, link, and network layers provides an optimal path while improving spectral efficiency for nodes.

In order to meet the rapidly changing environment of wireless spectrum, piggybacked status information collecting the change of spectrum on the path in real time can estimate transmission quality and judge timely avoiding routing interrupts on system performance. Simultaneously mixing routing will be the focus of study. Maintaining partial routing information can create routing quickly and avoid routing affected by the stale route information, while reducing the network load caused by routing establishment and maintenance.

Intelligent reasoning ability of CR itself is the most important characteristic of DSA networks so that its routing algorithms should also be intelligent. Routing algorithms can create reasoning models of network routing and provide evidence for various indicators of joint optimization such as latency, delay jitter, and package loss rate while providing support for the upper applications through the study of spectrum environment, information of path status, spectrum selection, and routing making.

Joint routing and SA is a key design issue for CRNs and many past efforts have proposed lots of techniques from different perspectives. In Section 2, we have provided the overall view of the development of SA in CRNs, routing schemes for CRNs, and joint routing and SA design. Nevertheless, there are still challenges and open problems for realizing effective and efficient joint routing and SA for CR communications.

2.7.1 Spectrum Characterization

Most proposed schemes also use a single target [i.e., fairness, Signal to Interference plus Noise Ratio (SINR), or throughput] to define the traffic load of the examined spectrum bands and choose the optimal channel based on either throughput requirements or interference threshold of the SUs. A famous issue is proposed that using the multiple QoS requirements to analyze the spectrum bands. According to each SU's service demand, there can be different QoS factors, such as delay, jitter, throughput, and bit error rate. Most researches consider only the throughput as a QoS requirement when jointing routing and SA algorithms, but this is not suitable for applications such as Voice-over-Internet Protocol (VoIP), which is sensitive to the delay. Since various applications are used into different performance/QoS requirements, which may run simultaneously on the SU's device, the joint routing and SA algorithms may be unable to be optimal methods because these schemes need to consider all the QoS parameters for the spectrum channels, so it is necessary to find optimal solutions to solve the problem.

2.7.2 Robust Route for SA

The presentation of PUs plays more serious on e2e connectivity for multihop communications than that in the single-hop case. The source nodes and destination nodes can only consider the local channel availability when this situation is in single-hop communications. However, in the multihop communications, each pair of nodes should be taken into the channel availability of the entire route. Therefore, it is important to find a robust route in CRNs, which is ignored in the most work. In traditional networks, the robust routing without considering the impact of PUs, so they only pay attention to the problem of node mobility, even the strong interference from the neighboring nodes. The existing work on routing and SA in CRNs ignores both these issues and single-flow scenarios.

2.7.3 Hidden User Effect

Detecting a primary transmitter signal is a major misconception in CR literature, which is equivalent to finding spectrum opportunities.

However, joint routing and spectrum opportunistic access is affected by the following three main problems when the primary signals can be perfectly detected: (1) the hidden transmitter, (2) the exposed transmitter, and (3) the hidden receiver. For the first one, a hidden transmitter senses the range of the SU sender outside but it is placed close to the SU receiver. For the second one, an exposed transmitter is a PU sender that is located near the cognitive transmitter; at the same time, the PU receiver is also outside the SU transmitter power interfering range. For the third one, a hidden receiver is a PU receiver that is located in the interfering range of the SU transmitter, whereas the PU transmitter is outside the detection range of the SU users. While the spectrum sensing solved the hidden transmitter problem at both the transmitter and the receiver side, there are still no optimal solutions for the latter problems. In order to solve these problems, an SU should be able to detect the presence of a neighbor PU receiver.

2.7.4 Heterogeneity-Induced Challenges

The heterogeneity of the CRNs that exhibited in a multiuser environment with coexistence of SUs and PUs introduces new challenges to the joint routing and SA schemes. Besides, the heterogeneous spectrum comprising both licensed and unlicensed bands for different purposes imposes additional challenges in CR techniques. For example, future heterogeneous CRNs exist multiple PNs and a PU that has subscription to all the PNs, but is also equipped with a CRN device. An efficient joint routing and SA algorithm should consider the possibility to access any network in order to maximize the network performance.

2.7.5 Spectrum Mobility Challenge

Each time or space an SU changes its frequency, the operation parameters should be modified of the network protocols. When an SU captures the best spectrum band, the activity of PU on the selected spectrum can necessitate so that it can change its operating spectrum band(s); this scheme is referred to as spectrum mobility. This is a big challenge that must be solved in the jointed routing and SA schemes. The open research issues that are solved in the joint routing and SA schemes, which can improve the efficient spectrum mobility in CRNs, are as follows:

- *The spectrum mobility in the time domain*: According to the information of available bands, CRNs adapt to the wireless spectrum accurately. Because these available channels also change over time, the QoS in this environment is challenging.
- *The spectrum mobility in the space domain*: As PUs or SUs move from one place to another, the available bands also change. Therefore, continuous allocation of spectrum is a major challenge.

2.7.6 Routing for SA in Multihop CRNs

It is true that CRNs present some resemblance to multihop multichannel networks, but CR technologies have to deal with transmissions over parallel channels and handle PU-to-SU interference. Moreover, SUs possess physical capabilities that, if efficiently exploited, allow them to sense, switch, and transmit over many bands of the spectrum, thus removing some physical constraints considered in the previous CRNs. In fact, the routing problem for SA becomes very similar to the one defined and resolved in a multihop multichannel network. Besides, if the environment imposed by the primary nodes' behavior gets more dynamic, new cognitive-specific approaches need to be proposed. Such dynamic routing approaches shall be used until the sporadic availability of primary bands. Practically, this last dynamic environment requires per-packet routing solutions since a path cannot be considered for whole flow duration. Therefore, in such cases an opportunistic forwarding approach based on the instantaneously available primary bands is a candidate to replace e2e routing approaches.

2.7.7 Selfish Users in Joint Distributed Routing and SA Schemes

The most common assumption in distributed routing and SA schemes is that the SUs are willing to follow a strict MAC protocol and exchange information with other SUs for taking a joint decision regarding spectrum allocation. However, in [128], this assumption is not valid when selfish users exist in the CRNs, which want to maximize their profits regardless of the impact on the other SUs and may avoid sending information or send false or misleading information. The investigation of incentives for the participation of selfish users into the SA schemes and how they can benefit (in terms of profit or performance) from the cooperation with other SUs in the SA process is not covered in the literature, which may degrade the performance in the route path selection and MAC protocol too.

2.7.8 Security Challenges

Cognitive communication is based on a *vertical* SA rights paradigm. In this context, CR nodes are secondary spectrum users authorized to access frequency channels only on a no- or limited-interference manner with respect to the PUs of the band. This SA status makes CRNs particularly vulnerable to attacks aiming to deny the CR nodes from the SA; the security considerations in a CRN setting are still in jointed routing and SA, and require a more thorough analysis by the research community. A number of threats specifically pertain to cognitive communications, most notably to secure the spectrum sensing process against exogenous and insider malicious nodes. The far-reaching effect of isolated attacks against CRNs, due to the learning-based interaction CR nodes with their RF environment, is another key issue to be more vigorously investigated [125].

2.7.9 Adaptive Algorithms

Another key challenge for the development of efficient joint routing and SA algorithms is that the algorithms should be adaptive to varying conditions or scenarios in order to meet the requirements of a highly dynamic cognitive environment. All algorithms so far focus on a static scenario and network and try to find an optimal solution according to some criteria. In mobile environments, though, this is not the case because the environmental characteristics change significantly as a user moves through space (and time). For example, a user may at one time be a part of a centralized CRN, but as the user moves, he or she may leave the area of that CRN and enter the area of a distributed CRN. Another case can be when a user is in a high contention area and after a while enters a low contention area. Developing adaptive route path selection and SA algorithms remains a key challenge for future research to provide efficient solutions for varying conditions and scenarios.

Acknowledgments

We sincerely thank anonymous reviewers for their constructive comments and suggestions to improve the quality of this chapter. This work was supported by the National Nature Science Foundation of China under contract numbers 61271259 and 61301123; the Chongqing Nature Science Foundation under contract number CTSC2011jjA40006; the Research Project of Chongqing Education Commission under contract numbers KJ120501, KJ120502, and KJ130536; the special fund of Chongqing key laboratory (CSTC); the Program for Changjiang Scholars and Innovative Research Team in University (IRT2129); and the project of Chongqing Municipal Education Commission under contract number Kjzh11206.

References

1. Federal Communications Commission (FCC), "ET Docket No. 03-108: Facilitating opportunities for flexible, efficient and reliable spectrum use employing cognitive radio technologies: Notice of proposed rule making and order," FCC, December 2003. http://citeseerx.ist.psu.edu/showciting?cid=4712477.
2. J. Mitola, "Cognitive radio: An integrated agent architecture for software defined radio," PhD dissertation, KTH Royal Institute of Technology, Stockholm, Sweden, 2000.
3. J. Mitola and G. Q. Maguire, "Cognitive radio: Making software radios more personal," *IEEE Personal Communications*, vol. 6, no. 4, pp. 13–18, 1999.
4. C.-T. Chou, N. Sai Shankar, H. Kim, and K.G. Shin, "What and how much to gain by spectrum agility?" *IEEE Journal on Selected Areas in Communications*, vol. 25, no. 3, pp. 576, 588, 2007.
5. K. R. Chowdhury and M. Di Felice, "SEARCH: A routing protocol for mobile cognitive radio ad-hoc networks," *IEEE Sarnoff Symposium*, Princeton, NJ, March 30–April 1, 2009, pp. 1, 6.
6. Z. Yuan, J. B. Song, and Z. Han, "Interference minimization routing and scheduling in cognitive radio wireless mesh networks," *2010 IEEE Wireless Communications and Networking Conference*, April 18–21, 2010, pp. 1, 6.
7. L. Ding, T. Melodia, S. N. Batalama, J. D. Matyjas, and M. J. Medley, "Cross-layer routing and dynamic spectrum allocation in cognitive radio ad hoc networks," *IEEE Transactions on Vehicular Technology*, vol. 59, no. 4, pp. 1969, 1979, 2010.

8. J. Chen, H. Li, J. Wu, and R. Zhang, "STARP: A novel routing protocol for multihop dynamic spectrum access networks," *Proceedings of the 1st ACM Workshop MICNET*, Beijing, China, 2009, pp. 49–54.

9. L. Yang, L. Cao, H. Zheng, and B. Zhao, "High-throughput spectrum aware routing for cognitive-radio-based ad hoc networks," *Proceedings of the 3rd International Conference on CrownCom*, Singapore, April 2008.

10. J. So and N. H. Vaidya, "Multi-channel MAC for ad hoc networks: Handling multi-channel hidden terminals using a single transceiver," *Proceedings of the 5th ACM International Symposium on Mobile Ad Hoc Networking and Computing*, Roppongi Hills, Tokyo, Japan, May 2004, pp. 222–233.

11. S. Haykin, "Cognitive radio: Brain-empowered wireless communications," *IEEE JSAC*, vol. 23, no. 2, pp. 201–220, February 2005.

12. I. F. Akyildiz, W.-Y. Lee, M. C. Vuran, and S. Mohanty, "Next generation/dynamic spectrum access/cognitive radio wireless networks: A survey," *Computer Networks Journal*, vol. 50, pp. 2127–2159, September 2006.

13. F. K. Jondral, "Software-defined radio—Basic and evolution to cognitive radio," *EURASIP Journal on Wireless Communications and Networking*, vol. 15, no. 8, pp. 37–48, 2005.

14. N. Devroye, M. Vu, and V. Tarokh, "Cognitive radio networks: Information theory limits, models and design," *IEEE Signal Processing Magazine*, vol. 25, no. 6, pp. 12–23, 2008.

15. C. Cordeiro, K. Challapali, D. Birru, and N. Sai Shankar, "IEEE 802.22: An introduction to the first wireless standard based on cognitive radios," *Journal of Communications*, vol. 1, no. 1, pp. 38–47, 2006.

16. I. F. Akyildiz, W.-Y. Lee, and K. R. Chowdhury, "CRAHNs: Cognitive radio ad hoc networks," *Ad Hoc Networks*, vol. 7, no. 5, pp. 810–836, 2009.

17. U. Berthold, F. Fu, M. van der Schaar, and F. Jondral, "Detection of spectral resources in cognitive radios using reinforcement learning," *Proceedings of the 3rd IEEE Symposium on New Frontiers in Dynamic Spectrum Access Networks*, Chicago, IL, October 2008, pp. 1–5.

18. Q. Zhao, L. Tong, A. Swami, and Y. Chen, "Decentralized cognitive MAC for opportunistic spectrum access in ad hoc networks: A POMDP framework," *IEEE Journal on Selected Areas in Communications*, vol. 25, no. 3, pp. 589–600, 2007.

19. Y. Chen, Q. Zhao, and A. Swami, "Joint design and separation principle for opportunistic spectrum access in the presence of sensing errors," *IEEE Transactions on Information Theory*, vol. 54, no. 5, pp. 2053–2071, 2008.

20. S. Geirhofer, L. Tong, and B. Sadler, "Cognitive medium access: Constraining interference based on experimental models," *IEEE Journal on Selected Areas in Communications*, vol. 26, no. 1, pp. 95–105, 2008.

21. Y. Chen, Q. Zhao, and A. Swami, "Distributed spectrum sensing and access in cognitive radio networks with energy constraint," *IEEE Transactions on Signal Process*, vol. 57, no. 2, pp. 783–797, 2009.

22. X. Wang, W. Chen, and Z. Cao, "Partially observable Markov decision process-based MAC-layer sensing optimisation for cognitive radios exploiting rateless-coded spectrum aggregation," *IET Communications*, vol. 6, no. 8, pp. 828, 835, 2012.

23. S. Wang and J. Zhang, "Opportunistic spectrum scheduling by jointly exploiting channel correlation and PU traffic memory," *IEEE Journal on Selected Areas in Communications*, vol. 31, no. 3, pp. 394, 405, 2013.

24. K. Prabhjot, K. Arun, and U. Moin, "Spectrum management models for cognitive radios," *Journal of Communications and Networks*, vol. 15, no. 2, pp. 222, 227, 201, 2013.

25. P. Kaur and A. Khosla, "Markovian queuing model for dynamic spectrum allocation in centralized architecture for cognitive radios," *IACSIT International Journal of Engineering and Technology*, vol. 3, no. 1, pp. 1–4, 2011.

26. L. R. Rabiner, "A tutorial on hidden Markov models and selected applications in speech recognition," *Proceedings of the IEEE*, vol. 77, no. 2, pp. 257–286, February 1989.

27. I. A. Akbar and W. H. Tranter, "Dynamic spectrum allocation in cognitive radio using hidden Markov models: Poisson distributed case," *Proceedings of the IEEE SoutheastCon*, Richmond, VA, March 22–25, 2007, pp. 196–201.

28. C.-H. Park, S.-W. Kim, S.-M. Lim, and M.-S. Song, "HMM based channel status predictor for cognitive radio," *Proceedings of the Asia-Pacific Microwave Conference*, Bangkok, Thailand, December 11–14, 2007, pp. 1–4.

29. Z. Chen, N. Guo, Z. Hu, and R. C. Qiu, "Experimental validation of channel state prediction considering delays in practical cognitive radio," *IEEE Transactions on Vehicular Technology*, vol. 60, no. 4, pp. 1314, 1325, 2011.

30. Y. Li, S. K. Jayaweera, M. Bkassiny, and K. A. Avery, "Optimal myopic sensing and dynamic spectrum access in cognitive radio networks with low-complexity implementations," *IEEE Transactions on Wireless Communications*, vol. 11, no. 7, pp. 2412, 2423, 2012.

31. N. Nie, C. Comaniciu, and P. Agrawal, "A game theoretic approach to interference management in cognitive networks," *IEEE Transactions on Wireless Communications*, vol. 07030, pp. 199–219, 2007.

32. Z. Ji and K. J. R. Liu, "Dynamic spectrum sharing: A game theoretical overview," *IEEE Communications Magazine*, vol. 45, no. 5, pp. 88–94, May 2007.

33. G. Owen, *Game Theory*, Academic Press, UK, 1995.

34. N. Nie and C. Comaniciu, "Adaptive channel allocation spectrum etiquette for cognitive radio networks," *Mobile Network Applications*, vol. 11, no. 6, pp. 779–797, 2006.

35. R. Etkin, A. Parekh, and D. Tse, "Spectrum sharing for unlicensed bands," *IEEE Journal on Selected Areas in Communications*, vol. 25, no. 4, pp. 517–528, 2007.

36. I. Malanchini, M. Cesana, and N. Gatti, "On spectrum selection games in cognitive radio networks," *IEEE Global Telecommunications Conference*, 2009, pp. 1–7.

37. Y. Cho and F. A. Tobagi, "Cooperative and Non-Cooperative Aloha Games with Channel Capture," *Global Telecommunications Conference, IEEE GLOBECOM 2008*. IEEE, New Orleans, LO, November 30–December 4, 2008.

38. R. Komali, R. Thomas, L. DaSilva, and A. MacKenzie, "The price of ignorance: Distributed topology control in cognitive networks," *IEEE Transactions on Wireless Communications*, vol. 9, no. 4, pp. 1434–1445, 2010.

39. D. Niyato and E. Hossain, "A game-theoretic approach to competitive spectrum sharing in cognitive radio networks," *Proceedings of the IEEE Wireless Communications and Networking Conference*, 2007, pp. 16–20.

40. D. Niyato and E. Hossain, "Competitive pricing for spectrum sharing in cognitive radio networks: Dynamic game, inefficiency of Nash equilibrium, and collusion," *IEEE Journal on Selected Areas in Communications*, vol. 26, no. 1, pp. 192–202, January 2008.

41. Z. Han, C. Pandana, and K. J. R. Liu, "Distributive opportunistic spectrum access for cognitive radio using correlated equilibrium and no-regret learning," *Proceedings of the IEEE Wireless Communications and Networking Conference*, 2007, pp. 11–15.

42. G. Liu, L. Zhou, K. Xiao, B. Yu, G. Zhou, B. Wang, and X. Zhu, "Receiver-centric channel assignment model and algorithm in cognitive radio network," *IEEE 4th International Conference on Wireless Communications, Networking and Mobile Computing*, Dalian, People's Republic of China, October 12–14 2008, pp. 1–4.

43. R. Thomas, R. Komali, A. MacKenzie, and L. DaSilva, "Joint power and channel minimization in topology control: A cognitive network approach," *IEEE International Conference on Communications*, 2007, pp. 6538–6543.

44. D. Xu, X. Liu, and Z. Han, "A two-tier market for decentralized dynamic spectrum access in cognitive radio networks," *Proceedings of the IEEE Society Conference on Sensor Mesh and Ad Hoc Communications and Networks*, Boston, MA, June 21–25, 2010, pp. 1–9.

45. Y. Wu, B. Wang, K. Liu, and T. Clancy, "Repeated open spectrum sharing game with cheat-proof strategies," *IEEE Transactions on Wireless Communications*, vol. 8, no. 4, pp. 1922–1933, 2009.

46. B. Wang, Y. Wu, Z. Ji, K. Liu, and T. Clancy, "Game theoretical mechanism design methods," *IEEE Signal Processing Magazine*, vol. 25, no. 6, pp. 74–84, November 2008.

47. Y. Leshem, "Stable matching for channel access control in cognitive radio systems," *2010 2nd International Workshop on Cognitive Information Processing*, 2010, pp. 470–475.

48. E. Maasland and S. Onderstal, "Auction theory," *Medium Econometrische Toepassingen*, vol. 13, no. 4, pp. 4–8, 2005.

49. S. Gandhi, C. Buragohain, L. Cao, H. Zheng, and S. Suri, "A general framework for wireless spectrum auctions," *Proceedings of the 2nd IEEE International Symposium on New Frontiers in Dynamic Spectrum Access Networks*, 2007, pp. 22–33.

50. Z. Ji and K. J. R. Liu, "Belief-assisted pricing for dynamic spectrum allocation in wireless networks with selfish users," *Proceedings of the IEEE Communications Society Conference on Sensor Mesh and Ad Hoc Communications and Networks*, Reston, VA, September 2006, pp. 119–127.

51. Z. Ji and K. J. R. Liu, "Multi-stage pricing game for collusion-resistant dynamic spectrum allocation," *IEEE Journal on Selected Areas in Communications*, vol. 26, no. 1, pp. 182–191, 2008.

52. L. Chen, S. Iellamo, M. Coupechoux, and P. Godlewski, "An auction framework for spectrum allocation with interference constraint in cognitive radio networks," *2010 Proceedings of the IEEE on INFOCOM*, 2010, pp. 1–9.

53. Y. Wu, B. Wang, K. J. R. Liu, and T. C. Clancy, "A scalable collusion resistant multi-winner cognitive spectrum auction game," *IEEE Transactions on Communications*, vol. 57, no. 12, pp. 3805–3816, 2009.

54. E. Z. Tragos, G. T. Karetsos, S. A. Kyriazakos, and K. Vlahodimitropoulos, "Dynamic segmentation of cellular networks for improved handover performance," *Wireless Communications and Mobile Computing*, vol. 8, no. 7, pp. 907–919, 2008.

55. A. Plummer and S. Biswas, "Distributed spectrum assignment for cognitive networks with heterogeneous spectrum opportunities," *Wireless Communications and Mobile Computing*, vol. 11, no. 9, pp. 1239–1253, 2011.

56. C. Peng, H. Zheng, and B. Y. Zhao, "Utilization and fairness in spectrum assignment for opportunistic spectrum access," *Mobile Networks and Applications*, vol. 11, no. 4, pp. 555–576, 2006.

57. H. Zheng and C. Peng, "Collaboration and fairness in opportunistic spectrum access," *Proceedings of the IEEE International Conference on Communications*, vol. 5, pp. 3132–3136, 2005.

58. D. Willkomm, M. Bohge, D. Holl'os, J. Gross, and A. Wolisz, "Double hopping: A new approach for dynamic frequency hopping in cognitive radio networks," *Proceedings on the IEEE International Symposium on Personal, Indoor and Mobile Radio Communications*, Cannes, France, September 15–18, 2008, pp. 1–6.

59. L. Zhu and H. Zhou, "A new architecture for cognitive radio networks platform," *IEEE 4th International Conference on Wireless Communications, Networking and Mobile Computing*, 2008, pp. 1–4.

60. L. Yang, X. Xie, and Y. Zheng, "A historical-information-based algorithm in dynamic spectrum allocation," *IEEE International Conference on Communication Software and Networks*, Macau, People's Republic of China, February 27–28, 2009, pp. 731–736.

61. M.-V. Nguyen and H.-S. Lee, "Effective scheduling in infrastructure-based cognitive radio networks," *IEEE Transactions on Mobile Computing*, vol. 10, no. 6, pp. 853, 867, 2011.

62. W. Wang and X. Liu, "List-coloring based channel allocation for open-spectrum wireless networks," *Proceedings of the IEEE on VTC*, 2005, pp. 690–694.

63. M. M. Halldorsson, J. Y. Halpern, L. E. Li, and V. S. Mirrokni, "On spectrum sharing games," *Proceedings of the 23rd Annual ACM Symposium on Principles Distributed Computing*, St. John's, NF, Canada, 2004, pp. 107–114.

64. J. Tang, S. Misra, and G. Xue, "Joint spectrum allocation and scheduling for fair spectrum sharing in cognitive radio wireless networks," *Computer Networks*, vol. 52, no. 11, pp. 2148–2158, 2008.

65. J. Zhao, H. Zheng, and G. Yang, "Distributed coordination in dynamic spectrum allocation networks," *Proceedings of the IEEE International Symposium on New Frontiers in Dynamic Spectrum Access Networks*, Baltimore, MD, IEEE Press, Baltimore, MD, November 8–11, 2005, pp. 259–268.

66. M. Hasegawa, T. Ha Nguyen, G. Miyamoto, Y. Murata, and S. Kato, "Distributed optimization based on neurodynamics for cognitive wireless clouds," *Proceedings of the IEEE 18th International Symposium on PIMRC*, Athens, Greece, September 3–7, 2007, pp. 1–5.

67. Z. Zhang and X. Xie, "Intelligent cognitive radio: Research on learning and evaluation of CR based on neural network," *Proceedings of the ITI 5th International Conference on ICICT*, Dhaka, Bangladesh, December 16–18, 2007, pp. 33–37.

68. Y. Liang, Y. SiXing, H. Weijun, and L. ShuFang, "Spectrum behavior learning in cognitive radio based on artificial neural network," *Military Communications Conference*, Baltimore, MD, November 7–10, 2011, pp. 25, 30.

69. S. K. Park, Y. Shin, and W. C. Lee, "Goal-Pareto based NSGA for optimal reconfiguration of cognitive radio systems," *Proceedings of the 2nd International Conference on CrownCom*, Orlando, FL, August 1–3, 2007, pp. 147–153.

70. D. Thilakawardana and K. Moessner, "A genetic approach to cell-by-cell dynamic spectrum allocation for optimising spectral efficiency in wireless mobile systems," *Proceedings of the 2nd International Conference on CrownCom*, Orlando, FL, August 1–3, 2007, pp. 367–372.

71. J. M. Kim, S. H. Sohn, N. Han, G. Zheng, Y. M. Kim, and J. K. Lee, "Cognitive radio software testbed using dual optimization in genetic algorithm," *Proceedings of the 3rd International Conference on CrownCom*, Singapore, May 15–17, 2008, pp. 1–6.

72. Z. Changchang and Z. Yunxiao, "Cognitive radio resource allocation based on niche adaptive genetic algorithm," *IET International Conference on Communication Technology and Application*, October 14–16, 2011, pp. 566, 571.

73. M. Xu and J. Hong, "Biologically-inspired distributed spectrum access for cognitive radio network," *2010 6th International Conference on Wireless Communications Networking and Mobile Computing*, September 23–25, 2010, pp. 1, 4.

74. H. Salehinejad, S. Talebi, and F. Pouladi, "A metaheuristic approach to spectrum assignment for opportunistic spectrum access," *2010 IEEE 17th International Conference on Telecommunications*, April 4–7, 2010, pp. 234, 238.

75. W. Fei and Y. Zhen, "An ACO algorithm for sum-rate maximization problem in opportunistic spectrum allocation," *4th International Conference on Wireless Communications, Networking and Mobile Computing*, October 12–14, 2008, pp. 1, 4.

76. H.-S. T. Le and H. D. Ly, "Opportunistic spectrum access using fuzzy logic for cognitive radio networks," *2008 2nd International Conference on Communications and Electronics*, June 4–6, 2008, pp. 240, 245.

77. L. Xuchen and Z. Wenzhu, "A novel dynamic spectrum access strategy applied to cognitive radio network," *2011 7th International Conference on Wireless Communications, Networking and Mobile Computing*, September 23–25, 2011, pp. 1, 5.

78. Z. Wenzhu and L. Xuchen, "Centralized dynamic spectrum allocation in cognitive radio networks based on fuzzy logic and Q-learning," *China Communications*, vol. 8, no. 7, pp. 64–78, 2011.

79. M. Cesana, E. Ekici, and Y. Bar-Ness, "Networking over multi-hop cognitive networks," *IEEE Network*, vol. 23, no. 4, pp. 4–5, 2009.

80. H. Khalife, S. Ahuja, N. Malouch, and M. Krunz, "Probabilistic path selection in opportunistic cognitive radio networks," *Proceedings of the IEEE Global Telecommunications Conference*, New Orleans, LA, November 30–December 4, 2008, pp. 1–5.

81. M. G. Khoshkholgh, K. Navaie, and H. Yanikomeroglu, "Access strategies for spectrum sharing in fading environment: Overlay, underlay, and mixed," *IEEE Transactions on Mobile Computing*, vol. 9, no. 12, pp. 1780–1793, 2010.

82. M. Xie, W. Zhang, and K.-K. Wong, "A geometric approach to improve spectrum efficiency for cognitive relay networks," *IEEE Transactions on Wireless Communications*, vol. 9, no. 1, pp. 268–281, 2010.

83. X.-L. Huang, G. Wang, F. Hu, and S. Kumar, "Stability-capacity-adaptive routing for high-mobility multihop cognitive radio networks," *IEEE Transactions on Vehicular Technology*, vol. 60, no. 6, pp. 2714, 2729, 2011.

84. G. Cheng, W. Liu, Y. Liu, and W. Cheng, "Spectrum aware on-demand routing in cognitive radio networks," *Proceedings of the 2nd IEEE on Dynamic Spectrum Access Networks*, Dublin, Ireland, April 17–20, 2007, pp. 571–574.

85. G. Zhu, I. F. Akyildiz, and G. Kuo, "STOD: A spectrum-tree based on demand routing protocol for multi-hop cognitive radio networks," *Proceedings of the IEEE Global Telecommunications Conference*, New Orleans, LA, November 8–11, 2008, pp. 1–5.

86. J. Cho and Z. J. Haas, "On the throughput enhancement of the downstream channel in cellular radio networks through multihop relaying," *IEEE Journal on Selected Areas in Communications*, vol. 22, no. 7, pp. 1206–1219, 2004.

87. A. Abbagnale and F. Cuomo, "Leveraging the algebraic connectivity of a cognitive network for routing design," *IEEE Transactions on Mobile Computing*, vol. 11, no. 7, pp. 1163, 1178, 2012.

88. F. Li, K. Wu, and A. Lippman, "Energy-efficient cooperative routing in multi-hop wireless ad hoc networks," *Proceedings of the IEEE International Performance, Computing, and Communications Conference*, April 2006, pp. 215–222.

89. C. Pandana, W. P. Siriwongpairat, T. Himsoon, and K. J. R. Liu, "Distributed cooperative routing algorithms for maximizing network lifetime," *Proceedings of the IEEE Wireless Communications and Networking Conference*, vol. 1, Las Vegas, NV, pp. 451–456, 2006.

90. I.-T. Lin and I. Sasase, "A primary traffic based multihop routing algorithm using cooperative transmission in cognitive radio ad hoc networks," *Consumer Communications and Networking Conference*, Las Vegas, NV, January 14–17, 2012, pp. 418, 423.

91. Y. T. Hou, Y. Shi, and H. D. Sherali, "Spectrum sharing for multi-hop networking with cognitive radios," *IEEE Journal on Selected Areas in Communications*, vol. 26, no. 1, pp. 146, 155, 2008.

92. M. Caleffi, I. F. Akyildiz, and L. Paura, "OPERA: Optimal routing metric for cognitive radio ad hoc networks," *IEEE Transactions on Wireless Communications*, vol. 11, no. 8, pp. 2884, 2894, 2012.

93. P. B. Nagaraju, L. Ding, T. Melodia, S. N. Batalama, D. A. Pados, and J. D. Matyjas, "Implementation of a distributed joint routing and dynamic spectrum allocation algorithm on USRP2 radios," *2010 7th Annual IEEE Communications Society Conference on Sensor Mesh and Ad Hoc Communications and Networks*, June 21–25, 2010, pp. 1, 2.

94. A. Sampath, L. Yang, L. Cao, H. Zheng, and B. Y. Zhao, "High throughput spectrum-aware routing for cognitive radio based ad-hoc networks," *Proceedings of the 3rd International Conference on Cognitive Radio Oriented Wireless Networks and Communications*, Singapore, 2008.

95. J. Sobrinho, "An algebraic theory of dynamic network routing," *IEEE/ACM Transactions on Networks*, vol. 13, no. 5, pp. 1160–1173, 2005.

96. I. Filippini, E. Ekici, and M. Cesana, "Minimum maintenance cost routing in cognitive radio networks," *Proceedings of the 2009 IEEE International Conference on Mobile Adhoc Sensor Systems*, 2009, pp. 284–293.

97. L. Ding, T. Melodia, S. Batalama, and M. J. Medley, "Rosa: Distributed joint routing and dynamic spectrum allocation in cognitive radio ad hoc networks," *Proceedings of the 2009 ACM International Conference on Modeling, Analysis Simulation Wireless Mobile Systems*, 2009, pp. 13–20.

98. Z. Wang, Y. Sagduyu, J. Li, and J. Zhang, "Capacity and delay scaling laws for cognitive radio networks with routing and network coding," *Proceedings of the 2010 Military Communications Conference*, 2010, pp. 1375–1380.

99. S.-C. Lin and K.-C. Chen, "Spectrum aware opportunistic routing in cognitive radio networks," *IEEE 2010 Global Telecommunication Conference*, Miami, FL, December 2010, pp. 1–6.

100. A. Abbagnale and F. Cuomo, "Gymkhana: A connectivity-based routing scheme for cognitive radio ad hoc networks," *2010 INFOCOM IEEE Conference on Computer Communications Workshops*, March 15–19, 2010, pp. 1, 5.

101. K. R. Chowdhury and I. F. Akyildiz, "CRP: A routing protocol for cognitive radio ad hoc networks," *IEEE Journal on Selected Areas in Communications*, vol. 29, no. 4, pp. 794–804, 2011.

102. S. Soltani and M. Mutka, "On transitional probabilistic routing in cognitive radio mesh networks," *2011 IEEE International Symposium on a World of Wireless, Mobile and Multimedia Networks*, June 20–24, 2011, pp. 1, 9.

103. Y. Song, C. Zhang, and Y. Fang, "Stochastic traffic engineering in multihop cognitive wireless mesh networks," *IEEE Transactions on Mobile Computing*, vol. 9, pp. 305–316, 2010.

104. Y. Cui, W. Hu, S. Tarkoma, and A. Yla-Jaaski, "Probabilistic routing for multiple flows in wireless multihop networks," *2009 IEEE 34th Conference on Local Computer Networks*, 2009, pp. 261–264.

105. A. Cagatay Talay and D. T. Altilar, "ROPCORN: Routing protocol for cognitive radio ad hoc networks," *International Conference on Ultra Modern Telecommunications & Workshops*, October 12–14, 2009, pp. 1, 6.

106. A. Abbagnale and F. Cuomo, "Connectivity-driven routing for cognitive radio ad-hoc networks," *2010 7th Annual IEEE Communications Society Conference on Sensor Mesh and Ad Hoc Communications and Networks*, June 21–25, 2010, pp. 1, 9.

107. L. Jun, Y. Zhou, L. Lamont, and F. Gagnon, "A novel routing algorithm in cognitive radio ad hoc networks," *IEEE Global Telecommunications Conference*, December 5–9, 2011, pp. 1, 5.

108. C. Xin, B. Xie, and C.C. Shen, "A novel layered graph model for topology formation and routing in dynamic spectrum access networks," *Proceedings of the 2005 First IEEE International Symposium on New Frontiers in Dynamic Spectrum Access Networks*, Baltimore, MD, November 2005, pp. 308–317.

109. X. Zhou, L. Lin, J. Wang, and X. Zhang, "Cross-layer routing design in cognitive radio networks by colored multigraph model," *Wireless Personal Communications*, vol. 49, no. 1, pp. 123–131, 2009.

110. A. B. McDonald and T. F. Znati, "A mobility-based framework for adaptive clustering in wireless ad hoc networks," *IEEE Journal on Selected Areas in Communications*, vol. 17, no. 8, pp. 1466–1487, 1999.

111. T. Yamamoto, T. Ueda, and S. Obana, "A proposal of adaptive traffic route control scheme in QoS provisioning for cognitive radio technology with heterogeneous wireless systems," *2009 4th International Conference on Cognitive Radio Oriented Wireless Networks and Communications*, Hannover, Germany, June 22–24, 2009, pp. 1, 6.

112. P. C. Ng, D. Edwards, and S. C. Liew, "Coloring link-directional interference graphs in wireless ad hoc networks," *Proceedings of the IEEE on Global Telecommunications Conference*, November 2007, pp. 859–863.

113. K. Duffy, N. O'Connell, and A. Sapozhnikov, "Complexity analysis of a decentralized graph coloring algorithm," *Information Processing Letters*, vol. 107, no. 2, pp. 60–63, 2008.

114. C. S. R. Murthy and B. S. Manoj, *Ad Hoc Wireless Networks: Architectures and Protocols*. Upper Saddle River, NJ: Prentice Hall, 2004.

115. M. H. Manshaei, M. Felegyhazi, J. Freudiger, J.-P. Hubaux, and P. Marbach, "Spectrum sharing games of network operators and cognitive radios," in *Cognitive Wireless Networks: Concepts, Methodologies and Visions—Inspiring the Age of Enlightenment of Wireless Communications*, F. Gagnon, ed. New York: Springer-Verlag, 2007.

116. I. Pefkianakis, S. H. Y. Wong, and S. Lu, "SAMER: Spectrum-aware mesh routing in cognitive radio networks," *Proceedings of the 3rd IEEE International Symposium on Dynamic Spectrum Access Networks*, Chicago, IL, October 2008, pp. 1–5.

117. K. Hendling, T. Losert, W. Huber, and M. Jandl, "Interference minimizing bandwidth guaranteed online routing algorithm for traffic engineering," *Proceedings of the 12th IEEE on ICON*, vol. 2, pp. 497–503, November 2004.

118. R. S. Komali, A. B. MacKenzie, and P. Mähönen, "On selfishness, local information, and network optimality: A topology control example," *Proceedings of the International Conference on Computer Communications and Networks*, San Francisco, CA, August 3–6, 2009, pp. 1–7.

119. G. Cheng, W. Liu, Y. Z. Li, and W. Cheng, "Joint on-demand routing and spectrum assignment in cognitive radio networks," *Proceedings of the IEEE International Conference on Communications*, Glasgow, Scotland, 2007.

120. H. S. Ma, L. L. Zheng, X. Ma, and Y. Luo, "Spectrum aware routing for multi-hop cognitive radio networks with a single transceiver," *Proceedings of the 3rd International Conference on Cognitive Radio Oriented Wireless Networks and Communications*, Singapore, May 15–17, 2008.

121. Q. W. Wang and H. T. Zheng, "Route and spectrum selection in dynamic spectrum networks," *Proceedings of the 3rd IEEE Consumer Communications and Networking Conference*, Munich, Germany, January 8–10, 2006.

122. K. G. Bian and P. Jung-Min, "Segment-based channel assignment in cognitive radio ad-hoc networks," *Proceedings of the 2nd International Conference on Cognitive Radio Oriented Wireless Networks and Communications*, Orlando, FL, August 1–3, 2007.

123. T. Chen, H. Zhang, G. M. Maggio, and I. Chlamtac, "CogMesh: A cluster-based cognitive radio network," *Proceedings of the 2nd IEEE International Symposium on New Frontiers in Dynamic Spectrum Access Networks*, Dublin, Ireland, April 17–20, 2007.

124. T. Fujii, Y. Kamiya, and Y. Suzuki, "Multi-band ad-hoc cognitive radio for reducing inter-system interference," *Proceedings of the IEEE 17th International Symposium on Personal, Indoor and Mobile Radio Communications*, Helsinki, Finland, September 11–14, 2006.

125. T. Fujii and Y. Suzuki, "Ad-hoc cognitive radio development to frequency sharing system by using multi-hop network," *Proceedings of the 1st IEEE International Symposium on New Frontiers in Dynamic Spectrum Access Networks*, Baltimore, MD, 2005.

126. H. M. Almasaeid, "Spectrum allocation algorithms for cognitive radio mesh networks," PhD dissertation, Iowa State University, Ames, IA, 2011.

127. A. Attar, H. Tang, A. V. Vasilakos, F. R. Yu, and V. C. M. Leung, "A survey of security challenges in cognitive radio networks: Solutions and future research directions," *Proceedings of the IEEE*, vol. 100, no. 12, pp. 3172, 3186, December 2012.

128. F. Wu, S. Zhong, and C. Qiao, "Strong-incentive, high-throughput channel assignment for noncooperative wireless networks," *IEEE Transactions on Parallel and Distributed Systems*, vol. 21, no. 12, pp. 1808–1821, 2010.

Biographical Sketches

Xianzhong Xie (xiexzh@cqupt.edu.cn) received his PhD degree in communication and information systems from Xi'dian University, Xi'an, People's Republic of China, in 2000. He is currently a professor of the School of Computer Science and Technology at Chongqing University of Posts and Telecommunications, Chongqing, People's Republic of China, and a director of the Institute of Broadband Access Technologies, Chongqing, People's Republic of China. His research interests include cognitive radio (CR) networks, MIMO precoding, and signal processing for wireless communications. He is the principal author of five books on cooperative communications, 3G, multiple-input multiple-output (MIMO), CR, and time division duplexing (TDD) technology. He has published more than 80 papers in journals and 30 papers in international conferences.

Helin Yang (yhelincqupt@163.com) received his BS degree from Chongqing University of Posts and Telecommunications, Chongqing, People's Republic of China, in 2013. He is a postgraduate student of Chongqing University of Posts and Telecommunications, People's Republic of China, and will receive his MS degree in communication and information systems in 2016. His research interests include cognitive radio networks, cooperative communications, and wireless communications.

Athanasios V. Vasilakos received his PhD degree in computer engineering from the University of Patras, Patras, Greece, in 1988. He is currently a professor at the department of computer and telecommunications engineering, University of Western Macedonia, Kozani, Greece, and a visiting professor of the graduate program of the department of electrical and computer engineering, National Technical University of Athens (NTUA), Athens, Greece. He is a senior member of the Institute of Electrical and Electronics Engineers (IEEE). He has authored or coauthored over 200 technical papers in major international journals and conferences. He is the author or coauthor of 5 books and 20 book chapters in the areas of communications. Dr. Vasilakos served as a general chair, technical program committee chair, and symposium chair for many international conferences. He served or is serving as an editor and/or guest editor for many technical journals, that is, the *IEEE Transactions on Network and Service Management*, the *IEEE Transactions on Systems, Man, and Cybernetics Part B: Cyber-Netics*, the *IEEE Transactions on Information Technology in Biomedicine*, and the *IEEE Journal on Selected Areas in Communications*. He is the founding editor-in-chief of the *International Journal of Adaptive and Autonomous Communications Systems* (*IJAACS*, http://www.inderscience.com/ijaacs) and the *International Journal of Arts and Technology* (*IJART*, http://www.inderscience.com/ijart). He is the chairman of the European Alliance for Innovation.

3

NEIGHBOR DISCOVERY FOR COGNITIVE RADIO NETWORKS

ATHAR ALI KHAN, MUBASHIR HUSAIN REHMANI, AND YASIR SALEEM

Contents

3.1 Introduction

In recent years, cognitive radio (CR) technology has gained a lot of attention by the research community to alleviate the spectrum shortage problem faced by traditional wireless systems. Equipped with highly flexible spectrum sensing capabilities, CR technology enables radio nodes to change their transmission parameters, so that they can dynamically adapt to spectrum availability in their geographical region. CR nodes scan and identify the unused spectrum resources in the licensed bands without causing any interference to the legitimate users, assigned to these frequency bands [1,2].

A *CR* node maintains a set of locally available channels after scanning the spectrum usage. This set of channels may be different for different nodes, leading to the following layer-2 autoconfiguration (L2AC) issues:

- Without any central authority, how do nodes know about their potential neighbors?
- What is the set of available channels to form a communication infrastructure [1,3]?

Hence, the problem of neighbor discovery (ND) comes forward.

To ensure successful communication in CR networks (CRNs), CR nodes have to adapt to certain mechanisms, such as information sharing about neighborhood and spectrum availability. However, it is possible only if nodes are aware of each other and a network of trusted users is formed. The first step in this regard is the successful completion of ND phase and the knowledge of available channels that can be used for communication among neighbors. The dynamic and challenging wireless environment requires that this phase must be as quick as possible [2]. In ND, information retrieval at the local level enables every node to determine the set of nodes that can help for reliable communication. Neighbor discovery is defined as follows: "Two radio nodes are considered as neighbors, whether or not they may be geographically located near to each other, if they are within transmission range of each other for a considerable amount of time and have at least one common channel between them" [1,3,4]. Figure 3.1 shows a general illustration of the concept of neighborhood in which CR nodes are involved in an ND process. In this figure, CR users 1, 2,

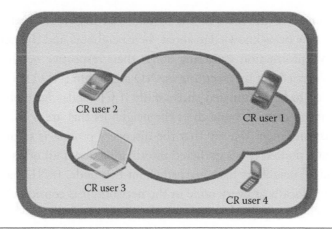

Figure 3.1   The concept of neighborhood.

and 3 are neighbors because these are within the transmission range of each other and are connected with at least one common channel. But CR user 4 is not a neighbor because although it resides in the transmission range of CR users 1 and 3, but it does not have a common channel.

An efficient discovery algorithm should have certain features as depicted in Figure 3.2 [1]:

- The algorithm must complete in a minimum time discovering maximum neighbors.
- There is no false discovery, that is, it must be secure.
- All of the neighbors are eventually discovered by each node.
- It should have strong termination guarantees, which mean that after successful completion, the discovery process eventually terminates at every node.

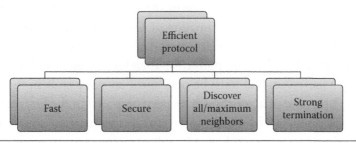

Figure 3.2   Features of an efficient ND protocol.

It is assumed at the start of the system that individual nodes may not have any prior knowledge about their neighbors and the network topology. Information gathering about their respective neighboring nodes is carried out by executing an ND algorithm (NDA). The ND is believed to be terminated successfully if CR nodes have obtained sufficient information on each of the neighbors that are within their transmission range. We say that the discovery process is terminated locally at a node if it has gathered information about all of its neighbors and all neighbors have learned about this node. If ND has been completed locally at every node in the network, it is considered to be completed globally [1,2].

The discovery process is easily implemented in traditional ad hoc networks, as all nodes have the same availability of channels and a precise wireless standard is followed. But this is somehow very difficult in CRN due to primary users' (PUs) activities and heterogeneity in the available channels, such as spatial variations in hardware and frequency usage. These characteristics create a complex situation for the ND process in CRN. Different CR nodes may have different available channels, which may vary over time and space due to the arrival and departure of legitimate users and other CR nodes at different geographical locations. Furthermore, all CR nodes may not have the time synchronization. Therefore, these issues must be addressed in order to develop efficient NDAs [3]. ND is a fundamental step because it can be used as a basis for solving other important communication problems, such as broadcasting or gossiping a message, finding a globally common control channel (CCC) that can be used by all neighbors for control information exchange, and computing a deterministic transmission schedule [1,5]. These important features are also depicted in Figure 3.3.

In a scenario where each node in the network has the knowledge about its neighbors, then estimation of linear time leader and linear time depth-first search (DFS) traversal becomes easy, which in turn can solve the broadcasting problems. Therefore, in order to broadcast multiple messages, ND must be performed on every node in the network. A number of algorithms have been proposed in the literature to solve the problem of ND [1]. Successful communication among two channels can be accomplished only if they are tuned to the same channel at the same time, so determination of globally common channel

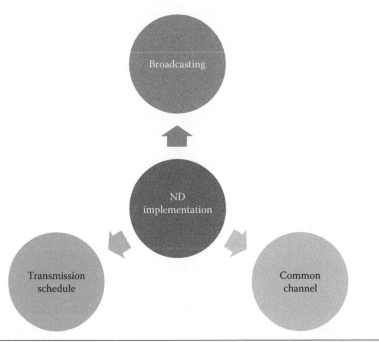

Figure 3.3   Use of ND in a CR network.

is required. Furthermore, due to dynamic channel availability sets on different nodes, nodes may be needed to perform ND on multiple channels based on their available channels. Therefore, timely discovery of neighbors and time synchronization can facilitate the nodes to detect the ND completion at the same time [5,6].

The goal of this chapter is to provide and consolidate information about ND protocols in CRNs. We provide the features and characteristics of various NDAs in order to promote the development of NDAs for CRNs.

The remainder of this chapter is organized as follows: Section 3.2 discusses the research work carried out for ND in traditional wireless networks. Moving one step ahead, Section 3.3 gives a detailed information on NDAs in CRNs with their classification and Section 3.4 provides issues, challenges, and future research directions. Finally, Section 3.5 concludes the chapter.

3.2 ND in Traditional Wireless Networks

ND protocols are the fundamental building blocks of traditional wireless networks. In a dynamic environment, ND is critical, so it must be robust, reliable, and secure. Significant amount of work has

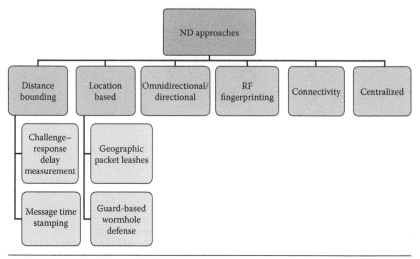

Figure 3.4   ND approaches in wireless networks.

been carried out in the literature on problems and definitions of types of ND. The *existing approaches* of ND protocol are classified as given in Figure 3.4.

In the distance-bounding approach, the signal round-trip time is measured and multiplied by the propagation speed in order to calculate the distance to a potential neighbor. These protocols employ the cryptographic features for secure discovery, and for radio-frequency (RF) signals, it is very difficult for an adversary node to decrease the estimated distance. The distance-bounding approaches have two variants: challenge–response delay measurement and message time stamping.

In the location-based approach, the location information is used as the basic feature for ND. The location-based protocols can be either geographic packet leashes or guard-based wormhole defense.

Omnidirectional antenna approach is used in a majority of wireless applications due to its simplicity. However, directional antenna approach has some limitations, but it is used to provide increased protection by focusing transmission energy in a certain direction.

There are specific signal patterns induced by radio transmitters. These are uniquely identified at the receiver and the identification of these patterns is the basis of RF fingerprinting approaches.

Connectivity approaches use local network connectivity information for a secure ND. These approaches do not require any specialized node hardware.

The discovery schemes that do not focus on the security are classified under centralized approaches. Operating in a long-range environment, these approaches are designed to detect the presence and approximate location of attackers.

The details about NDAs are provided in Table 3.1. The table describes the simulators used to check the performance of the protocols. Some of the features of the related work summarized here are the parameters, whether the coordination is required between a transmitter and a receiver, whether the algorithm is employed on a single channel or multichannels, the algorithm type, and the initial knowledge required. It is worth mentioning that people have used a variety of simulators for performance evaluation.

An interesting point observed from this table is that probabilistic ND protocols are employed in a single-channel network, whereas both probabilistic and random ND protocols are used for multichannel networks. The parameters discussed in passive algorithms are related to beacon interval. Energy detection or energy conservation is focused in probabilistic NDAs. Another notable thing is that almost all the NDAs require distinctive knowledge to start the process, whereas both single-channel algorithms need to know the number of nodes at the initial point.

3.3 Cognitive Radio Networks

With the increasing applications in wireless communication in recent times, there is a possibility of an acute shortage of bandwidth in the near future. The available spectrum is becoming extremely populated due to various applications, such as television transmission, microwave communication, and cellular communication. However, it has been observed that a significant portion of this allocated spectrum is not properly utilized [1]. CR is a promising technology likely to be adopted in the very near future to alleviate the spectrum underutilization problem [2]. CR technology enables the radio nodes to optimally exploit the available spectrum resources at a given geographical region

Table 3.1 ND in Traditional Wireless Networks

PROTOCOL NAME	SIMULATOR USED	PARAMETERS	NUMBER OF CHANNELS	COORDINATION	TYPE	REQUIRED KNOWLEDGE
OPT, SWOPT, SUBOPT [8]	Crossbow TelosB mote platform, TinyOS 2.1.1	Different beacon intervals	Multichannels	Tx, Rx	Passive	Beacon intervals
Birthday protocols [9]	–	Energy conservation environment	Single channel	Tx, Rx	Probabilistic	Number of nodes
ALOHA-like [10]	–	Coupon collector's problem, collision detection	Multichannels	Tx, Rx	Randomized	–
OPT, SWOPT [11]	OMNeT++ 3.3 with the mobility framework (MF), OMNeT++ IEEE 802.15.4 implementation developed at TKN	Linear programming optimization, different beacon intervals, multiple frequency bands	Multichannels	Tx, Rx	Passive	Beacon intervals
Hello [12]	Netsens	Interference considerations	–	Tx, Rx	Random	Node density
SEDINE [13]	MATLAB	Security against nonneighbor and malicious nodes	–	Tx, Rx	–	Pair-wise key management protocol
[14]	Setdest validation framework (NS-2)	Random waypoint mobility model	–	–	–	Mobility parameters
Two-hop MAC [15]	–	Multihop consideration for spectrum allocation	Multichannels	–	Distributed	Traffic demand, traffic dependency

Direct, gossip-based [16]	OPNET version 14.5	Direct and reflected beams, node density, antenna beamwidth	–	Tx, Rx	Random	Slot length
SAND [17]	Qualnet	Serial ND process, bounded time, reliable relay, no external global synchronization	–	Tx, Rx	–	Number of sectors
–[18]	MATLAB	Collision detection feedback at transmitters, fading	Single channel	Tx, Rx	Probabilistic	Max. no. of neighbors
–[19]	No simulation	Randomized algorithms for multihop, multi-channel, heterogeneous, wireless network	Multichannels	Tx, Rx	Randomized	Number of nodes, drift rate

ALOHA, Areal Locations of Hazardous Atmospheres; MAC, Medium Access Control; MATLAB, MATrix LABoratory; OPT, Optimized; Rx, Receiver; SAND, Sectored-Antenna Neighbor Discovery; SUBOPT, Suboptimal; SWOPT, Switched Optimized; TKN, The Telecommunication Networks Group, Faculty of Electrical Engineering and Computer Science at the Technische Universität, Berlin; Tx, Transmitter.

and dynamically adapt to spectrum availability without causing any disturbance to legitimate users or PUs. CR nodes opportunistically scan and identify the unused channels in the operating environment. A channel is believed to be available if the secondary user (SU) can communicate on this channel for a reasonable amount of time without causing any interference with the PU(s) [1]. A CRN is composed of CR-enabled SUs or secondary nodes scattered geographically and able to opportunistically perform spectrum access [5].

The users assigned to communicate on the licensed frequency band are called PUs and all other looking for opportunistic spectrum access are SUs. An SU scans the unused channels in the operating environment and, after identification of these channels, starts its communication over these channels. If the PU returns to the channel, the SU has to vacate the channel and switch to some other available channel for communication [3]. This mechanism is known as spectrum handoff.

A CR node scans the channels, and after finding available channels, it utilizes them to communicate with other CR nodes. But time and geographical location may vary this channel set. Two CR nodes may have different available channel sets due to uneven propagation of wireless signals making the network heterogeneous. Moreover, no information about the IDs of neighbors, diverse signal propagation characteristics, geographical scattering of nodes, collisions, and the unknown size of neighborhood makes the ND a noticeable problem due to which the termination of discovery process becomes difficult. ND is an important and initial step to set up a wireless ad hoc network. In case of multiple CR nodes, an ad hoc wireless network has a heterogeneous environment that increases the complexity of conducting ND in the available channel set across the network.

3.3.1 ND in CRNs

ND in CRNs may be based on certain characteristics and operational metrics. As shown in Figure 3.5, some protocols and algorithms are presented in the literature, which focus on various aspects such as the available channel set [1,3,4], CCC [7,20–22,25], jamming attacks [2], transmitter/receiver dependence [5], termination detection [6], algorithms/protocols [23], performance [24,25], and rendezvous problem

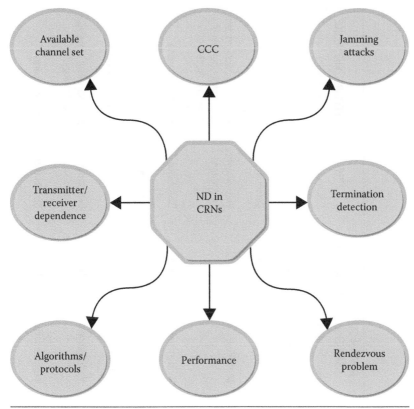

Figure 3.5 Different aspects of research work carried out for ND in CRNs.

[26,27]. A brief overview about ND protocols is given in Tables 3.2 and 3.3. In our work, we focus mainly on the following:

- Synchronous/asynchronous operational behavior
- Finding a global common channel
- Collision-free algorithm

In this section, we will use the notations as given in Table 3.3.

3.3.1.1 Synchronous Operation In synchronous operation, CR nodes operate in a slotted time and have access to synchronous clocks. This enables every CR node to detect the start and end of a timeslot. This may result in an improved efficiency as NDA may consume less time, but it requires a complex implementation. At the same time, this synchronous

Table 3.2 ND in CRNs

PROTOCOL NAME	SIMULATOR USED	PARAMETERS	NUMBER OF CHANNELS	COORDINATION	TYPE	REQUIRED KNOWLEDGE
NDA_SC, NDA_MC, NDA_BD, GA_SC [1]	Self-made in Java	Collision-aware, size-aware	Multichannels	Tx, Rx	Deterministic	Number of nodes, noise
JENNA [2]	–	Control packet dissemination	Multichannels	Tx, Rx	Distributed	Set of orthogonal channels
–[3]	No simulation	Leader election	Multichannels	Tx, Rx	Deterministic	Available channels, beacon duration
[5]	No simulation	TDMA-based collision-free algorithm, multiple receivers with single transmitter	Multichannels	Tx, Rx	Deterministic/ distributed	Number of nodes, number of channels, number of receivers, number of transmitters
FNDACC [4]	–	Available channel comparability	Multichannels	Tx, Rx	Distributed	Number of channels
ND_IDC [6]	–	Transmission concurrency, termination detection	Multichannels	Tx, Rx	Deterministic	Universal channel set
[23]	OMNeT++	ACROPOLIS joint research activities	Multichannels	Tx, Rx	–	Common channel
[28]	–	Handshake-based algorithm using quorum systems	Multichannels	Tx, Rx	Deterministic	Universal channel set, fixed node topology
[24]	–	Discovery failures	Multichannels	Tx, Rx	Distributed/ random	Common channel

[25]	No simulation	Set of common channels, lower bound, diameter-aware, diameter-unaware	Multichannels	Tx, Rx	Distributed	Node mobility, time synchronization, number of possible channels
[29]	–	UWB technology	Single channel	Tx, Rx	–	UWB interface/energy transmitted
[30]	–	Delay bound, energy consumption	–	Tx, Rx	–	Time interval
[31]	No simulation	Local cognitive pilot channel approach	Single channel	–	–	Set of neighbors
[32]	MATLAB	Sequences	Multichannels	Tx, Rx	–	Nonorthogonal sequences
Random, generated orthogonal sequence based, modular clock, modified modular clock [33]	Simulator not mentioned	Rendezvous problem	Multichannels	Tx, Rx	Deterministic/probabilistic	Available channels, orthogonal sequence
RAC²E [26]	Spectrum analyzer	Cooperative environments	Multichannels	Tx, Rx	Randomized	Local spectrum map

Table 3.3 Description of Notations

NOTATION	DESCRIPTION
k	Size of space used for assigning labels (identifiers)
M	Number of channels
N	Number of nodes
b	Difference of channel availability sets of two neighboring nodes
D	Diameter of the network
P	Co-prime
B	Delay bound
r	Number of receivers

condition must be maintained throughout the process. In this section, we will discuss some additional features of NDAs while operating in a synchronous mode.

In a network consisting of synchronous CR nodes, a new paradigm of the NDA is presented. This research work is different from other existing works in many ways, such as the following:

1. The information about the size of space used for assigning labels is available to the nodes (k). The information about the number of channels over which these nodes can operate (M) is also available, but nodes do not know the actual number of nodes present in the network (N).
2. There is no relationship between N and k, and $M \geq 1$.
3. Nodes may operate over multiple channels and different channel availability sets are available to different nodes. Moreover, nodes do not have directional antennas and the NDA has strong termination guarantees.

3.3.1.1.1 ND for Collision-Aware Nodes In undirected networks with collision-aware nodes, fast and deterministic algorithms have been described that satisfy the strong termination condition, that is, nodes not only detect the completion of ND but this process also terminates at the same time. The NDA for a single channel (NDA_SC) gives eventual but no false discovery with time complexity $O(N \log k)$. For multichannels, the same algorithm is modified when each timeslot is replaced by a frame in NDA_SC to give NDA_MC algorithm, with the same characteristics and a time complexity of $O(NM \log k)$. It has been assumed so far that there is an arbitrary change in the channel availability sets of CR nodes. But this is somehow different in practice, because the channel availability sets may be close to each other. Therefore, in this case if nodes know a small bound on divergence in channel availability sets of neighbors, then we can achieve a comparatively faster algorithm. This is done by changing the structure of a frame used when transforming the single-channel algorithm to the multichannel algorithm. The resulting new NDA with bounded channel divergence (NDA_BD) enhances the speed of ND when nodes are collision aware, and there is no dynamic change in channel availability sets of neighbors. NDA_BD has the same properties as in case of NDA_SC and NDA_MC. Its time complexity is $O(N(b + 1)^2 \log k)$ and it is much faster than NDA_MC if $b \ll \sqrt{M}$, where b is the difference of channel availability sets of two neighboring nodes.

3.3.1.1.2 ND for Size-Aware Nodes Now if nodes are size aware in undirected networks and they have a polynomial upper bound on the network size, then a fast *deterministic gossiping algorithm* for a single channel (GA_SC) is described with a strong termination condition and a time complexity of $(\phi N \log^2 N \log^2 k)$, for some known constant ϕ. For ND on all channels, the time complexity is m times that of ND on a single channel, that is, $O(Nm \log^2 N \log^2 k)$.

3.3.1.1.3 ND in Directed Networks In all of the above scenarios, network is believed to be undirected but a significant work has been carried out in directed networks also. Here, to discover neighbors over a single channel, the NDA has an overall time complexity of $O(ND \log k)$ timeslots, where D denotes the diameter of the network. In multiple channels, the algorithm has the general time complexity of $O(NDM \log k)$, and the bounded divergence time complexity of $O(ND(b + 1)^2 \log k)$. Therefore, the NDA has the improvement in the time complexity by a factor of $O(D)$ for a directed network than that for an undirected network [1].

3.3.1.1.4 ND with Available Channel Comparability In another slot-based scenario with synchronized nodes, a fast distributed NDA in CR ad hoc networks (CRAHNs) relies on the available channel comparability (ACC) among neighboring nodes and uses the data and control transceivers for a quick discovery process. The fast NDA based on ACC (FNDACC) algorithm is different from the previous work in the sense that it does not increase the hardware requirement for CR nodes. Due to distributed characteristics of the algorithm, CR nodes attempt to search the neighboring nodes individually and without any help of network infrastructure or CCC. The performance of the FNDACC is compared against two schemes: L2AC algorithm and NDA referred as spectrum-opportunity clustering (*SOC*). For this purpose, the topology and channel information are taken into account to calculate the time when more than 90% of nodes gain a complete knowledge of their neighbors. It is observed that SOC and FNDACC are not sensitive to nodes and do not depend on channel assignment for individual nodes. This results to the fact that both algorithms require less time compared to L2AC. Also, the time consumed by FNDACC is at most 29% less compared to SOC [4].

3.3.1.1.5 ND for Multihop Environment At the start of the ND process in a heterogeneous multihop CRN, a CR node knows only the universal channel set, the size of this set, the channel availability set for this CR node, the ID of CR node, and space size, and everything else is unknown. In such an environment, ND becomes a challenge. Researchers have worked in this regard and a new *deterministic* NDA *ND_IDC* is proposed. This algorithm allows multiple nodes to broadcast simultaneously without interfering with each other and has the termination detection capability. In fact, it has lightweight termination detection with 100% discovery rate. Based on *transmission concurrency*, it works in multichannel networks with unknown topology. During runtime, the nodes may join and/or leave the network and the available channel sets may change over time, creating dynamism in the network. To deal with this situation, ND_IDC is periodically executed at the predetermined time instants so that all nodes start executing the algorithm at the same time. A performance comparison between ND_IDC and another algorithm ND_SA (serial advertisement) shows that ND_IDC has a significant reduction in time complexity, that is, up to 97%–98%. ND_IDC gives a smoother time complexity, whereas in ND_SA, it is a step function. The improvement in timeliness is achieved with increased message complexity. We can say that ND_IDC will be beneficial for applications that need rapid configuration after deployment [6].

3.3.1.1.6 Neighbor and Network Discovery within ACROPOLIS Network of Excellence Framework Network discovery is also as important as ND. To perform both of these functions, ND and network discovery problems within the *ACROPOLIS* (advanced coexistence technologies for radio optimization in licensed and unlicensed spectrum) *network of excellence* (NoE) framework are considered. ND decides the functions of CRNs. This is similar to the situation with network discovery in which network wide-scale information is shared among cognitive networks and traditional wireless networks. Within ACROPOLIS NoE framework, not only algorithms and results of ND and *medium access control* are presented but the routing for underlay cognitive networks through network-wide protocols can also be described. This framework also enables to familiarize with centralized and distributed cooperative spectrum sensing schemes.

In order to evaluate this study, the development of a reliable simulation platform is required for an accurate modeling of all the relevant aspects in the CRN function. This includes channel modeling, interference estimation, packet error rate, terminal mobility, and primary system activity model. One possible solution is to introduce the reference generalized model in OMNeT++ simulator. All proposed modules and functions are included in this regard. Every ACROPOLIS partner provides its inputs to obtain these modules and functions. We can also study the implementation of cognitive network architecture in OMNeT++. The new architecture and the simulation platform development improve the performance of OMNeT++ compared to any other models. As stated above, packet error rate is an important aspect for a reliable simulation platform. Evaluation of the packet error rate in terms of packet collisions and corresponding interference real-time tracking of packet collisions can also be observed with resolution equal to the symbol time. The simulator estimates an average bit error probability for every bit region to determine the number of bit errors. A coding scheme is then selected to check whether the errors are recoverable or not [23].

3.3.1.1.7 Quorum-Based ND A reliable algorithm not only performs well in a synchronous environment but also works equally well in an asynchronous environment. To achieve this, a *deterministic handshake-based*, *quorum-based* algorithm is a suitable solution. Its operation in a multichannel network is the additional feature. Two techniques of this algorithm are found in the literature: grid and sync grid. This algorithm works by using *random backoff* time before each transmission and by employing *quorum* systems to discover neighbors, so that discovery procedure becomes independent of the number of nodes in the network. The quorum system ensures at least a common active interval for any two frequency channels. For each channel in the universal channel set, the quorum system allocates the active interval set without considering the node's available channel set and the quorum system of same size is used for tuning on each channel. If the nodes do not find any available channel, they go into sleep mode, skipping the broadcast intervals dedicated for these channels.

This quorum-based proposed algorithm is checked for performance improvement in random networks with constant network densities, assuming unchanged radio characteristics and nodes' available channel set for all the nodes. For a synchronous network, both grid and sync grid techniques are found to have the same behavior. However, it is observed that the consumption of energy in grid technique is noticeably greater than the transmission energy in sync grid system. In asynchronous networks, it is noticed that the discovery procedure starts just after the last node joins the network. These results are found to be almost the same as those of synchronous networks. However, it is analyzed that sync grid techniques are suitable for synchronous networks. To study the advertise message transmission rate for a typical node, the sync grid system consumes less energy [28].

3.3.1.1.8 Diameter-Aware and Diameter-Unaware NDAs If CR nodes have access to synchronized clocks and slotted time, an L2AC enables each participating node to determine a common set of channels, leading to a time-efficient distributed algorithm with its diameter-aware and diameter-unaware versions. The feature of operating in multi-hop, multichannel networks where CR nodes remain static for the duration of autoconfiguration makes this algorithm a unique one. Another important characteristic of this algorithm is the ability to compute a globally common channel set. Both diameter-aware and diameter-unaware versions use $2Mk + O(Dk)$ timeslots to find the globally common channel set, where a range $[1, 2,..., k]$ is used to assign a unique identifier to each node, M is the maximum number of channels available, and D denotes the diameter of the network.

Every node has the information about M and k but does not know about the neighborhood. The algorithm has two phases: phase 1 has $2k$ timeslots equally divided in ND and configuration procedure. The configuration procedure provides the information about a locally common broadcast channel. Phase 2 consists of $O(Dk)$ timeslots and it is used for the computation of a globally common channel set. This global common channel has no influence on the running time of the algorithms. Diameter-aware algorithm terminates in 8 s, whereas it takes 12 s for completion in case of diameter-unaware algorithm. Both algorithms provide a matching lower bound of $\Omega(Mk)$ timeslots in the worst case [25].

3.3.1.1.9 Rendezvous Problem for CR Nodes Rendezvous problem defines a situation in which two or more than two CR nodes try to find each other. In such a situation where CR nodes operate using fixed length timeslots, a noticeable work is carried out, focusing on rendezvous for CRs under *dynamic system capabilities*, *spectrum policies*, and *environmental conditions*. Here, we study the two types of rendezvous techniques: *sequence based* and *modular clock blind rendezvous*, that is, rendezvous without the presence of control channels or centralized controllers. It is observed that these algorithms have positive performance comparison with that of a random blind rendezvous and a bounded *time to rendezvous* (TTR) is achieved through a sequence-based algorithm. We can study the ability to set the priority of channels where rendezvous is more likely to occur. The research carried out so far presents four rendezvous algorithms with different performance characteristics based on the assumptions of the system model. These are as follows:

1. Random
2. Generated orthogonal sequence based
3. Modular clock
4. Modified modular clock

Each of these variants has certain characteristics. The random algorithm has the robust operation under all condition models. The generated orthogonal sequence-based algorithm has the ability to set the priority of channels under the shared model and a bounded TTR. When unbounded, the modular clock algorithms enhance the speed in expected TTR at less pre-coordination requirement and the flexibility to operate in both the shared and individual models.

In terms of maximum TTR, the sequence-based rendezvous has much greater performance than the random rendezvous. It supports preferred rendezvous channels at a higher expected TTR. It is observed that in case of more than three channels and under the shared model, the CR nodes must employ the modular clock rather than the random algorithm. The results give a crossover point for the expected TTR at approximately three channels. The CR nodes do not require the initiation synchronization of their rendezvous processes. It is also observed that almost in all scenarios, the modular clock algorithm consumes less than $p_1 p_2$ timeslots on average, where p_1 and p_2 are co-primes. If the observed channels are analyzed for a fixed ratio, the probability of the different prime numbers being selected increases when the number of observed channels by each radio increases [33].

3.3.1.2 Asynchronous Operation In asynchronous operation, channel access methods are adopted without using synchronous clocks, such as contention-based schemes. Asynchronous behavior is simple and beneficial when CR nodes are initiated at different times and remain connected for a relatively short time interval. Asynchronous mode operation is difficult to implement as CR nodes have different time intervals; therefore, CR nodes may have a little time to perform ND. Achieving a fast and reliable discovery process becomes challenging. A variety of algorithms have been adopted featuring specific attributes in this regard, as described in Sections 3.3.1.2.1 through 3.3.1.2.5.

3.3.1.2.1 NDA for Jamming Attacks Working in an asynchronous and distributed manner, a novel *jamming evasive network-coding neighbor-discovery algorithm* (JENNA) uses network coding for full ND with fast and reliable *control packet dissemination* in a single-hop network setting. The algorithm with the help of all CR nodes in the area provides a very robust solution to various jamming attacks. JENNA is compared with the state-of-the-art solutions such as random ND schemes. It has been observed that JENNA is useful for various wireless environments that may be *jamming free* or multiple *reactive* and *static jamming*. Its dissemination performance depends on the actual number of nodes in the network instead of the label space size, and it also operates without the need to know the number of nodes in advance.

The results obtained by employing JENNA show that using network coding gives faster control packet dissemination so that the ND finishes in less time. A 3–6 times improvement is obtained compared to selfish scheme in these settings. Considering the coding performance, the increase in the size of Galois fields (finite number of elements) does not provide improved gains in terms of dissemination delay, but this may not be true for small number of CR nodes where packet diversity is highly beneficial. The increase in dissemination delay is observed for all schemes as the number of free channels increases, and for an increase in the number of free channels, the number of transmitters per slot is also increased. For the number of CR nodes involved in the ND process, network coding is particularly robust in terms of dissemination delay for varying numbers of CR nodes compared to selfish and random message selection schemes. Moreover, improvement in the performance of the schemes using network coding is observed for the increased diversity in packet mixing. Network coding dissemination provides gain over random message selection and selfish schemes that are 6 and 4 times, respectively, and this gain is constant with the number of reactive jammers. It is also observed that the number of available channels decreases with the same number of reactive jammers. Also in a given channel, there is an increase in the probability to disrupt a packet transmission that results in higher ND delay. For multihop networks, network coding cannot provide a suitable solution for an increase in the decoding matrix size and the encoding vectors associated with the encoded packets. Therefore, this algorithm also provides some important problems that must be considered for the situations with high node densities [2].

3.3.1.2.2 Leader Election ND There are various NDAs. Another good work has been carried out to formulate a solution for ND and configuration problems in asynchronous CRNs using leader election method. The main attribute of this algorithm is the ability to find a leader node at the start of the discovery process. Then this leader node not only discovers its neighbors and their channel sets but also computes a global common channel set. Dealing with a situation where cognitive nodes are fully connected, an *asynchronous distributed* algorithm is proposed for single-hop CRNs, allowing neighbors and channel discovery that can be used to communicate among CR nodes. Here, the first process is to declare a *leader node* and then this leader performs the ND operations. Furthermore, it allows the leader node to provide information to other nodes about the presence of globally common channels. Then an upper bound is calculated for the time required to elect a leader and to discover neighbors. For a maximum number of nodes (N) and a maximum number of channels available (M), the upper bound is given by $O(NM^2)$ timeslots. The above-mentioned processes help to provide network scheduling and management [3].

3.3.1.2.3 Energy-Efficient ND   As we know that nodes in CRNs may be initiated asynchronously and remain connected for a short duration of time. For these cases and dealing with *one sender* and *one receiver*, a novel optimal NDA was proposed in [30]. It is optimal because without missing a contact, it consumes minimum energy and probabilistically ensures ND within a delay bound B. Its performance evaluations are checked against two wake-up schedule functions: *WSF-(7, 3, 1)* and *WSF-(73, 9, 1)*. Although WSF-(73, 9, 1) has the same energy saving at $B = 350$, yet its low success probability makes it inapplicable in real systems. Besides achieving energy efficiency, this research work has the success probability of almost 1 and remains higher than WSF-(7, 3, 1) and WSF-(73, 9, 1). Despite the fact that all success probabilities attain the value of 1 as B increases, the proposed algorithm has more energy consumption than the other two schemes.

3.3.1.2.4 Sequence-Based Rendezvous   An interesting feature of rendezvous when CR nodes do not require any synchronization is to use nonorthogonal sequences. This enables CR nodes to determine the order in which they attain rendezvous, that is, the ability to meet and establish a link on a common channel. The order in which radios search for potentially available channels is determined through a *sequence-based rendezvous*. This enables to mark an upper bound to the TTR and to set a priority order for channels. In sequence-based rendezvous, the expected TTR is minimized compared to random rendezvous. A number of sequences can be used to derive a *closed-form* expression for expected TTR, proving that it has an upper bound. This sequence-based rendezvous is also helpful to tackle the issue of PUs' presence. Using these sequences, we can derive expressions for the probability that rendezvous occurs in either the best or the worst channels. The same scenario can be generated to express the conditional expectation of TTR. There have been reports of dealing with *blind rendezvous* [32].

3.3.1.2.5 Rendezvous with Cooperative Environments   A novel *rendezvous protocol for asynchronous CRs in cooperative environments* (*RAC²E*), featuring *lightweight* design and *scalability*, is employed for dynamic control channel in a decentralized ad hoc environment without the requirement of any synchronization. Its functions are verified on a test-bed platform realized with four *universal software radio peripheral 2* (USRP2) nodes. As the control channel duration and the number of network nodes increase, the probability of missing the control channel messages decreases. Hence, it is believed that RAC²E will operate successfully in large networks. But a trade-off between backoff duration and performance occurs; as a result, a successful RAC²E is possible for high values of control channel duration. The RAC²E shows a fast reaction to varying environmental changes and gives acceptable performances in large node density areas [26].

3.3.1.3 Common Channel Set   We know that ND requires that CR nodes involved in the ND process must converge onto a CCC for distributed learning and knowledge sharing. A CCC defines a channel for transmitting elements of information necessary to manage and realize various operations. In CRNs, dynamicity in the network may result in the variation of potentially available channel set, and finding a CCC is an open challenge [24]. Also no specific frequency bands are reserved to set up a CCC in CRNs. Reserving frequency bands in CRNs may result in efficiency degradation. There are certain mechanisms employed to establish a CCC without reserving any space in the frequency spectrum. This control channel is used to exchange the context information [29]. Setting up a CCC is important, but this must be done with minimum cost. Minimum capital expenditure (CAPEX) and operational expenditure (OPEX) solutions are achieved by using minimum infrastructure [31]. These problems have been investigated by many researchers as described in Sections 3.3.1.3.1 and 3.3.1.3.2 respectively.

3.3.1.3.1 NDAs for Distributed Learning As discussed above, a CCC is responsible for distributed learning. The distributed learning schemes relying at the basis of the context information acquisition in multiple channels are applied on optimal network selection to discuss the relationship between the ND and the distributed learning process. This gives rise to a model for mapping the learning process on an ND problem. The distributed learning mechanisms are implemented to solve the problem of optimal selection. The optimal selection in terms of the achieved quality-of-service (QoS) levels by a device is performed in a set of candidate networks. An optimal network selection algorithm based on distributed learning focuses on the CCC. The CCC is used to exchange the context information. CR nodes have to converge to a common control or pilot channel for communication among them, but how they can do it remains a problem. The establishment of a common communication channel is the first and important step in ND and gives the efficiency of the distributed learning algorithm. There have been reports of a significant work carried out in this regard.

One such solution gives a thorough insight into the requirement for ND phase in order to establish a CCC. In this chapter, the impact of ND, especially discovery failures on algorithms for distributed learning, is critical if these algorithms do not set up a common communication channel. As the number of channels in a network increases, the average time required for successful ND also increases. Moreover, this time also increases if a node spends more time on each channel. Improved performance of the NDA can be achieved if each node spends minimum time in idle state, but this may consume more energy. The probability of false alarm considerably affects the time required to select the same signal, and if these probabilities become higher, the impacts are more dramatic. The worst failure rates are observed for low values of probability of false alarm even in case of low number of channels. This gives rise to a device failing to receive context information in the network selection problem and reduces the achievable performance [24].

3.3.1.3.2   CCC Allocation through Ultra Wideband Communication   It has been said earlier that one possible solution to establish a CCC is to reserve a specific portion of spectrum for exchanging control information. But it may not be feasible due to the following:

1. The reduction in the available bandwidth for data communication
2. Change in CCC to a different spectrum portion if PU returns.

One promising solution to overcome the above-mentioned two challenges is the use of *ultra wideband* (UWB) technology. This technology is preferred due to its negligible interference to narrowband transmissions, low complexity, and low power consumption. CR nodes equipped with UWB interface can discover each other through exchange of control information and hence can establish a communication link by exchanging the control information. A communication protocol is defined in this regard to allow CR nodes to discover each other and exchange control information so that a link can be set up. The coverage gap between the UWB and long-medium range technologies is solved through multihop communication.

UWB CCC has very good success probability and small latency. The performance of this UWB scheme is found to be better suited in terms of success probability of transmitted single messages on the UWB CCC and of the complete message handshake required to establish a communication link. The success probability of single message transmission is always less than for complete handshakes carried out through direct transmissions, whereas in multihop transmissions, the success probability for complete handshakes and high number of nodes is always lower than success probability for single message transmissions. In single-hop CCC, the average duration of a successful message handshake is found to be on the order of milliseconds regardless of the number of nodes. The number of nodes N has affect on the system performance since the receiver experiences a higher interference for higher number of nodes. The handshake success probability decreases with increasing number of network nodes. UWB implementation has been observed fine for high-order pulse position modulations (PPMs). The 16-PPM has almost half average delay compared to the 2-PPM and the average delay in single-hop has the duration of a successful message handshake on the order of tens of milliseconds whatever may be the number of nodes. Moreover, both 16-PPM and 2-PPM give almost the same average value. Hence, it is observed that the 16-PPM gives the best performance, whereas the 2-PPM gives the worst performance.

Some interesting characteristics of UWB-adopted communication protocol are observed in case of common codes. The probability of bit errors is higher than the probability of distinct codes. The results prove the Bose–Chaudhuri–Hocquenghem (BCH) code to be a suitable candidate and no significant improvement is observed with a BCH code having higher error correction capability. In terms of a comparison between UWB CCC and in-band signaling schemes, the UWB CCC has a considerable higher success probability as the number of users increases. Moreover, the UWB CCC has much improved performance in in-band signaling solution in terms of average handshake delay. The nodes as far as 100 m away successfully communicate among each other with high probability through a UWB CCC [29].

3.3.1.4 Collision-Free Algorithm   In a CRN, a successful communication among two neighbors requires that they must be tuned to the same channel at the same time. Besides this, there must be no collision between the transmission of sender and the transmission of another node when both of them are within the communication range of the receiver. This is particularly important if multiple CR nodes transmit on the same channel in a given timeslot. To address these challenges, a *deterministic, collision-free*, and *time division multiple access (TDMA) based distributed* algorithm for ND is proposed for one or more than one node, when nodes have multiple receivers but only a single transmitter. In this NDA, a CR node can have the following three states: transmitting on one channel, receiving on one or more channels (up to the number of receivers), or remaining turned off. The time complexity of this algorithm is given by

$$ M\left\lceil \frac{k}{r} \right\rceil + O(\max(M, k) \log r) $$

where:

M is the maximum number of channels on which a node can operate
k is the size of the space used to assign identifiers to nodes
r is the number of receivers at a node (with $1 \leq r \leq \min(M, k)$)

Here, the time complexity is related to the size of the space from which identifiers are selected and it does not depend on the actual number of nodes. The high time complexity is due to the following:

1. Node has the previous knowledge of the entire transmission schedule.
2. Node's operation does not depend on the knowledge of its neighbors.
3. Node does not have the information of the channels it could use to communicate.
4. The schedule is collision free.
5. There is no CCC.

But still this algorithm is proved to be *close* to optimal. For useful information, only up to M receivers can be used in a timeslot, if $r > M$. And if $r > k$, then from r receivers there must not be more than k number of receivers [5].

3.3.2 Classification of ND Protocols in CRNs

The NDAs may be classified on different features. Time complexity defines one major aspect in this manner. Time complexity analysis with special emphasis on its dependence on certain parameters has been focused by many researchers. These parameters include the size of space, the number of nodes, the number of available channel sets, and the network size [1,4,5,25].

Size-aware capability is another type of classification. It is discussed in [1] and [4] how to achieve a fast NDA when nodes know or do not know about the network size. Algorithms have been presented for directed and undirected networks. The network topology and the node density play an important role in this regard, and channel availability set may vary from node to node [1]. Some of the discovery schemes have the termination guarantees, that is, each node detects the completion of the discovery process and this process terminates simultaneously at each node [1,6]. Usually NDA executes on several nodes but a significant work has been carried out to elect a leader first. After this leader election, the leader node not only discovers its neighbors and channels set but also finds a global common channel set. This information is shared among other nodes by the leader node [3].

The NDAs may be classified into randomized [2] and deterministic [5]. In a randomized NDA, each node discovers its neighbors by transmitting at randomly chosen times. These algorithms have a high probability. In a deterministic NDA, each node has a predetermined time schedule for its transmission. Deterministic NDAs discover all its neighbors with probability one.

The time it takes to discover maximum number of neighbors is essential because a quick completion of the algorithms determines its efficiency [4]. Network discovery in which network wide-scale information is shared among CRNs decides the function of the ND [23]. The classification may be based upon the nature of operation, that is, either synchronous or asynchronous [28]. Some researchers have focused on rendezvous for CRN under dynamic system capabilities, spectrum policies, and environmental conditions [32,33]. Then the NDAs may also be differentiated according to the number of receivers and transmitters in the network [3,5]. Some of these have to first find a globally common channel [3,33], and then there are some that operate on multichannels [6].

Security describes an important attribute of the NDA. The algorithm must be prone to any threats from a malicious or selfish behavior. At the same time, these algorithms must be fast enough. A noticeable work has been carried out in order to deal with jamming attacks in [2].

3.4 Issues, Challenges, and Future Directions for ND Protocols in CRNs

In this section, we discuss some issues, challenges, and future directions of ND protocols in CRNs.

3.4.1 Coordination among Nodes

In order to perform an efficient ND, a protocol must be adopted to allow nodes to operate over multiple channels. Obviously this comes at the cost of additional level of complexity. It is essential that all the nodes must have the capability to detect the ND completion on one channel simultaneously and move on to the next channel in a coordinated manner. Therefore, any algorithm should be adaptive because future directions of a process may have to take the past information about the process.

3.4.2 Collision/Interference

Interference causes loss of packets. Therefore, if control messages exchanged for ND face interference, then nodes will not receive control messages from neighbors, and, subsequently, ND would not be achieved. This is why the consideration of interference is very important for an ND process. It is also difficult to develop an efficient discovery algorithm if nodes have neither the proper knowledge of network density nor the capability to detect collisions [1].

3.4.3 Minimum Delay

ND phase must be as quick as possible and CRs must develop a network of trusted neighbors, so that malicious attacks from false neighbors can be counteracted. Random hopping over a large number of available channels gives a large discovery delay. Hence, a fast and secure ND over multiple channels should have a minimum delay [2].

3.4.4 Finding a Global Common Channel Set

Finding a global common communication channel among neighboring nodes is an important problem. Without a handshaking mode and a control channel, there is a possibility that discovery algorithms may interfere with a PU(s) when it senses a vacant channel and returns to it [33].

3.4.5 Termination Detection

If the time complexity depends on the actual number of nodes, then the nodes can easily detect the termination of the NDA. Existing research uses maximum number of nodes to estimate the time complexity, so taking actual number of nodes into account to find a fast algorithm is a challenge [5,24].

3.4.6 Link Reliability

Diversity in signal propagation on different channels may cause a disruption of a link present between two nodes, even if these channels are present in the available channel sets of both nodes [6].

3.4.7 Security

The efficiency of discovery process depends upon the accurate and complete knowledge of neighbors among other features. Security of legitimate users against malicious or selfish nodes becomes an interesting challenge in a dynamic environment as faced by the nodes in a CRN [2].

3.4.8 Energy Consumption

The continuous sensing of the spectrum requires a considerable amount of energy. This issue becomes prominent when multiple CR nodes are present. Therefore, adopting energy-efficient NDAs can save considerable amount of energy [28,30].

3.5 Conclusion

This chapter presented an overview of NDAs for CRNs. First the ND was defined and then various ND protocols, starting from NDAs in traditional wireless networks, were discussed. Then a detailed analysis of different NDAs was presented for CRNs. A brief overview of the CRNs was presented before describing the areas of focus for ND process in wireless networks and CRNs. The discovery process might have a synchronous or asynchronous operation depending upon the environment. Some ND processes first established a CCC before starting this phase. Then neighbor discovery algorithms in collision-aware, size-aware and undirected networks was discussed. It was also mentioned that a good algorithm should have a strong termination condition, and the network topology and channel information could affect the ND protocols. It was deduced and discussed that algorithms adopted by various researchers had distinct features. Finally, some issues, challenges, and future directions for ND in CRNs were discussed.

References

1. N. Mittal, S. Krishnamurthy, R. Chandrasekaran, S. Venkatesan, and Y. Zeng, "On neighbor discovery in cognitive radionetworks," *Journal of Parallel and Distributed Computing*, Volume 69, Issue 7, July 2009, Pages 623–637.

2. A. Asterjadhi and M. Zorzi, "JENNA: A jamming evasive network-coding neighbor discovery algorithm for cognitive radio networks," *Journal of IEEE Wireless Communications*, Volume 17, Issue 4, August 2010, Pages 24–32.
3. C. J. L. Arachchige, S. Venkatesan, and N. Mittal, "An asynchronous neighbor discovery algorithm for cognitive radio networks," *IEEE 3rd Symposium on New Frontiers in Dynamic Spectrum Access Networks*, 2008, Pages 1–5.
4. Z. Jian-zhao, Z. Hang-sheng, Y. Fu-qiang, "A fast neighbor discovery algorithm for cognitive radio ad hoc networks," *12th IEEE International Conference on Communication Technology*, November 11–14, 2010, Pages 446–449, IEEE, Nanjing, People's Republic of China.
5. S. Krishnamurthy, N. Mittal, R. Chandrasekaran, and S. Venkatesan, "Neighbor discovery in multi-receiver cognitive radio networks," *International Journal of Computers and Applications*, Volume 31, Issue 1, 2009, Pages 50–57.
6. Y. Zeng, N. Mittal, S. Venkatesan, and R. Chandrasekaran, "Fast neighbor discovery with lightweight termination detection in heterogeneous cognitive radio networks," *Proceedings of the 9th International Symposium on Parallel and Distributed Computing*, July 7–9, 2010, Pages 149–156, IEEE, Istanbul, Turkey.
7. B. F. Lo, "A survey of common control channel design in cognitive radio networks," *Journal of Physical Communication*, Volume 4, Issue 1, March 2011, Pages 26–39.
8. N. Karowski and A. Carneiro Viana, "Optimized asynchronous multi-channel discovery of IEEE 802.15.4-based wireless personal area networks," *Journal of IEEE Transactions on Mobile Computing*, Volume 12, Issue 10, October 2013, Pages 1972–1985.
9. M. J. McGlynn and S. A. Borbash, "Birthday protocols for low energy deployment and flexible neighbor discovery in ad hoc wireless networks," *Proceedings of the 2nd ACM International Symposium on Mobile Ad hoc Networking and Computing*, October 1, 2001, Pages 137–145, ACM, New York.
10. S. Vasudevan, D. Towsley, and D. Goeckel, "Neighbor discovery in wireless networks and the coupon collector's problem," *Proceedings of the 15th Annual International Conference on Mobile Computing and Networking*, September 20, 2009, Pages 181–192, ACM, New York.
11. N. Karowski, A. Carneiro Viana, and A. Wolisz, "Optimized asynchronous multi-channel neighbor discovery," *IEEE Proceedings of the INFOCOM*, April 10–15, 2011, Pages 536–540, IEEE, Shanghai, People's Republic of China.
12. E. B. Hamida, G. Chelius, and E. Fleury, "Revisiting neighbor discovery with interferences consideration," *Proceedings of the 3rd ACM International Workshop on Performance Evaluation of Wireless Ad Hoc, Sensor and Ubiquitous Networks*, October 6, 2006, Pages 74–81, ACM, New York.
13. S. Hariharan, N. B. Shroff, and S. Bagchi, "Secure neighbor discovery through overhearing in static multihop wireless networks," *Journal of Computer Networks*, Volume 55, Issue 6, April 25, 2011, Pages 1229–1241.

14. R. Oliveira, M. Luis, L. Bernardo, R. Dinis, and P. Pinto, "The impact of node's mobility on link-detection based on routing hello messages," *IEEE Wireless Communications and Networking Conference*, April 18–21, 2010, Pages 1–6, IEEE, Sydney, NSW, Australia.

15. R. Huang and Y. Fang, "Utilizing multi-hop neighbor information in spectrum allocation for wireless networks," *Journal of IEEE Transactions on Wireless Communications*, Volume 8, Issue 8, August 2009, Pages 4360–4367.

16. J. Ning, T.-S. Kim, S. V. Krishnamurthy, and C. Cordeiro, "Directional neighbor discovery in 60 GHz indoor wireless networks," *Journal of Performance Evaluation*, Volume 68, Issue 9, September 2011, Pages 897–915.

17. R. Murawski, E. Felemban, E. Ekici, S. Park, S. Yoo, K. Lee, J. Park, and Z. Hameed Mir, "Neighbor discovery in wireless networks with sectored antennas," *Journal of Ad Hoc Networks*, Volume 10, Issue 1, January 2012, Pages 1–18.

18. R. Khalili, D. L. Goeckel, D. Towsley, and A. Swami, "Neighbor discovery with reception status feedback to transmitters," *Proceedings of the 29th Conference on Information Communications*, March 14–19, 2010, Pages 2375–2383, IEEE, San Diego, CA.

19. N. Mittal, Y. Zeng, S. Venkatesan, and R. Chandrasekaran, "Randomized distributed algorithms for neighbor discovery in multi-hop multi-channel heterogeneous wireless networks," *Proceedings of the 31st International Conference on Distributed Computing Systems*, June 20–24, 2011, Pages 57–66, IEEE, Minneapolis, MN.

20. A. Asterjadhi, N. Baldo, and M. Zorzi, "A distributed network coded control channel for multihop cognitive radio networks," *Journal of IEEE Network*, Volume 23, Issue 4, July/August 2009, Pages 26–32.

21. K. Dasgupta, M. Kukreja, and K. Kalpakis, "Topology-aware placement and role assignment for energy-efficient information gathering in sensor networks," *Proceedings of the 8th IEEE International Symposium on Computers and Communication*, June 30-July 3, 2003, Pages 341–348, IEEE.

22. C. Drula, C. Amza, F. Rousseau, and A. Duda, "Adaptive energy conserving algorithms for neighbor discovery in opportunistic Bluetooth networks," *IEEE Journal on Selected Areas in Communications*, Volume 25, Issue 1, January 2007, Pages 96–107.

23. L. De Nardis, M.-G. Di Benedetto, O. Holland, A. Akhtar, H. Aghvami, V. Rakovic, V. Atanasovski et al., "Neighbour and network discovery in cognitive radio networks: research activities and results in the ACROPOLIS network of excellence," *Proceedings of the 19th European Wireless Conference*, April 16–18, 2013, Pages 1–6, VDE, Guilford, UK.

24. L. DeNardis, M. G. Di Benedetto, V. Stavroulaki, A. Bantouna, Y. Kritikou, and P. Demestichas, "Role of neighbour discovery in distributed learning and knowledge sharing algorithms for cognitive wireless networks," *International Symposium on Wireless Communication Systems*, August 28–31, 2012, Pages 421–425, IEEE, Paris, France.

25. S. Krishnamurthy, M. Thoppian, S. Kuppa, R. Chandrasekaran, N. Mittal, S. Venkatesan, and R. Prakash, "Time-efficient distributed layer-2

auto-configuration for cognitive radio networks," *Journal of Computer Networks*, Volume 52, Issue 4, March 2008, Pages 831–849.

26. V. Pavlovska, D. Denkovski, V. Atanasovski, and L. Gavrilovska, "RAC²E: novel rendezvous protocol for asynchronous cognitive radios in cooperative environments," *IEEE 21st International Symposium on Personal, Indoor and Mobile Radio Communications*, September 26–30, 2010, Pages 1848–1853, IEEE, Istanbul, Turkey.

27. D. Liu, "Protecting neighbor discovery against node compromises in sensor networks," *Proceedings of the 29th IEEE International Conference on Distributed Computing Systems*, June 22–26, 2009, Pages 579–588, IEEE, Montreal, QC, Canada.

28. S. Khatibi and R. Rohani, "Quorum-based neighbor discovery in self-organized cognitive MANET," *IEEE 21st International Symposium on Personal, Indoor and Mobile Radio Communications*, September 26–30, 2010, Pages 2239–2243, IEEE, Istanbul, Turkey.

29. A. M. Masri, C.-F. Chiasserini, C. Casetti, and A. Perotti, "Common control channel allocation in cognitive radio networks through UWB communication," *Journal of Communications and Networks*, Volume 14, Issue 6, December 2012, Pages 710–718.

30. D. Yang, J. Shin, J. Kim, and C. Kim, "Asynchronous probing scheme for the optimal energy-efficient neighbor discovery in opportunistic networking," *IEEE International Conference on Pervasive Computing and Communications*, March 9–13, 2009, Pages 1–4, IEEE, Galveston, TX.

31. M. Mueck and A. Hayar, "A local cognitive pilot channel (LCPC) for neighbourhood discovery, relaying and cluster based local cognitive information management," *Proceedings of the 5th International Conference on Cognitive Radio Oriented Wireless Networks and Communications*, June 9–11, 2010, Pages 1–5, IEEE, Cannes, France.

32. L. A. DaSilva and I. Guerreiro, "Sequence-based rendezvous for dynamic spectrum access," *3rd IEEE Symposium on New Frontiers in Dynamic Spectrum Access Networks*, October 14–17, 2008, Pages 1–7, IEEE, Chicago, IL.

33. N. C. Theis, R. W. Thomas, and L. A. DaSilva, "Rendezvous for cognitive radios," *Journal of IEEE Transactions on Mobile Computing*, Volume 10, Issue 2, February 2011, Pages 216–227.

34. M. Kohvakka, J. Suhonen, M. Kuorilehto, V. Kaseva, M. Hännikäinen, and T. D. Hämäläinen, "Energy-efficient neighbor discovery protocol for mobile wireless sensor networks," *Journal of Ad Hoc Networks*, Volume 7, Issue 1, January 2009, Pages 24–41.

35. A. Meier, M. Weise, J. Beutel, and L. Thiele, "NoSE: Neighbor search and link estimation for a fast and energy efficient initialization of WSNs," *Proceedings of the 6th ACM Conference on Embedded Network Sensor Systems*, November 5, 2008, Pages 397–398, ACM, New York, NY.

36. P. Dutta and D. Culler, "Practical asynchronous neighbor discovery and rendezvous for mobile sensing applications," *Proceedings of the 6th ACM Conference on Embedded Network Sensor Systems*, November 5, 2008, Pages 71–84, ACM, New York, NY.

37. W. Cha and P.-S. Mah, "Reactive and reliable neighbor discovery protocols in wireless networks," *Proceedings of the 11th International Conference on Advanced Communication Technology*, February 15–18, 2009, Volume 3, Pages 1755–1758, IEEE, Phoenix Park, Dublin, Ireland.

38. A. Cornejo, S. Viqar, and J. L. Welch, "Reliable neighbor discovery for mobile ad hoc networks," *Journal of Ad Hoc Networks*, Volume 12, January 2014, Pages 259–277.

39. M. Younis and K. Akkaya, "Strategies and techniques for node placement in wireless sensor networks: A survey," *Journal of Ad Hoc Networks*, Volume 6, Issue 4, June 2008, Pages 621–655.

40. I. Khalil, S. Bagchi, C. N. Rotaru, and N. B. Shroff, "UNMASK: Utilizing neighbor monitoring for attack mitigation in multihop wireless sensor networks," *Journal of Ad Hoc Networks*, Volume 8, Issue 2, March 2010, Pages 148–164.

41. A. Kandhalu, K. Lakshmanan, and R. (Raj) Rajkumar, "U-Connect: A low-latency energy-efficient asynchronous neighbor discovery protocol," *Proceedings of the 9th ACM/IEEE International Conference on Information Processing in Sensor Networks*, April 12, 2010, Pages 350–361, ACM, New York, NY.

42. M. Poturalski, P. Papadimitratos, and J.-P. Hubaux, "Secure neighbor discovery in wireless networks: Formal investigation of possibility," *Proceedings of the ACM Symposium on Information, Computer and Communications Security*, March 18, 2008, Pages 189–200, ACM, New York, NY.

43. S. Gallo, L. Galluccio, G. Morabito, and S. Palazzo, "Rapid and energy efficient neighbor discovery for spontaneous networks," *Proceedings of the 7th ACM International Symposium on Modeling, Analysis and Simulation of Wireless and Mobile Systems*, October 4, 2004, Pages 8–11, ACM, New York, NY.

44. Z. Zhang, "Performance of neighbor discovery algorithms in mobile ad hoc self-configuring networks with directional antennas," *IEEE Military Communications Conference*, October 17–20, 2005, Pages 3162–3168, IEEE, Atlantic City, NJ.

45. A. Medina and S. Bohacek, "Performance modeling of neighbor discovery in proactive routing protocols," *Journal of Advanced Research*, Volume 2, Issue 3, July 2011, Pages 227–239.

46. R. C. Biradar and S. S. Manvi, "Neighbor supported reliable multipath multicast routing in MANETs," *Journal of Network and Computer Applications*, Volume 35, Issue 3, May 2012, Pages 1074–1085.

47. D. Angelosante, E. Biglieri, and M. Lops, "Neighbor discovery in wireless networks: A multiuser-detection approach," *Journal of Physical Communication*, Volume 3, Issue 1, March 2010, Pages 28–36.

48. E. B. Hamida, G. Chelius, A. Busson, and E. Fleury, "Neighbor discovery in multi-hop wireless networks: evaluation and dimensioning with interference considerations," *Journal of Discrete Mathematics and Theoretical Computer Science*, Volume 10, Issue 2, 2008, Pages 87–114.

49. S. Viqar and J. L. Welch, "Deterministic collision free communication despite continuous motion," *Journal of Ad Hoc Networks*, Volume 11, Issue 1, January 2013, Pages 508–521.

50. D. Liu, P. Ning, and W. Du, "Detecting malicious beacon nodes for secure location discovery in wireless sensor networks," *Proceedings of the 25th IEEE International Conference on Distributed Computing Systems*, October 10, 2005, Pages 609–619, IEEE, Columbus, OH.

51. F. Vázquez-Gallego, J. Alonso-Zarate, L. Alonso, and M. Dohler, "Analysis of energy efficient distributed neighbour discovery mechanisms for machine-to-machine networks," *Journal of Ad Hoc Networks*, Volume 18, July 2014, Pages 40–54.

52. G. Jakllari, W. Luo, and S. V. Krishnamurthy, "An integrated neighbor discovery and MAC protocol for ad hoc networks using directional antennas," *IEEE Transactions on Wireless Communications*, Volume 6, Issue 3, March 2007, Pages 1114–1124.

53. S. A. Borbash, A. Ephremides, and M. J. McGlynn, "An asynchronous neighbor discovery algorithm for wireless sensor networks," *Journal of Ad Hoc Networks*, Volume 5, Issue 7, September 2007, Pages 998–1016.

54. G. Pei, M. M. Albuquerque, J. H. Kim, D. P. Nast, and P. R. Norris, "A neighbor discovery protocol for directional antenna networks," *IEEE Military Communications Conference*, Volume 1, 2005, Pages 487–492.

55. P. Papadimitratos, M. Poturalski, P. Schaller, P. Lafourcade, D. Basin, S. Capkun, and J.-P. Hubaux, "Secure neighborhood discovery: a fundamental element for mobile ad hoc networking," *Journal of IEEE Communications Magazine*, Volume 46, Issue 2, February 2008, Pages 132–139.

56. R. Stoleru, H. Wu, and H. Chenji, "Secure neighbor discovery and wormhole localization in mobile ad hoc networks," *Journal of Ad Hoc Networks*, Volume 10, Issue 7, September 2012, Pages 1179–1190.

57. A. Willig, N. Karowski, and J.-H. Hauer, "Passive discovery of IEEE 802.15.4-based body sensor networks," *Journal of Ad Hoc Networks*, Volume 8, Issue 7, September 2010, Pages 742–754.

58. S. Vasudevan, J. Kurose, and D. Towsley, "On neighbor discovery in wireless networks with directional antennas," *Proceedings of the 24th Annual Joint Conference of the IEEE Computer and Communications Societies*, Volume 4, March 13–17, 2005, Pages 2502–2512.

59. Z. Deng, Y. Zhu, and M. Li, "On efficient neighbor sensing in vehicular networks," *Journal of Computer Communications*, Volume 35, Issue 13, July 2012, Pages 1639–1648.

60. K. Balachandran and J. H. Kang, "Neighbor discovery with dynamic spectrum access in adhoc networks," *IEEE 63rd Vehicular Technology Conference*, Volume 2, 2006, Pages 512–517.

61. J. Luo and D. Guo, "Neighbor discovery in wireless ad hoc networks based on group testing," *46th Annual Allerton Conference on Communication, Control, and Computing*, September 23–26, 2008, Pages 791–797, IEEE, Urbana-Champaign, IL.

62. Z. Zhang and B. Li, "Neighbor discovery in mobile ad hoc self-configuring networks with directional antennas: Algorithms and comparisons," *Journal of IEEE Transactions on Wireless Communications*, Volume 7, Issue 5, May 2008, Pages 1540–1549.

63. A. Cornejo, N. Lynch, S. Viqar, and J. L. Welch, "Neighbor discovery in mobile ad hoc networks using an abstract MAC layer," *Proceedings of the 47th Annual Allerton Conference on Communication, Control, and Computing*, September 30–October 2, 2009, Pages 1460–1467, IEEE, Monticello, IL.

64. L. Zhang, J. Luo, and D. Guo, "Neighbor discovery for wireless networks via compressed sensing," *Journal of Performance Evaluation*, Volume 70, Issues 7/8, July 2013, Pages 457–471.

65. A. Nasipuri and S. R. Das, "Multichannel CSMA with signal power-based channel selection for multihop wireless networks," *52nd Vehicular Technology Conference*, Volume 1, 2000, Pages 211–218.

66. I. Koutsopoulos and L. Tassiulas, "Fast neighbor positioning and medium access in wireless networks with directional antennas," *Journal of Ad Hoc Networks*, Volume 11, Issue 2, March 2013, Pages 614–624.

67. R. A. Santosa, B.-S. Lee, C. K. Yeo, and T. M. Lim, "Distributed neighbor discovery in ad hoc networks using directional antennas," *Proceedings of the 6th IEEE International Conference on Computer and Information Technology*, September 2006, Page 97, IEEE, Seoul, Republic of Korea.

68. K. Bian, J.-M. Park, and R. Chen, "Control channel establishment in cognitive radio networks using channel hopping," *IEEE Journal on Selected Areas in Communications*, Volume 29, Issue 4, April 2011, Pages 689–703.

69. V. Dyo and C. Mascolo, "A node discovery service for partially mobile sensor networks," *Proceedings of the 2nd International Workshop on Middleware for Sensor Networks*, November 30, 2007, Pages 13–18, ACM, New York, NY.

70. M. Sekido, M. Takata, M. Bandai, and T. Watanabe, "A directional hidden terminal problem in ad hoc network MAC protocols with smart antennas and its solutions," *IEEE Global Telecommunications Conference*, Volume 5, December 2, 2005, Pages 2579–2583, IEEE, St. Louis, MO.

71. F. Yildirim and H. Liu, "A cross-layer neighbor-discovery algorithm for directional 60-GHz networks," *IEEE Transactions on Vehicular Technology*, Volume 58, Issue 8, October 2009, Pages 4598–4604.

Biographical Sketches

Athar Ali Khan obtained his BSc in electrical (computer and communication) engineering with honors from CECOS University of IT and Emerging Sciences, Peshawar, Pakistan, in 2001. He received his MS degree in computer engineering from COMSATS Institute of Information Technology, Pakistan, in 2011. He is currently pursuing his PhD degree in electrical engineering at COMSATS Institute of Information Technology, Pakistan. His research interests include cognitive radio networks, wireless sensor networks, optical communication, and bandgap engineering.

Mubashir Husain Rehmani obtained his BE from Mehran UET, Jamshoro, Pakistan, in 2004. He received his MS and PhD from the University of Paris XI, Orsay, France, and the University of Paris VI, Paris, France, in 2008 and 2011, respectively. He was a postdoctoral fellow at the University of Paris Est, France, in 2012. He is an assistant professor at COMSATS Institute of Information Technology, Wah Cantt, Pakistan. His current research interests include cognitive radio networks and wireless sensor networks. He served in the technical program committee of Association for Computing Machinery (ACM) CoNEXT'13 SW, Institute of Electrical and Electronics Engineers (IEEE) ICC'14, and IEEE IWCMC'13. He currently serves as a reviewer of *IEEE* Journal on Selected Areas in Communications (*JSAC*), *IEEE* Transactions on Wireless Communications (*TWireless*), *IEEE* Transactions on Vehicular Technology (*TVT*), Computer Communications (*ComCom*; Elsevier), and *Computers and Electrical Engineering* (*CAEE*) journals.

Yasir Saleem received his 4-year BS in information technology from National University of Sciences and Technology, Islamabad, Pakistan, in 2012. He is currently doing MS in computer sciences (by research) under a joint program of Sunway University, Selangor, Malaysia, and Lancaster University, Lancaster, UK. He is also a reviewer of journals and conferences such as *Computers and Electrical Engineering, International Journal of Communication Networks and Information Security* (*IJCNIS*), and IEEE International Conference on Communications (ICC) 2013. His research interests include cognitive radio networks, cognitive radio sensor networks, cloud computing, and wireless sensor networks.

PART II
RADIO SPECTRUM SENSING

4

Time-Domain Cognitive Sensor Networking

STEFANO BUSANELLI, GIANLUIGI FERRARI, ALESSANDRO COLAZZO, AND JEAN-MICHEL DRICOT

Contents

4.1 Introduction

Because of the increasing number of low-cost wireless applications, the unlicensed spectrum is quickly becoming a scarce resource. In voice-oriented wireless networks, such as, cellular systems, it has been shown that a relevant portion of the licensed spectrum is underused [1], thus yielding significant inefficiencies. A better performance can be obtained using new techniques, such as dynamic spectrum access (DSA), that allow a secondary network to exploit the white spaces in the licensed spectrum of a primary network, owing to cognitive capabilities. Most of the research activity on cognitive systems focuses on efficient spectrum utilization, in the realm of cellular systems. However, in the case of wireless sensor networks (WSNs), a network typically generates bursty traffic over the entire available bandwidth.

In this chapter, we propose a cognitive sensor networking strategy such that a secondary WSN transmits in the inactivity periods of a primary WSN, using all the (common) shared bandwidth. Clearly, one of the main problems of the secondary WSN is to decide when to transmit its packets, in order to maximize its throughput while, yet, minimizing the interference with the primary WSN. In order to tackle this problem, we consider a cognitive system similar to the scheme presented in [2]. The reference scenario is given by a primary IEEE 802.15.4 WSN coexisting with a secondary WSN that tries to exploit the inactivity periods of the primary one. More precisely, both WSNs share the same bandwidth and the cognitive coexistence is carried out in the time domain. In particular, assuming synchronization between the nodes of the secondary WSN, upon waking up, they sense the channel and act accordingly: if the channel is busy, the secondary WSN defers any activity (namely, data collection, i.e., transmissions from the sensor nodes to the sink); if the channel is idle, the secondary WSN transmits for an *optimal* time interval. By relying on a rigorous queueing theoretic approach, the length of this interval is optimized in order to minimize the probability of interference with the primary WSN, yet maximizing the throughput of the secondary WSN.

This chapter is structured as follows. Section 4.2 presents an overview of related work. Section 4.3 derives an accurate queuing model for a single-hop IEEE 802.15.4 WSN. Section 4.4 discusses the exact distribution of idle times in a single-hop IEEE 802.15.4 WSN. By using the developed analytical models, Section 4.5 characterizes the performance trade-offs involved by time-domain cognitive sensor networking. Finally, Section 4.6 concludes this chapter.

4.2 Related Work

The interest in the application of cognitive principles to wireless networks dates back to the end of the 1990s, after the introduction of the basic concept of cognitive radio by Mitola [3]. In the last decade, the ever-increasing demand for data exchange (in particular, Internet access in the presence of mobility) has led to a higher and higher need to access the electromagnetic spectrum [in particular, some portions of it, e.g., the industrial, scientific, and medical (ISM) band] [4].

Cognitive networking is an approach that tries to exploit the *empty* (temporarily unused) spaces in the electromagnetic spectrum. In particular, a cognitive radio follows a cognitive cycle [5], which attributes to a node both the ability to perceive the surrounding environment and the intelligence required to identify spectral holes and exploit them efficiently.

The Federal Communications Commission (FCC) has attributed various characteristics to cognitive radios [6]. In particular, spectrum *sensing* and *sharing* play key roles. In [1], spectrum sensing techniques are classified into three groups: identification of a primary transmitter, identification of a primary receiver, and interference temperature measurement. In the same work, spectrum sharing is classified according to the used access technology: in the presence of *overlay* spectrum sharing, the secondary nodes access the spectrum using portions of the spectrum currently unused by the primary nodes, and with *underlay* spectrum sharing, spread spectrum techniques are used to make a primary user perceive the transmissions by the secondary users as noise.

Various approaches to the design of communication protocols able to exploit opportunistically the electromagnetic spectrum have been proposed. Sharma et al. [7] proposed a medium access control (MAC) protocol, relying on the theory of alternating renewal processes, for a cognitive (secondary) network, which opportunistically uses the channels of a primary network with the carrier sense multiple access with collision avoidance (CSMA/CA) MAC protocol (e.g., a WiFi network). According to this approach, a pair of secondary nodes estimates the duration of an idle period, and then try to maximize the secondary throughput by maximizing the number of frames transmitted during idle periods. In [8], a proactive channel access model was proposed, such that the secondary nodes, based on sensing, build a statistical model of the spectral availability for each channel. The proposed approach is validated considering a primary network given by a television broadcast network. Yang et al. [9] presented a similar proactive approach to the estimation of spectral availability by the secondary nodes, followed by intelligent commutation (between available channels) on the basis of a renewal theoretic prediction approach. Three statistical models of busy/idle times are considered: (1) the busy and idle times have both exponential distributions; (2) the busy and idle times have fixed durations; and (3) the idle time has an exponential distribution, whereas the busy time is fixed. Kim and Shin [10], exploiting the theory of alternate renewal processes, derived the optimized duration of the channel sensing phase in order to maximize the identification of spectral holes. Geirhofer et al. [11] investigated the coexistence of cognitive radios and WiFi nodes, modeled through a continuous-time Markov chain (CTMC)-based model. Geirhofer and Tong [12] investigated the busy and empty times in an IEEE 802.11 network experimentally, in order to identify accurate statistical models: in particular, it is concluded that a hyper-Erlang distribution provides the most accurate fit, but an exponential distribution allows to design MAC protocols more efficiently. In [13], with reference to IEEE 802.11 networks, the CSMA/CA protocol of secondary users is modified to be able to operate in an intermittent manner of spectrum pooling. Jeon et al. [14] proposed a cognitive scenario where secondary users adjust their communication protocols by taking into account the locations of the primary users. By introducing *preservation regions* around primary receivers, a modified multihop routing protocol is proposed for the cognitive users.

The features of cognitive radios can also be exploited by WSNs, which are typically designed to use fixed portions of the electromagnetic spectrum in a bursty manner and are formed by nodes with limited communication and processing capabilities. The focus of [15] is on the coexistence of IEEE 802.11 wireless local area networks (WLANs) and IEEE 802.15.4 WSNs in the ISM band. Distributed adaptation strategies, based on spectrum scanning and increased cognition through learning, are proposed for IEEE 802.15.4 nodes, in order to minimize the impact of the interference from IEEE 802.11 nodes. In [16], it is shown that a WSN, provided that the nodes are equipped with a cognitive radio interface, can have several benefits, such as DSA (which may avoid the acquisition of expensive licenses to transmit a very limited amount of data), opportunistic use of the available channels for a bursty traffic, adaptivity to channel conditions (leading to a reduced energy consumption), and feasibility of coexistence of competing WSNs (partially or totally sharing given spectrum portions). Liang et al. [17] formulated the sensing-throughput trade-off problem mathematically and used an energy detection sensing scheme to prove that the formulated problem allows to identify an optimal sensing time that yields the highest secondary network throughput. This optimal sensing time decreases when distributed spectrum sensing is applied. In [18], a cognitive radio sensor combines multiple sensing results obtained at different time points, that is, time diversity, to make an optimal decision on the existence of spectrum access opportunity.

Hu et al. [19,20] proposed a time-domain cooperative spectrum sensing framework, in which the time consumed by reporting for one cognitive user is also utilized for other cognitive users' sensing, that is, space diversity is exploited. The obtained results show that optimal sensing settings allow to maximize the throughput of the secondary network under the constraint that the primary users are sufficiently protected. In [21], a novel and comprehensive metric, denoted as the coexistence goodness factor (CGF), was introduced to accurately model the inherent trade-off between uninterfered primary users and unlicensed access efficiency (from secondary users) for time-domain DSA-based coexistence.

4.3 An Accurate Queuing Model of a Single-Hop IEEE 802.15.4 WSN

In order to derive the optimal transmission time that a secondary IEEE 802.15.4 WSN should adopt, we first develop an accurate queuing model for the primary IEEE 802.15.4 WSN. More precisely, by leveraging on the theory of renewal processes and discrete-time Markov chains (DTMCs) [22], the primary WSN is modeled as an $M/G/1/N$ queue, where N is the number of primary nodes and the service time distribution depends on the IEEE 802.15.4 standard. An illustrative scheme of the (general) queuing model is shown in Figure 4.1. More specifically, the $M/G/1/N$ queue models the overall number of packets in the system and is solved using its embedded DTMC. The time is discretized in minislots, and the length of each minislot coincides with the duration of the backoff unit of the IEEE

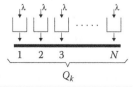

Figure 4.1 Queuing model for a multiple access scheme with *N* nodes.

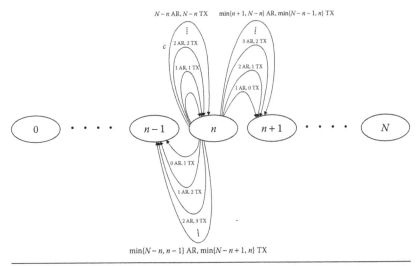

Figure 4.2 Transition diagram of the process $\{Q_k\}$ in a single-hop IEEE 802.15.4 network, depending on the number of arrivals (AR). TX, Transmitters.

802.15.4 MAC protocol. Each primary node acts as a Bernoulli traffic source with parameter p: more precisely, in each minislot a source node generates a packet with probability p. Each node transmits packets of fixed size D_{pck} (dimension of D_{pck}: [b/pck]) and, under the assumption of fixed transmit data rate, fixed duration (corresponding to a given number of minislots). The final corresponding transition diagram of the process $\{Q_k\}$ is shown in Figure 4.2. It is then possible to determine all transition probabilities $\{P_{i,j}\}$ [for all admissible pairs (i,j)] shown in the diagram in Figure 4.2, and then determine the stationary distribution $\pi = \{\pi_n\}_{n=0}^{N}$, such that

$$\pi = P\pi \qquad (4.1)$$

Solving the above equation corresponds to identifying the eigenvectors of P associated with the eigenvalue 1—note that P has for sure the unitary eigenvalue as the transition matrix is stochastic [23]. The existence and uniqueness of the steady-state distribution are guaranteed by the ergodicity of the considered DTMC. It can be shown that the obtained solution depends on the probability with which a packet, arriving in an idle minislot at a given node, is transmitted at the beginning of the next minislot. This probability, denoted as q, depends on the particular backoff algorithm under use. In order to determine the value of q, we apply the renewal theoretic approach proposed in [22],

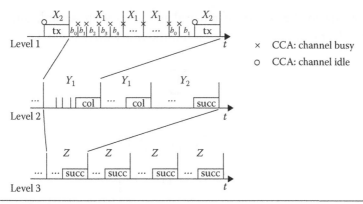

Figure 4.3 Three-level renewal process with maximum number of retransmissions equal to $M = 5$. col, collision; succ, successful transmission; tx, transmission.

where the three-level renewal process shown in Figure 4.3 (in the illustrative case with maximum number of retransmissions $M = 5$) is proposed:

- The first-level renewal cycle X is defined as the period between two consecutive instants at which the selected node starts with the backoff stage 0. In particular, two types of the first-level cycles can be observed: the X_1 cycle does not contain any transmission because of M consecutive busy CCAs, whereas the X_2 cycle contains one transmission (either successful or not), which is carried out after an idle CCA.

- The second-level renewal cycle Y is defined as the period between the end of the first-level cycle X_2 and the end of the consecutive first-level cycle X_2. Note that there could be $j(j \geq 0)X_1$ cycles before the X_2 cycle. The cycle Y can be of either type Y_1 (if the transmission reduces to a collision) or type Y_2 (if the transmission is successful).

- The third-level renewal cycle Z is defined as the period between the end of the second-level cycle Y_2 and the end of the consecutive second-level cycle Y_2. As for the previous case, there could be $k(k \geq 0)Y_1$ cycles before a Y_2 cycle. The successful transmissions carried out in the Z cycle can thus be considered as the *reward* for the third-level renewal process. The throughput of the selected node can thus be computed as the average reward in the Z cycle.

According to the theory of renewal processes with reward, one can derive the following two equations for the case with double CCAs [22]:

$$\tau = \frac{\bar{R}}{\bar{X}} = \frac{\sum_{m=0}^{M-1} \alpha^m}{\bar{X}} \qquad (4.2)$$

$$\alpha = \frac{L\left[1-(1-\tau)^N\right]}{1+L\left[1-(1-\tau)^N\right]} \qquad (4.3)$$

where:

τ is the sensing probability

α is the *failure* probability, that is, the probability of finding the channel busy in a minislot

N is the number of nodes

L is the packet length (in minislots)

M is the maximum number of backoff cycles

If L, N, and M are known, then Equations 4.2 and 4.3 are a set of fixed point equations and can be solved. In particular, the probability τ corresponds to the probability q introduced in our model.

In order to verify the validity of the proposed framework, we compare the predicted results with the Markov chain-based model proposed in [24]. In Figure 4.4, we compare the steady-state distribution predicted by our model with that predicted by the model in [24], in a scenario with $p = .0002$, $N = 12$ nodes, and $L = 10$.

4.4 Exact Idle Time Distribution in Single-Hop IEEE 802.15.4 WSN

Unlike the approximate model presented in [2], we propose an innovative analytical approach to numerically derive the exact distribution of the idle channel times of the primary IEEE 802.15.4 WSN, by using the queueing model presented in Section 4.3. The basic idea consists of the application of the total probability theorem for the evaluation of each term of the probability mass function (PMF) of the idle time, by conditioning on the starting states and assuming that they have the steady-state probabilities $\{\pi_n\}_{n=0}^N$ of the DTMC associated with the process $\{Q_k\}$. More precisely, one obtains

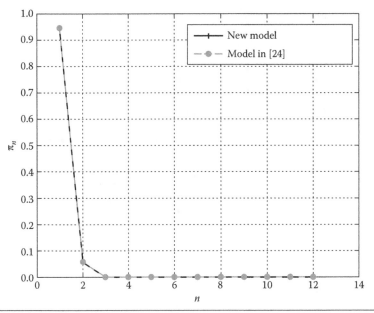

Figure 4.4 Distribution of the number of packets in the system in a scenario with $p = .0002$.

$$P\{I = i\} = P\{I = i|Q_k = 0\}\pi_0 + P\{I = i|Q_k = 1\}\pi_1$$

$$+\cdots+ P\{I = i|Q_k = N\}\pi_N \qquad \forall i > 0 \tag{4.4}$$

The terms $[P\{I = i|Q_k = n\}]_{n=0}^{N}$ are obtained through a recursive algorithm that *runs* over a particular trellis diagram, derived from the transition diagram in Figure 4.2—in networking theory, the transition diagram is commonly considered, as steady-state transition probabilities are of interest; the use of a trellis diagram, which takes into account the time evolution through the specific sequence of states (i.e., a *path*), is often used in communication theory [25]. Instead of considering the *full* trellis diagram (i.e., with all possible transitions), we consider a *reduced* trellis diagram only with the transitions associated with the idle events. Illustrative representations of full and reduced trellises are shown in Figures 4.5 and 4.6, respectively.

Assuming to start from the state n in the trellis diagram, the sum of the state probabilities at the ith step corresponds to the probability $P\{I \geq i|Q_k = n\}$. Therefore, it follows that

$$P\{I = i|Q_k = n\} = P\{I \geq i|Q_k = 1\} - P\{I \geq i+1|Q_k = n\} \tag{4.5}$$

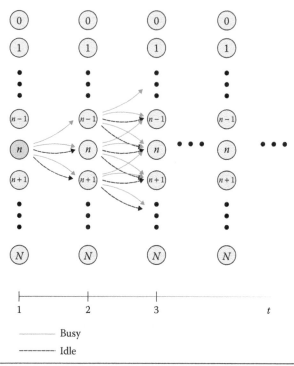

Figure 4.5   Full trellis diagram associated with the transition diagram in Figure 4.2.

At this point, the complete PMF (4.4) of the idle time duration (in minislots) can be derived—the accuracy of this calculation can be made as high as desired simply by considering a sufficiently large number of trellis steps, in order to correctly estimate the tail of the PMF.

We remark that in the derivation of the idle time distribution, we assume that the channel is busy immediately before becoming idle. Moreover, the distribution starts from one minislot as we decided to consider the two idle CCA minislots as belonging to a busy period.

In the remainder of this section, we investigate the idle time distribution considering a fixed network scenario with $N = 12$ nodes directly connected to the sink, fixed packet duration of $L = 10$ minislots (of data), and two supplementary CCA minislots related to the CSMA/CA MAC protocol. All remaining parameters of the CSMA/CA MAC protocol are set to their default values. The network load will be set to two values representative of low and high traffic situations. We recall that the traffic generation at each node has a Bernoulli distribution with parameter p (dimension: [pck/minislot]). In general, the aggregate and normalized network load can be expressed as [24]

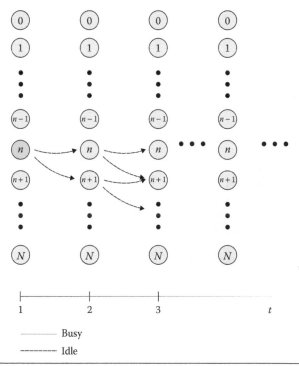

Figure 4.6 Reduced trellis diagram, with only branches associated with idle minislots, derived from the full trellis diagram in Figure 4.5.

$$G = NLp \qquad (4.6)$$

By fixing the values of N and L (as will be done in the following), G depends only on p.

In Figure 4.7, we show the idle time distribution associated with four values of p: (1) .0002, (2) .003, (3) .007, and (4) .01. In all cases, $N = 12$ and $L = 10$. From the results in Figure 4.7, it can be observed that the PMF of the idle time concentrates to smaller and smaller values for increasing values of p. However, it can also be observed that the *shape* of the PMF tends to remain the same. This suggests that there might exist a closed-form distribution that approximates the exact idle time distribution. Heuristically, the results in Figure 4.7 suggest that the shape of the PMF looks like a translated geometric with properly set parameter p_{idle}, that is,

$$P\{I = i\} = p_{\text{idle}}(1 - p_{\text{idle}})^{i-1} \qquad \forall i > 0 \qquad (4.7)$$

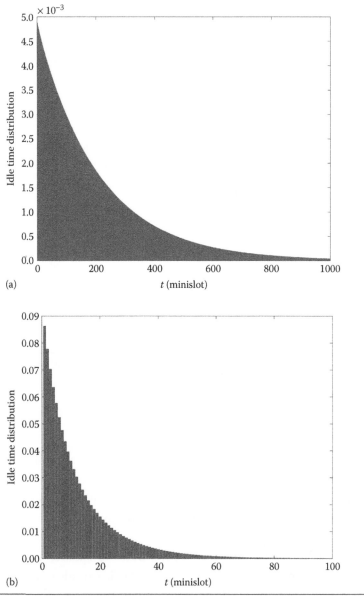

Figure 4.7 Idle time distributions corresponding to various values of p: (a) .0002; (b) .003; (c) .007; (d) .01. In all cases, $N = 12$ and $L = 10$.

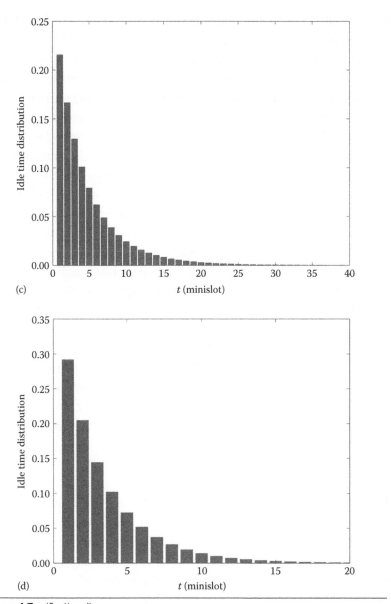

Figure 4.7 (Continued)

In particular, it can be concluded that $p_{idle} = P\{I = 1\}$ and this probability can be computed by applying the proposed trellis-based approach only in its first step. The average idle time duration is thus $1/p_{idle}$. In Figure 4.8, we compare directly the exact and approximate (geometric) distributions of the idle time considering various values of p: (1) .0002, (2) .003, and (3) .01. An excellent agreement can be observed. This result compares favorably with the results, relative to an IEEE 802.11 network, presented in [26], where the idle time distribution is assumed to be geometric. In [26], it is shown that the assumption of constant access probability per minislot is confirmed. Moreover, the geometric idle time distribution determined by our framework is in agreement with the results presented in [7,11], where the channel occupancy is modeled as exponential (continuous-time approach). We remark, however, that our recursive trellis-based approach is *exact*, and thus confirms the validity of the geometric approximation.

In Figure 4.9, the parameter p_{idle} is shown as a function of the aggregate (normalized) load G of the primary WSN. As mentioned earlier, $N = 12$ and $L = 10$. It can be observed that, as intuitively expected, p_{idle} is approximately a linearly increasing function of G. In Figure 4.10, the average idle time duration $\bar{I}$ is shown as a function of G. It can be observed that the average idle time $\bar{I}$ decreases very quickly for increasing values of the traffic load in the primary WSN.

4.5 Time-Domain Cognitive Sensor Networking

Even if not surprising, the fact that the idle time distribution is accurately modeled as geometric has a relevant consequence for the considered cognitive system. In fact, thanks to the memoryless property of the geometric distribution of the idle times of the primary WSN, when the secondary WSN wakes up in the middle of an idle interval, the elapsed portion of this idle time is irrelevant, as the distribution of the remaining portion of the idle time does not change. However, the optimal transmission strategy for the secondary WSN simply requires to estimate the average per-node traffic load of the primary network, together with its number of nodes. An illustrative representation of the overall cognitive sensor networking system is shown in Figure 4.11.

Depending on the adopted strategy, the estimation of the primary WSN activity behavior can be carried out in several ways: (1) by the

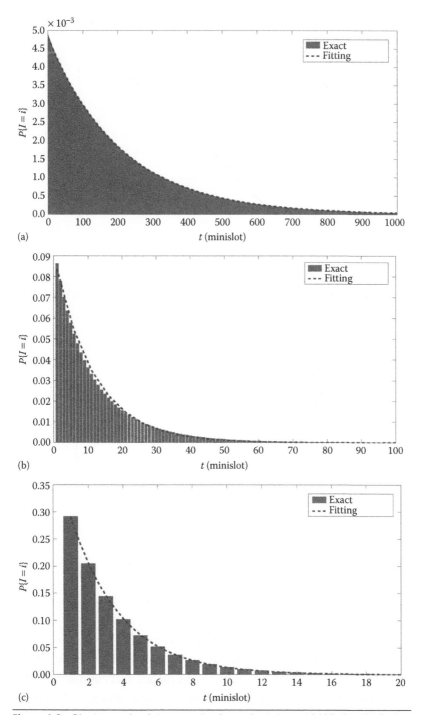

Figure 4.8 Direct comparison between exact and approximate (geometric) idle time distributions for various values of p: (a) .0002; (b) .003; (c) .01. In all cases, $N = 12$ and $L = 10$.

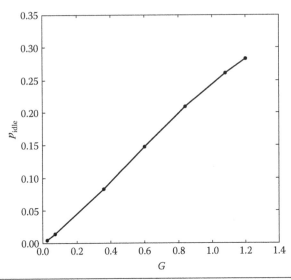

Figure 4.9 p_{idle} as a function of the aggregate (normalized) load G of the primary WSN.

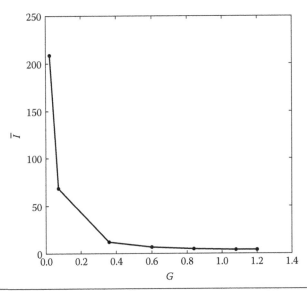

Figure 4.10   Average idle time duration $\bar{I}$ as a function of G.

primary node sink (in a very accurate way) and (2) by the secondary sink or a generic node of the secondary network. Clearly, these solutions lead to different levels of performance and complexity. The first approach implies cooperation between the primary and secondary sinks. The second approach seems more appealing for typical scenarios relative to coexisting sensor networks.

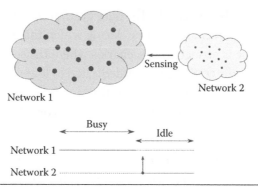

Network 1

Network 2

Figure 4.11 Cognitive sensor networking system.

In order to evaluate the performance of a cognitive system where the secondary nodes are aware of the geometric idle time distribution of the primary IEEE 802.15.4 WSN, we carry out simulations where the idle times of the primary network are generated according to the geometric distribution. The overall simulation time is $T_{sym} = 100,000$ minislots—the duration is such that the arithmetic average of the idle times is sufficiently close to $1/p_{idle}$. We assume that the secondary nodes have always packets to transmit but are allowed to transmit only by the access point (AP) of the secondary WSN (sort of polling mechanism). Once the channel is sensed idle by the secondary AP, the transmission in the secondary network, of duration equal to L minislots, can start: if, during the entire transmission, the channel remains idle, then the transmission is considered successful; otherwise (i.e., if the primary network starts its activity), there is a failure. In Figure 4.12, an illustrative example of the interaction between the two WSNs is shown.

In order to evaluate the performance of the proposed cognitive system, various parameters need to be considered: the load of the primary network, the packet duration L (in minislots), and the duty

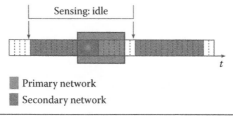

■ Primary network
■ Secondary network

Figure 4.12 Illustrative representation of channel utilization by the primary and secondary WSNs.

cycle of the secondary network. The duty cycle of the secondary WSN is defined as the percentage of time during which this WSN is active. A low duty cycle is attractive in scenarios where the secondary WSN generates a sporadic traffic (e.g., it is dedicated to low-rate background measurement of specific physical quantities).

The following two performance metrics are considered for performance analysis:

1. The probability of interference, denoted as P_i, of the secondary network on the primary network is directly related to the performance degradation of the primary network caused by the activity of the secondary network.
2. The throughput of the secondary network is defined as

$$S_2 = \frac{N_s L}{\sum_i T_i} \tag{4.8}$$

where:

N_s is the number of successful secondary transmissions (i.e., without interference with the primary network)

T_i is the duration of the ith idle period

In Figure 4.13, P_i is shown as a function of S_2 for various values of G and the *duty cycle*. The curves, parameterized with respect to L, show clearly the trade-off between the performance of primary and

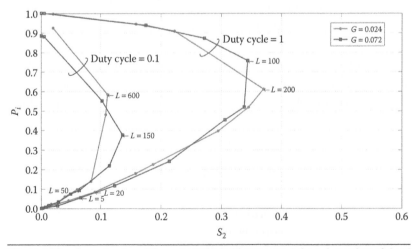

Figure 4.13 P_i as a function of S_2, parameterized with respect to L, for various values of G and the duty cycle.

secondary networks. In Figure 4.13, a few values of L of interest are indicated. For example, in order to have $P_i < 0.1$ with $G = 0.024$, the secondary network should transmit for $L = 20$ minislots, with a corresponding secondary throughput $S_2 = 0.1$.

4.6 Conclusions

In this chapter, we have considered the coexistence of a primary IEEE 802.15.4 WSN with a secondary WSN. The goal of the latter WSN is to transmit in the inactivity periods of the former WSN. First, we have proposed a rigorous DTMC-based queuing model of a single-hop IEEE 802.15.4 WSN (considering a minislotted approach to discretize the original continuous time $M/G/1/N$ WSN queue). On the basis of this model, we have derived the exact distribution of the idle times of the primary IEEE 802.15.4 WSN. Since these idle times can be accurately approximated as geometric, owing to the memoryless property of the geometric distribution the best transmission strategy for the secondary WSN simply requires to estimate the average per-node traffic load of the primary network, together with its number of nodes. Our results show clearly that there exists a trade-off between the throughput achievable in the secondary WSN and the probability of interference of the same WSN with the primary WSN.

References

1. I. Akyildiz, W.-Y. Lee, M. Vuran, and S. Mohanty, "A survey on spectrum management in cognitive radio networks," *IEEE Communications Magazine*, vol. 46, no. 4, pp. 40–48, 2008.
2. M. Sharma, A. Sahoo, and K. Nayak, "Model-based opportunistic channel access in dynamic spectrum access networks," *Proceedings of IEEE Global Telecommunications Conference*, pp. 1–6, Honolulu, HI, November/ December 2009.
3. J. Mitola and G. Q. Maguire, "Cognitive radio: Making software radios more personal," *IEEE Personal Communications*, vol. 6, no. 4, pp. 13–18, 1999.
4. Federal Communications Commission (FCC), "Mobile broadband: The benefits of additional spectrum," October 2010. Available at http://www. broadband.gov/.
5. K.-L. Yau, P. Komisarczuk, and P. Teal, "Cognitive radio-based wireless sensor networks: Conceptual design and open issues," *IEEE 34th Conference on Local Computer Networks*, pp. 955–962, Zurich, Switzerland, October 2009.

6. Federal Communications Commission (FCC), "ET Docket No. 03-322: Notice of proposed rule making and order," December 2003. Available at http://www.cs.ucdavis.edu/~liu/289I/Material/FCC-03-322A1.pdf.

7. M. Sharma, A. Sahoo, and K. Nayak, "Alternating renewal theory-based MAC protocol for cognitive radio networks," Department of Computer Science and Engineering, Indian Institute of Technology, Mumbai, India, Technical Report, 2008. Available at http://www.cse.iitb.ac.in/internal/techreports/reports/TR-CSE-2008-15.pdf.

8. P. A. K. Acharya, S. Singh, and H. Zheng, "Reliable open spectrum communications through proactive spectrum access," *The 1st International Workshop on Technology and Policy for Accessing Spectrum*, Boston, MA, August 2006.

9. L. Yang, L. Cao, and H. Zheng, "Proactive channel access in dynamic spectrum networks," *Physical Communication*, vol. 1, no. 2, pp. 103–111, 2008.

10. H. Kim and K. G. Shin, "Efficient discovery of spectrum opportunities with MAC-layer sensing in cognitive radio networks," *IEEE Transactions on Mobile Computing*, vol. 7, no. 5, pp. 533–545, 2008.

11. S. Geirhofer, L. Tong, and B. M. Sadler, "Cognitive medium access: Constraining interference based on experimental models," *IEEE Journal on Selected Areas in Communications*, vol. 26, no. 1, pp. 95–105, 2008.

12. S. Geirhofer and L. Tong, "Cognitive radios for dynamic spectrum access—Dynamic spectrum access in the time domain: Modeling and exploiting white space," *IEEE Communications Magazine*, vol. 45, no. 5, pp. 66–72, May 2007.

13. C. Zhao, J. Hu, and L. Shen, "A MAC protocol of cognitive networks based on IEEE 802.11," *The 12th IEEE International Conference on Communication Technology*, pp. 1133–1136, Nanjing, People's Republic of China, November 2010.

14. S.-W. Jeon, N. Devroye, M. Vu, S.-Y. Chung, and V. Tarokh, "Cognitive networks achieve throughput scaling of a homogeneous network," *IEEE Transactions on Information Theory*, vol. 57, no. 8, pp. 5103–5115, 2011.

15. S. Pollin, M. Ergen, M. Timmers, A. Dejonghe, L. V. der Perre, F. Catthoo, I. Moerman, and A. Bahai, "Distributed cognitive coexistence of 802.15.4 with 802.11," *The 1st International Conference on Cognitive Radio Oriented Wireless Networks and Communications*, pp. 1–5, Mykonos Island, Greece, June 2006.

16. O. B. Akan, O. Karli, and O. Ergul, "Cognitive radio sensor networks," *IEEE Network*, vol. 23, no. 4, pp. 34–40, 2009.

17. Y.-C. Liang, Y. Zeng, E. C. Y. Peh, and A. T. Hoang, "Sensing-throughput tradeoff for cognitive radio networks," *IEEE Transactions on Wireless Communications*, vol. 7, no. 4, pp. 1326–1337, 2008.

18. S. Lee and S.-L. Kim, "Optimization of time-domain spectrum sensing for cognitive radio systems," *IEEE Transactions on Vehicular Technology*, vol. 60, no. 4, pp. 1937–1943, 2011.

19. H. Hu, N. Li, L. Wu, Y. Xu, and Y. Chen, "Throughput maximization for cognitive radio networks with time-domain combining cooperative spectrum sensing," *IEEE 14th International Conference on Communication Technology*, pp. 235–239, Chengdu, People's Republic of China, November 2012.

20. H. Hu, Y. Xu, and N. Li, "Optimization of time-domain combining cooperative spectrum sensing in cognitive radio networks," *Wireless Personal Communications*, vol. 72, no. 4, pp. 2229–2249, 2013.

21. A. Kumar, K. G. Shin, Y.-J. Choi, and D. Niculescu, "On time-domain coexistence of unlicensed and licensed spectrum users," *IEEE International Symposium on Dynamic Spectrum Access Networks*, pp. 223–234, Bellevue, WA, October 2012.

22. X. Ling, Y. Cheng, J. Mark, and X. Shen, "A renewal theory based analytical model for the contention access period of IEEE 802.15.4 MAC," *IEEE Transactions on Wireless Communications*, vol. 7, no. 6, pp. 2340–2349, 2008.

23. A. Leon-Garcia, *Probability and Random Processes for Electrical Engineering*. Boston, MA: Addison Wesley, 1994.

24. M. Martalò, S. Busanelli, and G. Ferrari, "Markov chain-based performance analysis of multihop IEEE 802.15.4 wireless networks," *Performance Evaluation*, vol. 66, no. 12, pp. 722–741, 2009.

25. J. G. Proakis, *Digital Communications*, 4th Edition. New York: McGraw-Hill, 2001.

26. A. Motamedi and A. Bahai, "Optimal channel selection for spectrum-agile low-power wireless packet switched networks in unlicensed band," *EURASIP Journal on Wireless Communications and Networking*, vol. 2008, Article ID 896420, 10pp.

Biographical Sketches

Stefano Busanelli was born in Castelnovo né Monti, Reggio Emilia, Italy, in 1982. He received his *Laurea, Laurea Specialistica*, and PhD degrees in 2004, 2007, and 2011, respectively, from the University of Parma, Parma, Italy. From June 2008 to October 2008, he was an intern at Thales Communications (Colombes, France), working on cooperative wireless communications. Since March 2012, he has been leading the R&D department of Guglielmo Srl (Bibbiano di Barco, Reggio Emilia, Italy). As of today he has published more than 30 papers in leading international journals and conferences. His research interests include wireless ad hoc and sensor networking, vehicular ad hoc networks, digital signal processing, and vertical handover in heterogeneous networks. He participates in several research projects funded by public and private bodies. Dr. Busanelli is a corecipient of an award for the outstanding technical contributions at ITST-2011 and a member of the Wireless Ad Hoc and Sensor Networks Laboratory (WASN Lab) team that won the first Body Sensor Network (BSN) contest, held in conjunction with BSN 2011. He acts as a frequent reviewer for many international journals and conferences.

Gianluigi Ferrari (http://www.tlc.unipr.it/ferrari) is an associate professor of telecommunications at the University of Parma, Parma, Italy. He was a visiting researcher at the University of Southern California (USC), Los Angeles, California, during 2000–2001; Carnegie Mellon University (CMU), Pittsburgh, Pennsylvania, during 2002–2004; King Mongkut's Institute of Technology Ladkrabang (KMITL), Bangkok, Thailand, in 2007; and the Université libre de Bruxelles (ULB), Brussels, Belgium, in 2010. Since 2006, he has been the coordinator of the Wireless Sensor Networks Research Laboratory (WASN Lab; http://wasnlab.tlc.unipr.it/) at the department of information engineering. He has published extensively in the areas of wireless ad hoc and sensor networking, adaptive digital signal processing, and communication theory. Professor Ferrari is a corecipient of awards at the following international conferences: IWWAN'06, EMERGING'10, BSN 2011, ITST-2011, SENSORNETS 2012, and EvoCOMNET 2013 and BSN 2014. He is the editor of a few international journals. He was a guest editor of the 2010 European Association for Signal Processing (EURASIP) *Journal on Wireless Communications and Networking* (*JWCN*) special issue "Dynamic Spectrum Access: From the Concept to the Implementation" and is a guest editor of the 2014 Hindawi *International Journal of Distributed Sensor Networks* (*IJDSN*) special issue "Advanced Applications of Wireless Sensor Networks Using Sensor Cloud Infrastructure." He is a senior member of the Institute of Electrical and Electronics Engineers (IEEE).

Alessandro Colazzo was born in Galatina, Lecce, Italy, on February 29, 1984. He received the *Laurea* degree (3-year program, equivalent to a bachelor) in telecommunications engineering *summa cum laude* in March 2007, discussing a thesis titled "Study of the Correlation of UHF Antennas in Spatial Diversity" (in Italian). On December

2010, he received the *Laurea* degree (3 + 2-year program, equivalent to a master) in telecommunications engineering *summa cum laude*, discussing a thesis titled "Time-Domain Cognitive Wireless Systems with Primary IEEE 802.15.4 Network" (in Italian). Since January 2011, he has started to work in the information and communications technology (ICT) industry. Currently he works in the field of mobile communications, 4G long-term evolution (LTE), and 4G LTE-Advanced standards, focused on medium access control and physical layers.

Jean-Michel Dricot has more than 10 years of experience in the field of mobile, wireless, and sensor networks. He earned his MSc and PhD in electrical engineering at the Université Libre de Bruxelles (ULB), Brussels, Belgium, in 2001 and 2007, respectively. In 2007, he joined France Télécom R&D, before returning to the ULB as an Fonds de la Recherche Scientifique (FNRS) research associate. In 2010, he has been appointed as an associate professor in the Wireless Communications Group of the Brussels School of Engineering, Brussels, Belgium. Dricot is a member of the COST IC 0902 action and the ISO/TC215 standardization committee. He has regular collaborations with the University of Parma, Parma, Italy, and the French institute Institut National de Recherche en Informatique et en Automatique (INRIA). His current research interests include protocols and architecture design for 4G/5G mobile networks and the performance evaluation of body sensor networks and cognitive radio networks, with a strong focus on the impact of the physical layer.

RADIO SPECTRUM SENSING IN COGNITIVE WIRELESS NETWORKS

DANDA B. RAWAT AND CHANDRA BAJRACHARYA

Contents

5.1 Introduction

The future wireless systems are emerging toward ad hoc networks with uncertain topologies and the main idea in future wireless networks is to design networks that are self-configuring, self-organizing, self-optimizing, and self-protecting. Therefore, such cognitive wireless networks (CWNs) should be able to learn and adapt instantaneously to their operating environment depending on their operating radio-frequency (RF) environment, thus providing much needed flexibility and functional scalability. Moreover, in order to adapt to their operating parameters automatically and wisely, signal processing for spectrum sensing is regarded as a fundamental step in these types of networks. Currently existing wireless communication systems and networks are operating based on fixed spectrum assignment to the service providers and their users for exclusive use on a long-term basis and over a vast geographic area. The exclusive RF spectrum assignment, which is licensed by government regulatory bodies, such as the Federal Communications Commission (FCC) in the United States, was an efficient way for interference mitigation among adjacent bands. However, the fixed spectrum assignment leads to inefficient use of spectrum creating *spectrum drought* [1] since most of the channels actively transmit the information only for short duration while a certain portion of the spectrum is idle when and where the licensed users are not transmitting [2–5]. This implies that the inefficient radio spectrum usage has been a serious bottleneck for deployment of larger density of wireless devices. We note that the scarcity of RF spectrum is not a result of lack of spectrum but a result of wasteful static spectrum allocations. In order to alleviate the spectrum scarcity, cognitive radio (CR) for secondary user (SU) [6–13] has been introduced to facilitate the spectrum sharing [14–20] to increase the spectrum efficiency so that the larger density of wireless users can be accommodated without creating a new RF spectrum band. For SUs with software-defined radios along with some intelligence to work automatically according to their operating environment, the radio components are implemented in software rather than in hardware; therefore, it is possible to adapt system configurations to any frequency to transmit and receive the data. It is important to note that the primary users (PUs) are the authorized users of the licensed frequency band, and SUs, who are not the PUs but want to use the licensed frequency band, are the cognitive users in CWNs. It is also worth mentioning that allowing an SU to access the licensed spectrum (imposing some constraints on SUs) improves the spectrum utilization. In CWNs, devices detect each other's presence as interference and try to avoid the interference autonomously by changing their behavior accordingly. In dynamic spectrum sharing, SUs are not allowed to cause harmful interference to the incumbent PUs. It is worth noting that the SUs are essentially invisible to the PUs in CWNs; hence, possibly no changes are needed for licensed users/devices. In such a scenario, SUs can either be allowed to transmit at low power as in the ultra wideband (UWB) system or be allowed to use spectrum opportunities dynamically to transmit without causing the harmful interference to PUs. In the latter case, the CWN autonomously detects and exploits the idle spectrum where and when the PUs are not active. This helps to increase the system capacity and efficiency, and the dynamic spectral access implies that the SU be able to work in multiband, different wireless channels, and support multimedia services and/or applications.

As conventional and existing wireless communication networks are operating based on fixed spectrum assignment to the service providers and their users for exclusive use on a long-term basis result in spectrum scarcity, the CR technology uses the radio spectrum opportunities dynamically without creating harmful interference to the licensed users. In order to fully realize the CR system, the detection of PU signal is the most important and fundamental step. Therefore, the CR system requires a signal processing for radio spectrum sensing implementation to detect both the interference and the absence or presence of PUs.

To find a radio spectrum opportunity, an individual SU should undergo through a cognitive cycle starting from *sensing* (observe phase), analysis (reasoning phase), adaptation (switching to the best transmit parameter phase), and acting (act phase, i.e., SU communication phase

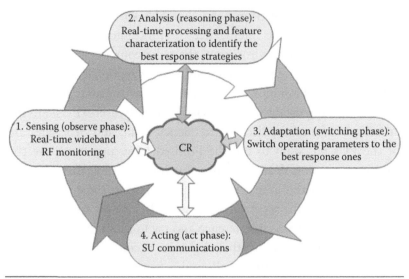

Figure 5.1   Cognitive cycle for CR network.

based on newly adapted parameters) [3,5] as shown in Figure 5.1. Among the four different phases of cognitive cycle, sensing (observe) phase is important since the remaining phases are based on the sensed information. In this chapter, we focus on radio spectrum sensing for dynamic spectrum access.

The rest of the chapter is organized as follows: Section 5.2 presents the background of spectrum sensing and access. Section 5.3 consists of different methods of signal detection for spectrum sensing. Section 5.4 presents a comparison of radio spectrum sensing methods. Section 5.5 concludes the chapter.

5.2 Radio Spectrum Access and Sensing: Background

In CR networks, the radio spectrum sensing plays a major role to realize its full potential of spectrum utilization in a real environment since the sensed information is used to decide the operations of SUs. The SUs perform the sensing of radio spectrum, and then the sensed information is analyzed to make wise decision in timely and accurate manner. A significant number of projects have addressed CR all over the world during the recent years. In Section 5.2.1, we first review the radio spectrum sharing models of dynamic spectrum access [21] for CWNs and capabilities of CR users.

5.2.1 Radio Spectrum Sharing Models of Dynamic Spectrum Access

The dynamic spectrum access is the new concept opposite of static frequency spectrum assignment and its management. The dynamic spectrum access strategies can be broadly categorized into three groups: (1) dynamic exclusive use model, (2) open radio spectrum sharing model, and (3) hierarchical radio spectrum access model.

5.2.1.1 Dynamic Exclusive Use Model   This spectrum access model is like a traditional model in which the radio spectrum bands are licensed to service providers for exclusive use. However, in order to introduce the flexibility to improve the radio spectrum efficiency, this model has two approaches: (1) *spectrum property rights* in which licensees have exclusive rights to freely choose technologies, and sell and trade their spectrum, and (2) *dynamic spectrum allocation* in which the spectrum is allocated exclusively to service providers for a given region and time.

5.2.1.2 Open Radio Spectrum Sharing Model   This model consists of wireless services operating in unlicensed industry, scientific, and medical (ISM) radio band such as in wireless local area network (LAN) or WiFi, where all users have equal opportunities to access the radio spectrum. However, the SUs can prefer the channels, which have sparse or moderate traffic (or users) than the channels with heavy traffic (or users).

5.2.1.3 Hierarchical Radio Spectrum Access Model   This model consists of a hierarchy between the licensed PUs and the secondary unlicensed SUs. In this sharing model, unlicensed SUs can access the spectrum dynamically, which is not licensed to them, but by making sure that the interference created to the licensed PUs is within the tolerable range or using the idle spectrum opportunistically without interfering the PU transmissions. This model has two basic approaches: *spectrum overlay* and *spectrum underlay* [21,22] as shown in Figure 5.2.

In the spectrum overlay model, the SUs will have to identify the idle spectrum bands, which are not used by the licensed system at a given time and location, and use those idle bands dynamically. This model is presented in Figure 5.2a. In the spectrum underlay approach, the SUs are allowed to transmit with low transmit power as in the UWB technology. This model is presented in Figure 5.2b.

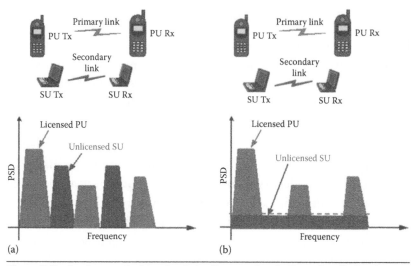

Figure 5.2   Spectrum overlay (a) and underlay (b) approaches. Tx, Transmitter, Rx, Receiver.

Both methods have their advantages and disadvantages. For instance, in the spectrum overlay approach, the SUs can transmit with high power to increase their rates for given spectrum opportunities; however, they have to identify the idle frequency bands that are not used by the PUs. Similarly, in the spectrum underlay approach, the SUs do not need to identify the spectrum opportunities and can transmit simultaneously coexisting with PUs; however, they are not allowed to transmit with high transmission power even if the entire RF band is idle (i.e., the entire RF band is not used by the PUs).

Moreover, the dynamic spectrum sharing by the SUs in a given spectrum band can be categorized as follows (Figure 5.3):

- *Horizontal sharing*, where SUs and PUs have equal opportunities to access the spectrum such as in wireless LAN operating in the ISM band at 2.4 GHz, and in order to improve the overall system performance, SUs can choose the channels that have less traffic or less number of users. In this approach, SUs and PUs coexist in the system and use the bands simultaneously.
- *Vertical sharing*, where SUs have less preference over the PUs, and, thus, SUs must vacate the spectrum as fast as possible once the licensed PUs are detected in the band. However, SUs can use the spectrum with potential whenever they detect the

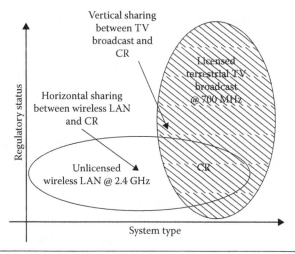

Figure 5.3 Spectrum sharing by SUs with different types of systems depending on regulatory status of incumbent radio systems: vertical and horizontal sharing of spectrum.

idle spectrum band. Moreover, in this approach, the CR system needs the operator's assistance.

5.2.2 Capabilities of CRs

CR technology is regarded as an emerging technology for its efficient spectrum use in a dynamic fashion. The main capabilities of CRs can be categorized according to their functionality based on the definition of the CR in [4] as follows:

- *Sense the environment,* which is cognitive capability, where the SUs sense the spectrum either to identify the frequency band not used by the licensed PUs or to make sure that the CR is not creating harmful interference to the PUs. In order to sense the environment, it will first discover the network around it. Furthermore, the CR will identify its location in order to choose the transmission parameters according to its position.

- *Analyze and learn sensed information,* which is self-organized capability, in which the SUs should be able to self-organize their communication based on sensed information.

- *Adapt to the operating environment,* which is reconfigurable capability, in which the SUs will choose the best transmission parameters such as operating frequency, modulation, and transmission power.

The main capabilities of SUs depend on the sensed information from which they analyze, learn, and then adapt their own operating parameters accordingly. Therefore, spectrum sensing is a fundamental step in the CR system, which is the major subject matter of Section 5.2.3.

5.2.3 Spectrum Sensing Methods for CRs

As we noted in Section 5.2.2, most of the functionalities of SUs depend on the information obtained from spectrum sensing. Furthermore, in order to access the available unused spectrum opportunistically and/ or dynamically, the CWN requires a spectrum sensing or estimation to detect both the interference and the presence of PUs. Several techniques have been proposed in the literature [23–37] to detect the PU signal in spectrum sensing.

We note that the signal processing for spectrum sensing or estimation can be categorized into *direct* and *indirect* methods. The *direct* method is recognized as a frequency-domain approach where the estimation is carried out directly from the signal, whereas the *indirect* method is known as a time-domain approach where the estimation is performed using autocorrelation of the signal.

Furthermore, the other way of categorizing the spectrum estimation and sensing techniques is by grouping them into model-based *parametric* methods and periodogram-based *nonparametric* methods [38].

However, in this chapter, we present signal detection techniques for spectrum sensing by categorizing into five groups as follows:

- *Primary transmitter detection*: Detection of PUs or signal is performed based on the received signal at the SU receiver. This approach includes matched filter (MF) detection [27,39], energy-based detection [25,29,30], covariance-based detection [23], waveform-based detection [26], cyclostationarity-based detection [28], radio identification-based detection [40], and random Hough transform-based detection [31].
- *Primary receiver detection*: In this method, spectrum opportunities are detected based on the PU receiver's local oscillator leakage power [33].
- *Cooperative detection*: In this approach, the PU signal for spectrum opportunities is detected reliably by interacting or cooperating with other users [24,35,36], and the method can

be implemented as either a centralized access to spectrum coordinated by a spectrum server [41] in the distributed approach implied by the spectrum load smoothing algorithm [42] or an external detection [37].

- *Interference temperature management*: In this approach, the CR system works as a UWB technology where the SUs coexist with the PUs and are allowed to transmit with low power and restricted by the interference temperature level so as not to cause harmful interference to the PUs [43,44].
- *Other advanced approaches*: There are some methods, which do not fit into the above categories, such as wavelet-based detection [32], multitaper spectrum sensing or estimation [4,45], and filter bank-based spectrum sensing [34] in this group.

Our main goal in this chapter is to present the comprehensive state-of-the-art research result of signal detection techniques for spectrum sensing CWNs in which the sensed information is used to make automatic and wise decision by the SUs so that they operate according to their RF environments. In Section 5.3, we present the system description and signal processing methods for CR networking.

5.3 Signal Detection Methods for Spectrum Sensing

In this section, we present the details of different methods for signal processing which are applicable in spectrum sensing for SUs.

5.3.1 Primary Transmitter Detection

This approach contains several methods in which we process the received signal (actually transmitted from the primary transmitter to the primary receiver) at the SU receiver that wants to use the spectrum opportunities locally for a given time and location. In this section, we present the most common techniques of signal detection for spectrum sensing in CR systems.

In order to detect the PU signal in the system to find the spectrum opportunity, we consider the received signal at the SU receiver in continuous time as

$$y(t) = gs(t) + w(t) \qquad (5.1)$$

where:

   $y(t)$ is the received signal

   g is the channel gain between the primary transmitter and the
      SU receiver

   $s(t)$ is the PU signal (to be detected)

   $w(t)$ is the additive white Gaussian noise (AWGN)

In order to use the signal processing algorithms for spectrum sensing, we consider the signal in the frequency band with a center frequency f_c and a bandwidth W, and sample the received signal at a sampling rate f_s, where $f_s > W$, and $T_s = 1/f_s$ is the sampling period. Then we can define $y(n) = y(nT_s)$ as the received signal samples, $s(n) = s(nT_s)$ as the primary signal samples, and $w(n) = w(nT_s)$ as the noise samples, and can write the sampled received signal as

$$y(n) = gs(n) + w(n) \qquad (5.2)$$

If we consider the channel gain $g = 1$ (i.e., ideal case) between the terminals, then Equation 5.2 becomes

$$y(n) = s(n) + w(n) \qquad (5.3)$$

For the received signal, two possible hypotheses considered for PU detection can be written as follows: H_0 to denote that the signal $s(n)$ is not present and H_1 to denote that the signal $s(n)$ is present. That is, the received signal samples under the two hypotheses are

$$y(n) = \begin{cases} w(n) & H_0 \\ s(n) + w(n) \quad \text{or} \quad gs(n) + w(n) & H_1 \end{cases} \qquad (5.4)$$

In this section, we deal with the system model given in Equation 5.2 or 5.3 with given two hypotheses in Equation 5.4 for primary transmitter detection methods, and then proceed with detection techniques to identify the presence of the PU signal. If the signal component $s(n) = 0$ in Equation 5.3, the particular frequency spectrum band may be idle (if the detection is error free). When the signal is present (i.e., $s(n) \neq 0$), the particular frequency spectrum/band is in use and there is no spectrum hole for a given time and location.

5.3.1.1 MF-Based Signal Detection When the transmitted signal is known at the receiver, MF is known as the optimal method for detection of PUs [39] since it maximizes the received signal-to-noise ratio (SNR), and the SNR corresponding to Equation 5.3 is

$$\gamma = \frac{|s(n)|^2}{E[w^2(n)]} \tag{5.5}$$

where:

   γ is SNR

   E is expected operator

A simple MF-based detection can be implemented as shown in Figure 5.4, where threshold is used to estimate the signal. Cabric et al. [27] used MF for pilot signal and MF-based detection where the method assumes that the PU sends a pilot signal along with the data. The process is depicted in Figure 5.5. The MF performs best when the signaling features to be received are known at the receiver.

In spite of its best performance, the MF has more disadvantages than its advantages: First, MF requires perfect knowledge of the PU signaling features (such as modulation type and operating frequency), which are supposed to be detected at the CR. As we know, CR will use wideband spectrum wherever it finds the spectrum opportunities. Therefore, it is almost impossible to have MF implemented in the CR for all types

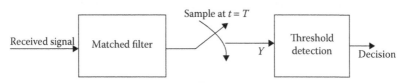

Figure 5.4 MF for signal detection.

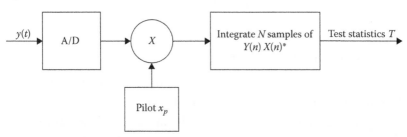

Figure 5.5 Pilot signal and MF-based detection. (Data from D. Cabric, et al., "Implementation Issues in Spectrum Sensing for Cognitive Radios," *Asilomar Conference on Signals, Systems and Computers*, Pacific Grove, CA, June 2004, pp. 772–776.)

of signals in the wideband regime. Second, MF implementation complexity of detection unit in SU devices is very high [27] because the CR system needs receivers for all signal types of wideband regime. Last, large power will be consumed to execute such several detection processes as SU devices sense the wideband regime. Therefore, the disadvantages outweigh the advantages of MF-based detection. It is important to note that the MF-based technique might not be a good choice for real CR system because of its above-mentioned disadvantages.

5.3.1.2 Covariance-Based Signal Detection   This is another method to detect the primary signal by SUs. Zeng and Liang [23] proposed the covariance-based signal detection technique whose main idea is to exploit the covariance of signal and noise since the statistical covariance of signal and noise is usually different. These covariance properties of signal and noise are used to differentiate signal from noise where the sample covariance matrix of the received signal is computed based on the receiving filter. We consider the system model for a received signal as in Equation 5.2, and the received signal in a vector channel form can be written as [23]

$$y = G_s + w \qquad (5.6)$$

where:

G_s is the channel matrix through which the signal travels

The covariance matrices corresponding to the signal and noise can be written as

$$\left. \begin{aligned} R_y &= E[yy^T] \\ R_s &= E[ss^T] \\ R_n &= E[ww^T] \end{aligned} \right\} \qquad (5.7)$$

where:

$E[.]$ is the expected value of $[.]$

If there is no signal ($s = 0$), then $R_s = 0$, and, therefore, the off-diagonal elements of R_y are all zeros. If there is signal ($s \neq 0$) and the signal samples are correlated, R_s is no more a diagonal matrix. Therefore, some of the off-diagonal elements of R_y should not be zeros. Hence, this method detects the presence of signals with the help of covariance matrix of the received signal. That is, if all the off-diagonal values of the matrix R_y are zeros, then the PU is not using the band at that time and location, and otherwise, the band is not idle.

5.3.1.3 Waveform-Based Detection   This is another approach for primary signal detection. In this approach, the patterns corresponding to the signal, such as preambles, midambles, regularly transmitted pilot patterns, and spreading sequences, are usually utilized in wireless systems to assist synchronization or detect the presence of signal. When a known pattern of the signal is present, the detection method can be applied by correlating the received signal with a known copy of itself [26] and the method is known as waveform-based detection. Tang [26] showed that waveform-based detection is better than energy-based detection (presented in Section 5.3.1.4) in terms of reliability and convergence time, and also showed that the performance of the algorithm increases as the length of the known signal pattern increases.

In order to perform waveform-based signal detection, we consider the received signal in Equation 5.3 and compute the detection metric as follows [26]:

$$D = \mathrm{Re}\left[\sum_{n=1}^{N} y(n)s \times (n) \right]$$

$$= \sum_{n=1}^{N} |s(n)|^2 + \mathrm{Re}\left[\sum_{n=1}^{N} w(n)s \times (n) \right]$$

(5.8)

where:

N is the length of the known pattern

The detection metric D for waveform-based detection in this equation consists of two terms: the first term $\sum_{n=1}^{N} |s(n)|^2$ of the second equality is related to a signal and the second term $\mathrm{Re}[\sum_{n=1}^{N} w(n)s \times (n)]$ of the second equality consists of a noise component. Therefore, we can conclude that when the PU is idle [i.e., $s(n) = 0$)], the detection metric D will have only the second term of the second equality in Equation 5.8 that is only noise, and when $s(n) \neq 0$, then D will have both terms of the second equality in the same equation. In order to detect the signal, the metric D value can be compared with some threshold value λ and the detection of the signal can be formulated as

$$P_T = \Pr(D > \lambda \mid H_1)$$

$$P_F = \Pr(D > \lambda \mid H_0)$$

$$(5.9)$$

where:

P_T is the probability of true detection, that is, when a signal is present in the frequency band, the detection is successful

P_F is the probability of false alarm, that is, the detection algorithm shows that the frequency is occupied; however, actually it is not

The basic idea is to reduce the probability of false alarm P_F. We note that the choice of threshold λ plays a major role in this detection approach and can be estimated or predicted based on noise variance. We also note that the measurement results presented by Cabric et al. [24] show that waveform-based detection requires short measurements time; however, it is susceptible to synchronization errors.

5.3.1.4 Energy-Based Detection   Another approach of PU detection for spectrum sensing is energy-based detection. This method is regarded as the most common way of signal detection due to its low computational and implementation complexities [27]. Unlike in MFs and other approaches, in this method, the receivers do not need any kind of knowledge of the PUs' signals.

In this method, the signal detection is performed by comparing the output of energy detector with a given threshold value [29], and the threshold value as in waveform-based approach depends on the noise floor and can be estimated based on it. Figure 5.6a and b show the digital implementation of energy-based detection. In the periodogram approach as in Figure 5.6a, first of all the signal is converted from analog to digital and then fast Fourier transform (FFT) is applied. The output of the FFT process is squared, which is then averaged to get test statistics. Based on the test statistics, the absence or presence of the signal in the particular band is identified. In the analog prefilter approach as depicted in Figure 5.6b, the signal is prefiltered before converting from analog to digital. Then the signal is converted to digital followed by squaring and averaging to get the test statistics. In both implementations, the statistics are compared with given thresholds and then decision is made about the presence or absence of the signal.

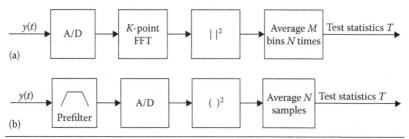

Figure 5.6   Digital implementation of energy-based detection (a) with periodogram: FFT magnitude squared and averaged (b) with analog prefilter and square-law device. (Data from D. Cabric, et al., "Spectrum Sensing Measurements of Pilot, Energy, and Collaborative Detection," *Proceedings of the IEEE Military Communication Conference,* October 2006, pp. 1–7.)

For energy-based detection method, we can consider the system model given in Equation 5.3 and compute the decision metric as

$$D = \sum_{n=1}^{N} |\, y(n)\,|^2 \qquad (5.10)$$

Assuming the variance of AWGN σ_n and the variance of the signal σ_s, the decision metric D follows chi-square distribution with $2N$ degrees of freedom $(\chi_{2N})^2$ [29] and can be modeled two hypotheses as follows:

$$D = \begin{cases} \left(\dfrac{\sigma_w^2}{2} \right) \chi_{2N}^2 & H_0 \\[4mm] \left(\dfrac{\sigma_s^2 + \sigma_w^2}{2} \right) \chi_{2N}^2 & H_1 \end{cases} \qquad (5.11)$$

In this approach, the false alarm probability P_F and the true detection probability P_T can be calculated using two hypotheses making comparison with the chosen threshold value as in Equation 5.9. Again we note that the method has some disadvantages such as without proper choice of threshold value results in undesirable probability of true detection and false alarm, poor performance under low SNR value [26], and inability to differentiate between interference to licensed users and noise that might limit the performance of this approach. In addition, this approach does not work optimally for detecting spread spectrum such as code division multiple access (CDMA) signals [27].

5.3.1.5 Cyclostationarity-Based Detection   This method is also regarded as a good candidate for spectrum sensing in CR systems. This method

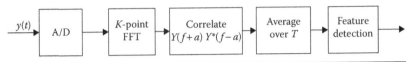

Figure 5.7 Digital implementation of cyclostationarity-based feature detection. (Data from D. Cabric, et al., "Implementation Issues in Spectrum Sensing for Cognitive Radios," *Asilomar Conference on Signals, Systems and Computers*, Pacific Grove, CA, June 2004, pp. 772–776.)

takes advantage of cyclostationarity properties of the received signals [28,46] to detect PU transmissions. The digital implementation of this approach is depicted in Figure 5.7.

The basic idea of this method is to use the cyclostationarity features of the signals. In general, the transmitted signals are stationary random process; furthermore, the cyclostationarity features, which are the periodicity in signal statistics such as mean and autocorrelation, are induced because of modulation of signals with sinusoid carriers, cyclic prefix in orthogonal frequency division multiplexing (OFDM), and code sequence in CDMA. However, the noise is considered as wide-sense stationary (WSS) with no correlation. Therefore, this method can differentiate PUs' signals from noise [28]. In this method, cyclic spectral correlation function (SCF) is used for detecting signals present in a given frequency band and the cyclic SCF of the received signal in Equation 5.3 can be calculated as [28,46]

$$S_{yy}^{\alpha} = \sum_{\tau=-\infty}^{\infty} R_{yy}^{\alpha}(\tau) e^{-j2\Pi f} \qquad (5.12)$$

where:

$R_{yy}^{\alpha}(\tau)$ is the cyclic autocorrelation function which is obtained from the conjugate time-varying autocorrelation function of $s(n)$, which is periodic in n

α is the cyclic frequency

We note that when the parameter $\alpha = 0$, the SCF becomes the power spectral density (PSD). When the signal is present in the given frequency spectrum, this method gives the peak in cyclic SCF, implying that the PU is present. If there is no such peak, the method implies that the given spectrum band is idle or there is no more PU active at a given time and location. Based on this observation, SUs identify the status of absence or presence of PUs in the particular band in a given time and location.

5.3.1.6 Random Hough Transform-Based Detection This approach is borrowed from image processing field. Random Hough transform is applied to a received signal to identify the presence of PU transmission.

5.3.1.7 Centralized Server-Based Detection In this method, a central unit, which is a server, collects all the sensed information related to spectrum occupancy from SU devices, aggregates the available information centrally, and then disseminates or broadcasts the aggregated information related to the spectrum status to all SUs [41,47]. When the SUs receive the aggregated information related to spectrum occupancy, they adapt their transmission parameters according to the received information. Since the spectrum server gathers the information from all other users, this spectrum server is assumed to be just an information collector without having spectrum sensing capability built on it. It is noted that the central server acts like an information fusing device for CR systems and does not play any role in spectrum sensing. Furthermore, individual SUs sense the information and transmit it to their server. They participate in information collaboration. They decide to utilize the spectrum opportunistically and dynamically based on the aggregated information received from a centralized server instead of individually sensed information.

5.3.1.8 External Detection This technique can be considered as an alternative approach of centralized server-based cooperative detection. In this method, similar to centralized server-based approach, all SUs obtain the spectrum information from an external detection agent [37]. However, unlike the centralized server-based detection, an external agent performs the spectrum detection or sensing since it is equipped with sensing capability with spectrum sensors. Once the external agent senses the spectrum, it disseminates the sensed information of spectrum occupancy for SUs. It is important to point out here that the individual SU devices will not have spectrum sensing capability, unlike in centralized server-based detection. SUs do not have their own sensing unit; however, they decide which spectrum band is to be used and with what transmission parameters and technologies. Because of the external agent-based sensing, this approach also helps to overcome the hidden PU problem as well as the uncertainty due to shadowing and fading [37]. Furthermore, this method is also efficient in terms of time and spectrum and power consumptions from the perspective of CR systems since SUs do not spend time and power for signal detection [37] as the spectrum sensing is performed by the external agent. This approach can be seen as a good candidate for the CR system to overcome the technical problems; however, it is important to perform the cost–benefit analysis before recommending it for the implementation in the CR system.

5.3.1.9 Distributed Detection   In contrast to the centralized server-based and external detection, in this method, SUs make their own decision based on their individually sensed information and the information received from other interacting or cooperating users. Unlike in centralized server-based and external detection, in this approach individual SUs need to have installed individual sensing unit on them. Therefore, we do not need a high-capacity centralized backbone infrastructure (centralized spectrum server or external agent) in this approach. This can be seen as economically advantageous over other methods. Instead of deciding the utilization of spectrum opportunities based on individually sensed information, this approach considers the sensed information from other collaborating SUs who are also seeking the spectrum access dynamically and opportunistically to make decision regarding the dynamic spectrum utilization. This method increases the probability of true detection and decreases the probability of false alarm related to actual spectrum occupancy. Basically, the distributed approach can be implemented in SU devices by using spectrum load smoothing algorithms [42].

5.3.2 Interference Temperature Management

This approach considers the spectrum underlay approach for dynamic spectrum access or spectrum utilization. When PUs and SUs coexist and transmit their data simultaneously, the *interference temperature management* is the best approach to protect PUs from the interference caused by SUs by imposing some constraints (e.g., low transmit power) on SUs so as not to exceed the specified interference limit. This approach is illustrated in Figure 5.9.

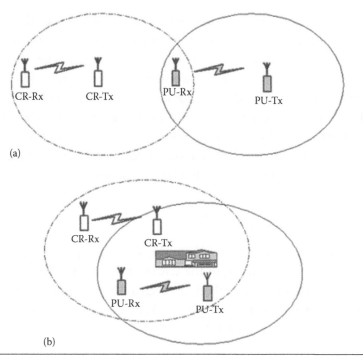

Figure 5.8   Hidden PU problem because of path loss (a) and shadowing/blocking (b).

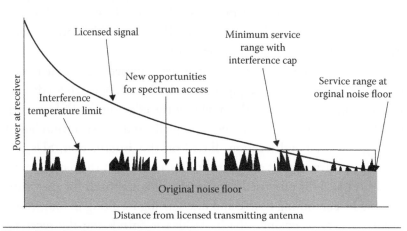

Figure 5.9   Interference temperature model.

The basic idea behind this approach is to set up an upper limit in interference power for a given frequency band in a specific geographic location such that SUs are not allowed to create harmful interference while using the specific band in the specific area [43,44]. This approach works as the UWB technology where the SUs are allowed to transmit simultaneously with the PUs using low transmit power and are restricted by the inter-ference temperature level so as not to cause harmful interference to PUs [43,44]. In this approach, SUs do not have to sense and wait for spectrum opportunities for their communications, but they are restricted by some operating constraints, as they have to respect the incumbent PUs.

However, SUs cannot transmit their data with higher power because of imposed low transmit power and interference temperature limit even if the licensed system is completely idle for a given time and location. This can be seen as a disadvantage of the approach.

5.3.3 Other Advanced Approaches

Some methods, which do not fit in the above groups and are appli-cable for spectrum sensing in the CR system, are considered in this category. Some of these methods are seen as more suitable candidates for future development of CR systems. We present such approaches in Sections 5.3.3.1 through 5.3.3.3.

5.3.3.1 Wavelet-Based Detection

This method is popular in image pro-cessing for edge or boundary detection. Tian and Giannakis [32] used wavelets for detecting the edges in the PSD of a wideband channel. This method is illustrated in Figure 5.10.

In this approach, signal spectrum is decomposed into smaller non-overlapping sub-bands to apply the wavelet-based approach for detect-ing the edges in the PSD. We note that the edges in the PSD are the divider of occupied bands and nonoccupied bands (or spectrum holes) for a given time and location. Based on this information, SUs can identify spectrum holes or opportunities and exploit them optimally.

Hur et al. [48] proposed another wavelet approach for spectrum sensing by combining coarse and fine sensing resulting in multireso-lution spectrum sensing. The basic idea is correlating the received signal with the modulated wavelet to obtain the spectral contents of the received signal around the carrier frequency in the given band

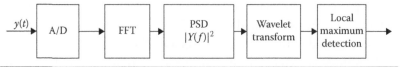

Figure 5.10   Digital implementation of a wavelet detector.

processed by the wavelet. By analyzing dilated versions of the wavelet and scaling functions, wavelet has the capability to dynamically tune time and frequency resolution [49]. In addition, time resolution can be compromised and traded-off with high-frequency resolution for segments of slow varying signal [49].

5.3.3.2 Multitaper Spectrum Sensing/Estimation Thomson [45] proposed the multitaper spectrum estimation (MTSE), in which the last N samples of the received signal are collected in a vector form and represented them as a set of Slepian base vectors. The Slepian base vectors are used to identify the spectrum opportunities in the targeted spectrum band. The main idea of this approach is to utilize its fundamental property, that is, the Fourier transforms of Slepian vectors have the maximal energy concentration in the bandwidth $f_c - W$ to $f_c + W$ under a finite sample size constraint [4,45]. After MTSE, by analyzing this feature, SUs can identify whether there is spectrum opportunity. This method is also regarded as an efficient method for small sample spaces [34].

5.3.3.3 Filter Bank-Based Spectrum Sensing Filter bank-based spectrum sensing (FBSE) is a simplified version of MTSE by introducing only one prototype filter for each band and was proposed for CR networks in [34]. The main idea of FBSE is to assume that the filters at the receiver and transmitter sides are a pair of matched root-Nyquist filters $H(z)$ as shown in Figure 5.11. Specifically, the FBSE was proposed for multicarrier modulation-based CR systems by using a pair of matched root-Nyquist filters [34]. The approach for demodulation of the received signal with ith subcarrier before it is processed through root-Nyquist filter is presented in Figure 5.11.

5.3.3.4 Compressive Radio Spectrum Sensing Generally speaking signals of interest are often sparse in a certain domain and the number of samples required for estimating the signal order can be much smaller than that for reconstructing the unknown sparse signal itself [50,51]. Thus, compressive spectrum sensing techniques can effectively reduce the acquisition costs of high-dimensional signals using compressive sensing for the sparse signals [50,51,52]. By estimating the sparsity order on the fly, compressed sensing can considerably reduce the sampling costs while achieving the desired sensing

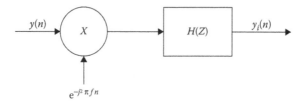

Figure 5.11 Demodulation of a received signal with ith subcarrier before it is processed through root-Nyquist filter. (Data from B. Farhang-Boroujeny and R. Kempter, *IEEE Communication Magazine*, vol. 48, no. 4, 2008.)

accuracy [50], whereas in conventional spectrum sensing, each and every channel should be sensed in a sequential manner, which consumes both battery life and time.

5.4 Comparison of Radio Spectrum Sensing Methods

We presented several signal processing methods for spectrum sensing applicable to CR systems. Among them some methods are suitable for one system consideration and technologies and others are suitable for different system consideration and technologies. Note that there is no such optimal and universal technique available yet, which is suitable for all kinds of technologies for CR systems in the wideband regime. In this section, we compare the main signal processing techniques for spectrum sensing in terms of sensing accuracies and complexities. The different methods of primary transmitter detection are presented in Figure 5.12. Among them, MF gives the highest accuracies with high complexity, which is due to implementation of many MFs in SU devices for spectrum sensing in the wideband regime. However, energy-based detector is least accurate and least complex since we do not need any special kind of filters and the detector uses the energy of the signal during the detection process. In terms of implementation complexity, this approach is suitable for the CR system; however, it is more prone to noise level and interference from close proximity. Others are in the kind of middle in terms of accuracy and complexities.

Sometimes the primary transmitter might be far away from SUs. In this case, if we apply the primary transmitter detection approach for spectrum sensing, low probability of true detection is because of fading and other blockings. In this case, primary receiver detection is

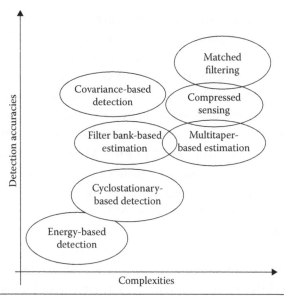

Figure 5.12 Comparison of different techniques for spectrum sensing methods for spectrum overlay in terms of sensing accuracies and implementation complexities.

suitable for spectrum sensing. However, in this system SUs should be close enough to listen LO power leakage in order to sense the spectrum. This might be seen as a disadvantage of this approach.

It is noted that in the spectrum overlay approach, SUs need to identify the spectrum opportunities or holes to utilize them opportunistically. If PUs use the spectrum all the time, SUs will not get a chance to access the channel for their transmission. In this case, the spectrum underlay approach is applicable in which interference temperature management is important where SUs do not need to sense for the spectrum but have to transmit with severe low transmit power so as not to create harmful interference to the primary systems. However, they are not allowed to transmit with high power when the primary system is idle leaving whole spectrum unutilized. This is the drawback of this method. However, there are many challenges to differentiate between the legitimate SUs and the malicious SUs whose intention is to disturb the PUs' transmissions.

In single SU-based spectrum sensing methods, an SU can come up with wrong decision of spectrum occupancy and create the interference to the PUs. For this type of problem, cooperative or collaborative spectrum sensing is more practical. In such collaborative approaches,

SUs rely on aggregated sensing information rather than only on their sensed information, the reliability and accuracy is increased significantly, and problems such as hidden PU problem (as shown in Figure 5.8) will be overcome. However, collaboration will create some network overhead both economically (to install external agent or spectrum server) and technically (network traffic overload).

Similarly, Farhang-Boroujeny and Kempter [34] noted that FBSE outperforms MTSE with low PSD since MTSE uses multiple prototype filters [4,45] for a given frequency band and the best response corresponds to the largest eigenvalue of the autocorrelation matrix of the observation vector. Moreover, MTSE is better for small sample spaces, whereas FBSE is better for large number of samples. In addition, MTSE approach increases the computational complexity and hence might not be suitable for CR system. Recently, FBSE has attracted the researcher from both academia and institutes [53]. However, FBSE assumes multicarrier modulation technology as the underlying communication technique, which can be considered as its drawback if we try to implement FBSE in other system with different communication technologies.

In order to realize the CR system in wireless communications for efficient utilization of underutilized scarce spectrum, the interference and spectrum sensing methods should be reliable and prompt so that the PUs will not be hampered or disturbed from the CR system to utilize their own spectrum. It is noted that the licensed PUs can claim their own frequency band at any time while the CR system operates on the band opportunistically in the spectrum overlay model and coexist in the spectrum underlay model. Whatever model and methods have been implemented, the CR system should be able to identify the presence of PUs as fast as possible in order to respect the PUs. However, the signal detection methods have limitations on the performance of detection in terms of time and frequency resolution. Since the CR system is still in its early stages of development, there are number of challenges for its implementation such as time for PU signal detection and hardware implementations and computational complexities in SU devices [53]. Furthermore, the *spread spectrum PU* (or CDMA system user) detection is also difficult for the user since the energy is spread all over wider frequency range.

Last but not least, the best signal processing technique for spectrum sensing in the CR system is the one that offers low sensing time, minimum hardware and computational complexities, and fine-tuned time and frequency resolution. In some sense, FBSE and wavelet-based spectrum estimation seem suitable for the time being. However, future research should be focused to address the existing problems in order to realize the full potential of the CR system for dynamic spectrum access.

5.5 Conclusions

This chapter has presented the state-of-the-art research results for radio spectrum sensing in the CWN for dynamic spectrum access. Spectrum sensing is one of the most important steps in the CWN and a crucial need to achieve satisfactory results in terms of efficient use of available radio spectrum while not causing harmful interference to the licensed PUs. As described in this chapter, the development of CWNs requires the involvement and interaction of many advanced techniques, including distributed spectrum sensing, interference management, CR reconfiguration management, and cooperative communications. Furthermore, in order to fully realize the CR systems in wireless communications for efficient utilization of scarce RF spectrum, the method used in identifying the interference and/or spectrum sensing should be reliable and prompt so that the PU will not suffer from the CR system to utilize their licensed spectrum. We have presented the different signal processing methods by grouping them into five basic groups and their details in turn. We have also presented the pros and cons of different spectrum sensing methods and performed the comparison in terms of operation, accuracies, complexities, and implementations. Since the CR system is still in its early stages of development, there are a number of challenges for its implementation such as time needed for PU signal detection, detection of spread spectrum PU signals, computational complexities, hardware implementation in SU devices, robust spectrum sharing policies/rules, and robust security for both licensed and licensed users in CWNs.

References

1. G. Staple and K. Werbach, "The End of Spectrum Scarcity," *IEEE spectrum*, vol. 41, no. 3, pp. 48–52, 2004.
2. M. A. McHenry, "NSF Spectrum Occupancy Measurements Project Summary," *Shared Spectrum Company Report*, Shared company, Washington, DC, 2005. http://www.sharedspectrum.com/wp-content/uploads/5_NSF_NRAO_Report.pdf.
3. I. F. Akyildiz, W.-Y. Lee, M. C. Vuran, and S. Mohanty, "NeXt Generation/ Dynamic Spectrum Access/Cognitive Radio Wireless Networks: A Survey," *Computer Networks*, vol. 50, no. 13, pp. 13–18, 2006.
4. S. Haykin, "Cognitive Radio: Brain-Empowered Wireless Communications," *IEEE Journal on Selected Areas in Communications*, vol. 23, no. 2, pp. 201–220, 2005.
5. R. W. Thomas, L. A. DaSilva, and A. B. MacKenzie, "Cognitive Networks," *Proceedings of the 1st IEEE International Symposium on New Frontiers in Dynamic Spectrum Access Networks*, Baltimore, MD, November 2005, pp. 352–360.
6. J. Mitola and G. Q. Maguire, "Cognitive Radio: Making Software Radios More Personal," *IEEE Personal Communications Magazine*, vol. 6, no. 6, pp. 13–18, 1999.
7. W. Krenik, A. M. Wyglinsky, and L. Doyle, "Cognitive Radios for Dynamic Spectrum Access," *IEEE Communications Magazine*, vol. 45, no. 5, pp. 64–65, 2007.
8. A. A. Abidi, "The Path to Software-Defined Radio Receiver," *IEEE Journal of Solid-State Circuits*, vol. 42, no. 5, pp. 954–966, 2007.
9. K.-C. Chen, R. Prasad, and H. V. Poor, "Special Issue on Software Radio," *IEEE Personal Communications Magazine*, vol. 6, no. 4, August 1999, pp. 1–3.
10. F. K. Jondral, "Software-Defined Radio—Basics and Evolution to Cognitive Radio," *EURASIP Journal on Wireless Communications and Networking*, vol. 2005, no. 3, pp. 275–283, 2005.
11. D. B. Rawat, Y. Zhao, G. Yan, and M. Song, "CRAVE: Cognitive Radio Enabled Vehicular Communications in Heterogeneous Networks," *Proceedings of the IEEE Radio and Wireless Symposium*, Austin, TX, January 20–23, 2013.
12. D. B. Rawat, B. B. Bista, and G. Yan, "CoR-VANETs: Game Theoretic Approach for Channel and Rate Selection in Cognitive Radio VANETs," *Proceedings of the 7th International Conference on Broadband and Wireless Computing, Communication and Applications*, Victoria, British Columbia, Canada, November 12–14, 2012.
13. D. B. Rawat and G. Yan, "Spectrum Sensing Methods and Dynamic Spectrum Sharing in Cognitive Radio Networks: A Survey," *International Journal of Research and Reviews in Wireless Sensor Networks*, vol. 1, no. 1, 2011, pp. 1–13.
14. R. Etkin and D. Tse, "Spectrum Sharing for Unlicenced Bands," *Proceedings of the IEEE International Symposium on New Fromtiers in Dynamic Access Networks*, Baltimore, MD, November 2005, pp. 251–258.

15. D. Cabric and R. W. Brodersen, "Physical Layer Design Issues Unique to Cognitive Radio Systems," *Proceedings of the 16th IEEE International Symposium on Personal, Indoor and Mobile Radio Communications*, Berlin, Germany, September 2005, pp. 759–763.

16. D. B. Rawat and D. C. Popescu, "Joint Precoder and Power Adaptation for Cognitive Radios in Interference Systems," *Proceedings of the 28th IEEE International Performance Computing and Communications Conference*, Phoenix, AZ, December 2009, pp. 425–430.

17. D. B. Rawat and S. Shetty, "Game Theoretic Approach to Dynamic Spectrum Access with Multi-radio and QoS Requirements," *IEEE Global Conference on Signal and Information Processing*, Austin, TX, 2013, pp. 1150–1153.

18. D. B. Rawat, S. Shetty, and K. Raza, "Geolocation-Aware Resource Management in Cloud Computing Based Cognitive Radio Networks," *International Journal of Cloud Computing*, Special Issue on Information Assurance and System Security in Cloud Computing, (in press).

19. D. B. Rawat, B. B. Bista, G. Yan, and S. Shetty, "Waiting Probability Analysis for Opportunistic Spectrum Access," *International Journal of Adaptive and Innovative Systems*, vol. 2, no. 1, pp. 15–28, 2014.

20. D. B. Rawat, S. Shetty, and K. Raza, "Game Theoretic Dynamic Spectrum Access in Cloud-Based Cognitive Radio Networks," *Proceedings of the IEEE International Conference on Cloud Engineering*, Boston, MA, March 10–14, 2014.

21. Q. Zhao and B. M. Sadler, "A Survey of Dynamic Spectrum Access," *IEEE Signal Processing Magazine*, vol. 24, no. 3, pp. 79–89, May 2007.

22. D. B. Rawat and D. Popescu, "Precoder Adaptation and Power Control for Cognitive Radios in Dynamic Spectrum Access Environments," *IET Communications*, vol. 6, no. 8, pp. 836–844, 2012.

23. Y. Zeng and Y.-C. Liang, "Spectrum-Sensing Algorithms for Cognitive Radio Based on Statistical Covariances," *IEEE Transactions on Vehicular Technology*, vol. 58, no. 4, pp. 1804–1815, 2009.

24. D. Cabric, A. Tkachenko, and R. Brodersen, "Spectrum Sensing Measurements of Pilot, Energy, and Collaborative Detection," *Proceedings of the IEEE Military Communication Conference*, Washington, DC, October 2006, pp. 1–7.

25. P. De and Y.-C. Liang, "Blind Sensing Algorithms for Cognitive Radio," *IEEE Radio and Wireless Symposium*, Long Beach, CA, January 2007, pp. 201–204.

26. H. Tang, "Some Physical Layer Issues of Wide-band Cognitive Radio Systems," *IEEE International Symposium on New Frontiers in Dynamic Spectrum Access Networks*, Baltimore, MD, June 2005, pp. 151–159.

27. D. Cabric, S. Mishra, and R. Brodersen, "Implementation Issues in Spectrum Sensing for Cognitive Radios," *Asilomar Conference on Signals, Systems and Computers*, Pacific Grove, CA, June 2004, pp. 772–776.

28. M. Öner and F. Jondral, "Air Interface Identification for Software Radio Systems," *AEÜ International Journal of Electronics and Communications*, vol. 61, no. 2, pp. 104–117, February 2007.

29. H. Urkowitz, "Energy Detection of Unknown Deterministic Signals," *Proceedings of the IEEE*, vol. 55, pp. 523–531, April 1967.

30. Y. Zhuan, G. Memik, and J. Grosspietsch, "PHY 28-1—Energy Detection Using Estimated Noise Variance for Spectrum Sensing in Cognitive Radio Networks," *IEEE Wireless Communications and Networking Conference*, Las Vegas, NV, April 2008, pp. 711–716.

31. K. Challapali, S. Mangold, and Z. Zhong, "Spectrum Agile Radio: Detecting Spectrum Opportunities," *Proceedings of the International Symposium on Advanced Radio Technologie*, Boulder, CO, March 2004.

32. Z. Tian and G. B. Giannakis, "A Wavelet Approach to Wideband Spectrum Sensing for Cognitive Radios," *Proceedings of the IEEE International Conference on Cognitive Radio Oriented Wireless Networks and Communications*, Mykonos, Greece, June 2006, pp. 1054–1059.

33. B. Wild and K. Ramchandran, "Detecting Primary Receivers for Cognitive Radio Applications," *Proceedings of the IEEE Dynamic Spectrum Access Networks*, Baltimore, MD, November 2005, pp. 124–130.

34. B. Farhang-Boroujeny and R. Kempter, "Multicarrier Communication Techniques for Spectrum Sensing and Communications in Cognitive Radios," *IEEE Communication Magazine*, vol. 48, no. 4, pp. 80–85, 2008.

35. G. Ganesan and Y. Li, "Cooperative Spectrum Sensing in Cognitive Radio, Part I: Two User Networks," *IEEE Transactions on Wireless Communications*, vol. 6, no. 6, pp. 2204–2213, 2007.

36. G. Ganesan and Y. Li, "Cooperative Spectrum Sensing in Cognitive Radio, Part II: Multiuser Networks," *IEEE Transactions on Wireless Communications*, vol. 6, no. 6, pp. 2214–2222, 2007.

37. Z. Han, R. Fan, and H. Jiang, "Replacement of Spectrum Sensing in Cognitive Radio," *IEEE Transactions on Wireless Communications*, vol. 8, no. 6, pp. 2819–2826, 2009.

38. J. Proakis and D. G. Manolakis, *Digital Signal Processing: Principles, Algorithms, and Applications*, 4th edn. Upper Saddle River, NJ: Prentice Hall, 2007.

39. J. G. Proakis, *Digital Communications*, 4th edn. Boston, MA: McGraw-Hill, 2000.

40. T. Farnham, G. Clemo, R. Haines, E. Seidel, A. Benamar, S. Billington, N. Greco, N. Drew, B. A. T. Le, and P. Mangold, "IST-TRUST: A Perspective on the Reconfiguration of Future Mobile Terminals using Software Download," *Proceedings of the IEEE International Symposium on Personal, Indoor and Mobile Radio Communications*, London, September 2000, pp. 1054–1059.

41. R. Yates, C. Raman, and N. Mandayam, "Fair and Efficient Scheduling of Variable Rate Links via a Spectrum Server," *Proceedings of the IEEE International Conference on Communications*, Istanbul, Turkey, June 2006, pp. 5246–5251.

42. L. Berlemann, S. Mangold, G. R. Hiertz, and B. H. Walke, "Spectrum Load Smoothing: Distributed Quality-of-Service Support for Cognitive Radios in Open Spectrum," *European Transactions on Telecommunications*, vol. 17, pp. 395–406, 2006.

43. Y. Xing, C. N. Mathur, M. Haleem, R. Chandramouli, and K. Subbalakshmi, "Dynamic Spectrum Access with QOS and Interference Temperature Constraints," *IEEE Transactions on Mobile Computing*, vol. 6, no. 4, pp. 423–433, 2007.

44. J. Bater, H.-P. Tan, K. Brown, and L. Doyle, "Modelling Interference Temperature Constraints for Spectrum Access in Cognitive Radio Networks," *Proceedings of the IEEE International Conference on Communications*, Glasgow, June 2007, pp. 6493–6498.

45. D. J. Thomson, "Spectrum Estimation and Harmonic Analysis," *Proceedings of the IEEE*, vol. 20, pp. 1055–1096, 1982.

46. W. Gardner, "Exploitation of Spectral Redundancy in Cyclostationary Signals," *IEEE Signal Processing Magazine*, vol. 8, no. 2, pp. 14–36, 1991.

47. T. A. Weiss and F. K. Jondral, "Spectrum Pooling: An Innovative Strategy for the Enhancement of Spectrum Efficiency," *IEEE Communications Magazine*, vol. 42, no. 3, pp. S8–S14, March 2004.

48. Y. Hur, J. Park, W. Woo, K. Lim, C. Lee, H. Kim, J. Laskar, S. Center, G. Tech, and G. Atlanta, "A Wideband Analog Multi-Resolution Spectrum Sensing (MRSS) Technique for Cognitive Radio (CR) Systems," *Proceedings of the IEEE International Symposium on Circuits and Systems*, Island of Kose, Greece, 2006, p. 4.

49. M. Lakshmanan and H. Nikookar, "A Review of Wavelets for Digital Wireless Communication," *Wireless Personal Communications*, vol. 37, no. 3, pp. 387–420, 2006.

50. Y. Wang, Z. Tian, and C. Feng, "Sparsity Order Estimation and Its Application in Compressed Spectrum Sensing for Cognitive Radios," *IEEE Transactions on Wireless Communications*, vol. 11, no. 6, pp. 2116–2125, 2012.

51. W. Yin, Z. Wen, S. Li, J. Meng, and Z. Han, "Dynamic Compressive Spectrum Sensing for Cognitive Radio Networks," *Proceedings of the IEEE CISS*, Baltimore, MD, March 23–25, 2011, pp. 1–6.

52. Y. Wang, Z. Tian, and C. Feng, "A Two-Step Compressed Spectrum Sensing Scheme for Wideband Cognitive Radios," *Proceedings of the IEEE Globecom Conference*, Miami, FL, December 2010.

53. D. B. Rawat and G. Yan, "Signal Processing Techniques for Spectrum Sensing in Cognitive Radio Systems: Challenges and Perspectives," *Proceedings of the IEEE/IFIP Asian Himalayas International Conference on Internet*, Kathmandu, Nepal, October 2009, pp. 1–5.

Biographical Sketches

Danda B. Rawat received his PhD in electrical and computer engineering from Old Dominion University, Norfolk, Virginia. He is currently an assistant professor in the department of electrical engineering at Georgia Southern University, Statesboro, Georgia. He is the founding director of Cybersecurity, Wireless Systems and Networking Innovations (CWiNs) Lab at Georgia Southern University. His research focuses on wireless communication systems and networks. His current research interests include design, analysis, and evaluation of cognitive radio networks, cyber physical systems, a vehicular/wireless ad hoc networks, OpenFlow-based networks, software-defined networks, wireless sensor networks, and wireless mesh networks. His research interests also include information theory, mobile computing, and network security. He has published over 70 scientific/technical papers on these topics. He has authored or edited 4 books and over 10 peer-reviewed book chapters related to his research areas. Dr. Rawat has been serving as an editor for over six international journals. He also served as a lead guest editor for a special issue "Recent Advances in Vehicular Communications and Networking" (Elsevier) published in *Ad Hoc Networks* (2012), a guest editor for a special issue "Network Protocols and Algorithms for Vehicular Ad Hoc Networks" (ACM/Springer) published in *Mobile Networks and Applications* (*MONET*) (2013), and a guest editor for a special issue "Recent Advances in Mobile Ad Hoc and Wireless Sensor Networks" (Inderscience) published in *International Journal of Wireless and Mobile Computing* (2012). He served as a program chair, a conference chair, and a session chair for numerous international conferences and workshops, and served as a technical program committee member for several international conferences including IEEE GLOBECOM, CCNC, GreenCom, AINA, WCNC, and VTC conferences. He has previously held an academic position at Eastern Kentucky University, Richmond, Kentucky; Old Dominion University; and Tribhuvan University, Kathmandu, Nepal. He is the recipient of the Best Paper Award at the International Conference on Broadband and Wireless Computing, Communication and Applications 2010 (BWCCA 2010) and the Outstanding Ph.D. Researcher Award 2009 in Electrical and Computer Engineering at Old Dominion University among others. Dr. Rawat is a senior member of IEEE and a member of ACM and ASEE. He has been serving as a vice chair of the Executive Committee of the IEEE Savannah Section since 2013.

Chandra Bajracharya received her PhD in electrical and computer engineering from Old Dominion University, Norfolk, Virginia. During her doctoral dissertation, she was associated with the Frank Reidy Research Center for Bioelectrics and worked on several projects funded by the US Air Force and the National Institutes of Health (NIH). Her PhD dissertation was on the characterization of near field focusing on impulse reflector antenna and its applications in target detection, imaging, and electromagnetic waves into biological targets. She also has an extensive teaching experience of over 5 years as a lecturer at the department of electrical engineering, Tribhuvan University, Kathmandu, Nepal. Her research interests include medical cyber physical systems, transportation cyber physical systems, numerical electromagnetics, biological effects of electromagnetic fields, ultra wideband antennas, and antenna design. Her research interests also include power electronics, alternative energy, communication systems, signal processing, and smart grid. She has published several scientific/technical papers on these topics. She is a member of the IEEE and has served as a technical program committee member and reviewer of several conferences.

6

COLLABORATIVE SPECTRUM SENSING TECHNIQUE

ASHISH BAGWARI AND
GEETAM SINGH TOMAR

Contents

6.1 Introduction

In the present era, wireless communication is going in a big way and cognitive radio network (CRN) is one of the future-based technologies in wireless communication system. The concept of cognitive radio (CR) was first proposed by Joseph Mitola III at the Royal Institute of Technology (KTH) in Stockholm, Sweden, in 1998. CR is an intelligent wireless communication system, which is aware of its surrounding environment, learns from the environment, and adapts its internal states to statistical variations in the incoming radio-frequency (RF) stimuli by making corresponding changes in certain operating parameters in real time. A CR comes under the Institute of Electrical and Electronics Engineers (IEEE) 802.22 wireless regional area network (WRAN) standard and has the ability to detect the channel usage, analyze the channel information, and make a decision whether and how to access the channel. The US Federal Communications Commission (FCC) uses a narrower definition for this concept: "Cognitive radio: A radio or system that senses its operational electromagnetic environment and can dynamically and autonomously adjust its radio operating parameters to modify system operation, such as maximize throughput, mitigate interference, facilitate interoperability, and access secondary markets." The primary objective of the CR is to provide highly reliable communication whenever and wherever needed and to utilize the radio spectrum efficiently. Static allocation of the frequency spectrum does not meet the needs of current wireless technology, so dynamic spectrum usage is required for wireless networks. CR is

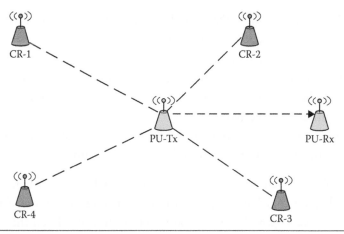

Figure 6.1   Cognitive radio network. Tx, Transmitter; Rx, Receiver.

considered as a promising candidate to be employed in such systems as they are aware of their operating environments and can adjust their parameters. It can sense the spectrum and detect the idle frequency bands; thus, secondary users (SUs) can be allocated in those bands when primary users (PUs) do not use those bands in order to avoid any interference to PUs by SUs. In the CRN literature, PU and SU are considered as shown in Figure 6.1. The PU is a licensed user that has an allocated band of spectrum for exclusive use. The SU is unlicensed user that does not have an allocated band of spectrum. We use spectrum sensing techniques to detect the presence of PU licensed signal at low signal-to-noise ratio (SNR).

6.2 CRN Functions

Basically, a CR should be able to quickly jump in and out of free spaces in spectrum bands, avoiding preexisting users, in order to transmit and receive signals. There are four basic functions of CRNs: spectrum sensing, spectrum sharing/allocation, spectrum mobility/ handoff, and spectrum decision/management.

- *Spectrum sensing*: It detects all the available spectrum holes in order to avoid interference. It determines which portion of the spectrum is available and senses the presence of licensed PUs.

- *Spectrum sharing allocation*: It shares the spectrum-related information between neighbor nodes.
- *Spectrum mobility handoff*: If the spectrum in use by a CR user is required for a PU, then CR leaves the present band and switches to another vacant spectrum band in order to provide seamless connectivity.
- *Spectrum decision management*: It captures the best available vacant spectrum holes from detected spectrum holes.

Many of the licensed air waves are too crowded. Some bands are so overloaded that long waits and interference are the norm. Other bands are used sporadically and are even underused. Even the FCC acknowledges the variability in licensed spectrum usage. According to the FCC report, 70% of the allocated PU licensed spectrum band remains unused called white space/spectrum hole at any one time as shown in Figure 6.2. This fluctuating utilization results from the current process of static allocation of spectrum, such as auctions and licensing, which is inefficient, slow, and expensive. This process cannot keep up with the swift pace of technology. In the past, a fixed spectrum assignment policy was more than adequate. However, today such rigid assignments cannot match the dramatic increase in access to limited spectrum for mobile devices. This increase strains the effectiveness of traditional, licensed spectrum policies. In fact, even unlicensed spectrum bands need an overhaul. Congestion resulting from

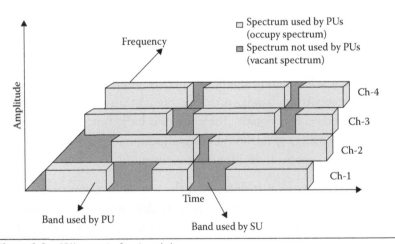

Figure 6.2   CRN concepts: Spectrum holes.

the coexistence of heterogeneous devices operating in these bands is on the rise. Consider the license-free industrial, scientific, and medical (ISM) radio band. It is crowded by wireless local area network equipment, Bluetooth devices, microwave ovens, cordless phones, and other users. Devices, which use unlicensed bands, need to have higher performance capabilities to have better job managing user quality of service. The limited availability of spectrum and the nonefficient use of existing RF resources necessitate a new communication paradigm to exploit wireless spectrum opportunistically and with greater efficiency. The new paradigm should support methods to work around spectrum availability traffic jams, make communications far more dependable, and of course reduce interference among users. The current shortage of radio spectrum can also be blamed in large part on the cost and performance limits of current and legacy hardware. Next-generation wireless technology such as software-defined radio (SDR) may well hold the key to promoting better spectrum usage from an underlying hardware/physical layer perspective. The SDR uses both embedded signal processing algorithms to sift out weak signals and reconfigurable code structures to receive and transmit new radio protocols. However, the system-wide solution is really CR.

In a typical CR scenario, users of a given frequency band are classified into PUs and SUs. PUs are licensed users of that frequency band. SUs are unlicensed users that opportunistically access the spectrum when no PUs operate on that frequency band. This scenario exploits the spectrum sensing attributes of CR. CRNs form when SUs utilize *holes* in licensed spectrum for communication. These spectrum holes are temporally unused sections of licensed spectrum that are free of PUs or partially occupied by low-power interferers. The holes are commonly referred to as white or gray spaces. Figure 6.2 shows a scenario of PUs and SUs utilizing a frequency band.

In the other cognitive scenario, there are no assigned PUs for unlicensed spectrum. Since there are no license holders, all network entities have the same right to access the spectrum. Multiple CRs coexist and communicate using the same portion of spectrum. The objective of the CR in these scenarios is more intelligent and fair spectrum sharing to make open spectrum usage much more efficient. It will help in utilizing the unused channels and also using spectrum efficiently, and also include the better channel assignment and management policy.

6.3 Spectrum Sensing Techniques

CR attempts to discern the areas of used or unused spectrum by determining if a PU is transmitting in its vicinity. This approach is predicated on detecting not the strongest transmitted signal from a PU but the weakest. The idea is that the weakest signal producing primary transmitter would ideally be the one furthest away from the CR, but still susceptible to RF interference from the radio.

The aim of the CR is to use the natural resources efficiently including the frequency, time, and transmitted energy. CR technologies can be used in lower priority secondary systems that improve spectral efficiency by sensing the environment and then filling the discovered gaps of unused licensed spectrum with their own transmissions. Unused frequencies can be thought of as a spectrum pool from which frequencies can be allocated to SUs and SUs can also directly use frequencies discovered to be free without gathering these frequencies into a common pool. In addition, CR techniques can be used internally within a licensed network to improve the efficiency of spectrum use. In the CRN, the CR users monitor the radio spectrum periodically and opportunistically communicate over the spectrum holes.

As shown in Figure 6.3 there are basically three types of spectrum sensing techniques for detecting PU licensed spectrum band [1–3]:

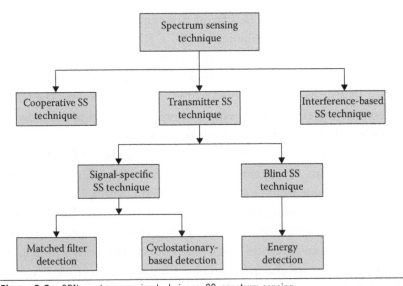

Figure 6.3   CRN spectrum sensing techniques. SS, spectrum sensing.

1. Cooperative or collaborative spectrum sensing (CSS) technique
2. Transmitter spectrum sensing technique
3. Interference-based spectrum sensing technique

6.3.1 Cooperative Spectrum Sensing Technique

In cooperative detection, multiple CRs work together to supply information to detect a PU. This technique exploits the spatial diversity intrinsic to a multiuser network. It can be accomplished in a centralized or distributed fashion. In a centralized manner, each radio reports its spectrum observations to a central controller that processes the information and creates a spectrum occupancy map of the overall network. In a distributed fashion, the CRs exchange spectrum observations among themselves and each individually develops a spectrum occupancy map.

Cooperative detection is advantageous because it helps to mitigate multipath fading and shadowing RF pathologies that increase the probability of PU detection. Additionally, it helps to combat the dreaded hidden node problem that often exists in ad hoc wireless networks. The hidden node problem, in this context, occurs when a CR has good line of sight to a receiving radio, but may not be able to detect a second transmitting radio also in the locality of the receiving radio due to shadowing or because the second transmitter is geographically distanced from it. Cooperation between several CRs alleviates this hidden node problem because the combined local sensing data can make up for individual CR errors made in determining spectrum occupancy. Sensing information from others results in an optimal global decision.

6.3.2 Transmitter Spectrum Sensing Technique

In this technique, the CR attempts to discern the areas of used or unused spectrum by determining if a PU is transmitting in its vicinity. This approach is predicated on detecting not the strongest transmitted signal from a PU but the weakest. The idea is that the weakest signal producing primary transmitter would ideally be the one furthest away from the CR, but still susceptible to RF interference from the radio. To detect the PU signal, there is a mathematical hypothesis expression for a received signal given as

$$x(n) = \begin{cases} w(n), & H_0 \\ s(n)h(n) + w(n), & H_1 \end{cases} \qquad (6.1)$$

where:

$x(n)$ denotes the signal received by each CR user

$s(n)$ is the PU licensed signal

$w(n) \sim N(0, \sigma_w^2)$ is additive white Gaussian noise with zero mean and variance σ_w^2

$h(n)$ denotes the Rayleigh fading channel gain of the sensing channel between the PU and the CR user

H_0 known as the null hypothesis shows the absence of PU

H_1 known as the alternative hypothesis shows the presence of PU

The channel considered between the PU and the CR user is the Rayleigh channel. Further, transmitter spectrum sensing technique is divided into two categories: signal-specific sensing technique and blind sensing technique.

6.3.2.1 Signal-Specific Spectrum Sensing Technique It requires prior knowledge of PU signal. The examples are matched filter detection and cyclostationary-based detection.

6.3.2.1.1 Matched Filter Detection This technique sometimes called coherent detection, which is an optimum spectrum detection method, requires prior information of PUs and increases the SNR. In other words, when PU signal information, such as modulation type, pulse shape, and packet format, is known to a CR, the optimal detector in stationary Gaussian noise is the matched filter since it maximizes the received SNR. The matched filter works by correlating a known signal, or template, with an unknown signal to detect the presence of the template in the unknown signal. Figure 6.4 provides a graphical representation of this process. Because most wireless network systems have pilots, preambles, synchronization words, or spreading codes, these can be used for coherent (matched filter) detection. A big plus in favor of the matched filter detection is that it requires less time to achieve a high processing gain due to coherency. Its main shortcoming is that it requires a priori knowledge of the

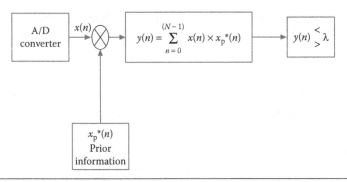

Figure 6.4   Matched filter detector. A/D, Analog to digital converter.

PU signal, which in a real-world situation may not be available, and implementation is complex.

6.3.2.1.2 Cyclostationary-Based Detection   In this method, the signal is seen to be cyclostationary if its statistics, that is, mean or autocorrelation, is a periodic function over a certain period of time. Because modulated signals (i.e., messages being transmitted over RF) are coupled with sine-wave carriers, repeating spreading code sequences, or cyclic prefixes, all of which have a built-in periodicity, their mean and autocorrelation exhibit periodicity that is characterized as being cyclostationary. Noise, however, is a wide-sense stationary signal with no correlation. Using a spectral correlation function, it is possible to differentiate noise energy from modulated signal energy and therefore detect if a PU is present. The cyclostationary detection has several advantages. It can differentiate noise power from signal power, more robust to noise uncertainty, and can work with lower SNR. But it requires partial information of PU that makes it computationally complex, and long observation time is required. Figure 6.5 shows the block diagram of cyclostationary-based detector.

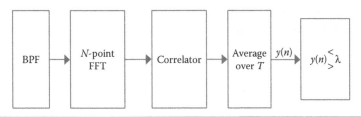

Figure 6.5   Cyclostationary-based detector. FFT, fast Fourier transform; BPF, Band pass filter.

6.3.2.2 Blind Spectrum Sensing Technique   It does not require prior knowledge of PU signal. Energy detector (ED) is the example of this kind of sensing technique.

6.3.2.2.1 Energy Detection   In energy detector, if a receiver cannot gather sufficient information about the PU signal, such as in the case that only the power of random Gaussian noise is known to the receiver, the optimal detector is an ED. Energy detection implementation and computation are easier than others. However, there are some limitations: At low SNR, its performance degrades; it cannot distinguish interference from a user signal; and it is not effective for signals whose signal power has been spread over a wide band. Figure 6.6 shows the block diagram of ED.

Now, there are some important parameters related to spectrum sensing performance: the probability of detection (P_d), the probability of false alarm (P_f), and the probability of missed detection (P_m). The probability of detection is the probability of accurately deciding the presence of the PU signal. The probability of false alarm refers to the probability that the SU incorrectly decides that the channel is idle when the PU is actually transmitting. The probability of missed detection refers to the probability that the SU misses the PU signal when the PU is transmitting.

6.3.3 Interference-Based Spectrum Sensing Technique

This method differs from the typical study of interference that is usually transmitter centric. Typically, a transmitter controls its interference by regulating its output transmission power and its out-of-band emissions, based on its location with respect to other users. Cognitive interference-based detection concentrates on measuring interference at the receiver. The FCC introduced a new model of measuring interference referred to as interference temperature. The model manages

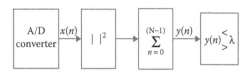

Figure 6.6   Energy detector.

interference at the receiver through the interference temperature limit, which is the amount of new interference that the receiver can tolerate. The model accounts for the cumulative RF energy from multiple transmissions and sets a maximum cap on their aggregate level. As long as the transmissions of CR users do not exceed this limit, they can use a particular spectrum band. The major hurdle with this method is that unless the CR user is aware of the precise location of the nearby PU, interference cannot be measured with this method. An even bigger problem associated with this method is that it still allows an unlicensed CR user to deprive a licensee (PU) access to his licensed spectrum. This situation can occur if a CR transmits at high-power levels while the existing PUs of the channel are quite far away from a receiver and are transmitting at a lower power level.

6.4 Issues in CRN

CRN is a future-based wireless communication technology. Due to this, there are various challenges or issues related to CRNs. In this chapter, we deal with two major problems: spectrum sensing failure and fading and shadowing problems.

6.4.1 Spectrum Sensing Failure Problem

In the ED-based spectrum sensing technique, noise uncertainty [4] arises the difficulty in setting the ideal threshold for a CR and therefore reduces its spectrum sensing reliability [5]. Moreover, this may not be optimum under low SNRs where the performance of a fixed threshold (λ_1)-based ED can fluctuate from the desired targeted performance metrics significantly.

In Figure 6.7, the x-axis shows the power level of signals and the y-axis shows the signal probability. There are two curves: PU signal and noise. According to the CRN scheme, it is very easy to detect PU signal and noise if both are separated from each other. If an ED gets a PU signal, then it shows H_1, that is, the channel is occupied; if it gets a noise signal, it shows H_0, that is, the channel is unoccupied. However, if both PU signal and noise intersect with each other, then it is very difficult to sense desired signals. In Figure 6.7, the area between the PU signal and noise or under the upper bound (λ_1) and the lower

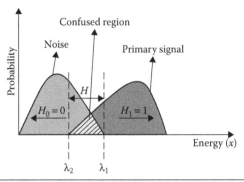

Figure 6.7   Energy distribution of PU signal and noise.

bound (λ_2) is known as a confused region. In this region, detection of noise and PU signal is very difficult using single threshold.

6.4.2 Fading and Shadowing Problem

Multipath fading and shadowing is one of the reasons of arising hidden node problem in carrier sense multiple access. Figure 6.8 depicts an illustration of a hidden node problem where the dashed circles show the operating ranges of the PU and the CR device. Here, the CR device causes unwanted interference to the PU (receiver) as the primary transmitter's signal could not be detected because of the locations of devices. Cooperative sensing is proposed in this chapter for handling multipath fading and shadowing problem.

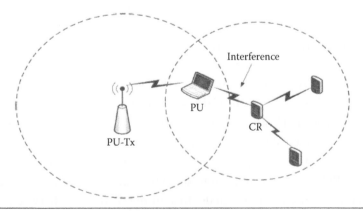

Figure 6.8   Illustration of hidden PU problem in CRNs. Tx, Transmitter.

6.5 Related Works

6.5.1 Adaptive Spectrum Sensing

Spectrum sensing is the most crucial part in the successful implementation of CRs. The main focus of our current research in CR is divided into two main streams: (1) to improve local sensing and (2) to focus on cooperative spectrum sensing for better data fusion results. In CRs during cooperative spectrum sensing, many SUs cooperate to achieve better data fusion results. In infrastructure-based networks, all the observations made by SUs are reported to a fusion center (FC) and a final decision about the presence or absence of PU is conducted at the FC [7]. In local spectrum sensing, all the SUs make observations individually and a final decision is made individually. In literature, many improvements for local spectrum sensing are proposed, but still there is room for improvement. Sensing a wideband spectrum is significant in CR; however, only a few researchers have worked on the wideband spectrum sensing in CRs. Two-stage spectrum sensing is considered as one of the techniques to deal with this issue. Hur et al. [8] proposed a two-stage wideband spectrum sensing technique, which combines coarse sensing and fine sensing. In the first stage, coarse sensing is performed over the entire frequency range with a wide bandwidth. A wavelet transform-based multiresolution spectrum sensing technique is presented as a coarse sensing. In the first stage, the occupied and candidate spectrum segments are identified. In the second stage, fine sensing is applied on candidate spectrum segments to detect unique features of modulated signals. Confirmation of an unoccupied segment is done by careful fine sensing. Luo et al. [9] presented a two-stage dynamic spectrum access approach that consists of preliminary coarse resolution sensing (CRS) followed by fine resolution sensing (FRS). In CRS, the whole spectrum is divided into equal-sized coarse sensing blocks (CSBs) of equal bandwidth. One of the CSBs is selected randomly and checked for at least one idle channel by applying ED of bandwidth equal to that of CSB. FRS is then applied on that CSB, using ED of bandwidth equal to that of the channel to determine the idle channel. Further, Maleki et al. [10] presented another two-stage sensing scheme in which ED is used in coarse sensing and, if required, cyclostationary detection is used in fine sensing. Only if a channel is declared as unoccupied in the coarse stage, the fine stage is used for the final decision. Otherwise, coarse sensing will give the final decision. Yue et al. [11] proposed a two-stage spectrum sensing scheme in which coarse detection is based on energy detection. Based on the power in each channel, it sorts the channels in ascending order. In the fine stage, a one-order cyclostationary technique is applied on the channel with the lowest power to detect weak signals. In all the above-mentioned techniques, both stages perform spectrum sensing and hence increase mean detection time. In this scheme, only one of the detection techniques will run during the two stages, based on the estimated SNR. Under the worst case, mean detection time is equal to one-order cyclostationary detection. Although two stages are running in this scheme, the SNR of the channel can be estimated in advance and the history of the channel SNR can be maintained to further reduce the mean sensing time.

6.5.1.1 Limitations The *adaptive spectrum sensing technique* [6] presents a new scheme, where out of two stages only one of the detection techniques is running at a time based on the estimated SNR [5]. Although this scheme reduces the mean sensing time, it does not consider the spectrum sensing failure problem [12].

6.5.2 Hierarchical Spectrum Sensing Technique

For achieving good spectrum sensing performance, many methods based on single CR user are investigated. Due to the multipath and shadow effect, their sensing performances are not perfect compared with the cooperative spectrum sensing based on multiuser diversity. In cooperative detection, a number of CR users coordinate to perform spectrum sensing in order to achieve more accurate detection performance. In [14], a censoring method using two thresholds in energy detection was proposed. This method can be used to decrease the average number of sensing bits to the common receiver at the expense of some sensing performance loss. Based on this method, some measurements are taken to promote the detection performance [15–18]. Srivastava and Banerjee [15] used *n*-ratio logic cooperative sensing to improve the sensing performance. There is a problem that when all the detected values fall between the two thresholds, the CR users will not send their local decisions, which would cause sensing failure problem. In this scheme, they considered hierarchical cooperative spectrum sensing based on two-threshold energy detection in CRNs to resolve sensing failure problem. Resolving the sensing failure problem in conventional two-threshold energy detection, soft combination of the observed energy value from different CR users is investigated. In this method, all the CR users perform local observation by using ED with double thresholds. If the collected energy value falls between the two thresholds, it will be sent to the FC. Otherwise, the local decision will be made and reported to the FC. The FC will make a final decision to determine whether the PU is present or absent based on the received information.

In conventional energy detection [19], each CR user makes their local decision depending only on a single threshold λ_0 as illustrated in Figure 6.9a, where "Decision H_0" and "Decision H_1" represent the absence and the presence of PU, respectively. A two-threshold method was introduced in [14] to decrease the average number of sensing bits to the common receiver. However, the decreases are achieved at the expense of some sensing performance loss, and a two-threshold

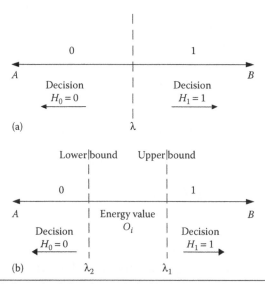

Figure 6.9 (a) Conventional detection method with single-threshold scheme; (b) hierarchical spectrum sensing method with double-threshold scheme.

method has sensing failure problem. In order to eliminate these problems effectively, two thresholds λ_1 and λ_2 are used to help the decision of the CR user, as shown in Figure 6.9b.

6.5.2.1 Limitations The *hierarchical with quantization scheme* [13] overcomes the sensing failure problem. In this method, if the collected energy value falls between the two thresholds, it requires multiple CR users with an FC to make a final decision; thus, this scheme increases system requirements and complexity.

6.6 Propose Scheme

6.6.1 Spectrum Sensing Method with Adaptive Threshold Scheme

In ED-based spectrum sensing [19], noise uncertainty arises the difficulty in setting the ideal threshold for a CR and therefore reduces its spectrum sensing reliability [20]. Moreover, this may not be optimum under low SNRs where the performance of fixed threshold (λ)-based ED can fluctuate from the desired targeted performance metrics significantly.

Figure 6.7 illustrates the energy distribution curve of PU signal and noise. Using single threshold, detection of noise and PU signal is difficult in the confused region. To overcome this problem, we consider the ADT scheme to define the local decision at the CR user using the following logic function rule (LR):

$$LR = \begin{cases} H_0 = 0, & X \leq \lambda_2 \\ H = M, & \lambda_2 < X < \lambda_1 \\ H_1 = 1, & \lambda_1 \leq X \end{cases} \tag{6.2}$$

where:
   M is the quantization decision
   X denotes the received signal energy by the CR user

To overcome the confused region problem effectively, two thresholds λ_1 and λ_2 are used to help the decision of the CR user, as shown in

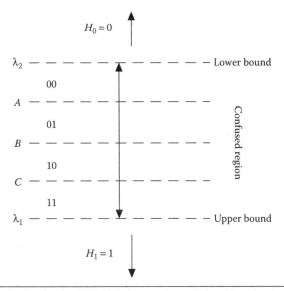

Figure 6.10 Confused region divided into four equal quantization intervals using two-bit quantization method.

Figure 6.10. In the figure, a two-bit quantization method divides the confused region into four equal quantization intervals ($\lambda_2 A$–AB–BC–$C\lambda_1$), where λ_2, A, B, C, and λ_1 are subthresholds (STs) and their values are chosen as follows:

$$\text{ST} = \begin{cases} A = \lambda_2 + D \\ B = A + D \\ C = B + D \\ \lambda_1 = C + D \end{cases} \tag{6.3}$$

$$D = \frac{(\text{Upper bound} - \text{Lower bound})}{\text{Number of quantization intervals}} = \frac{(\lambda_1 - \lambda_2)}{4} \tag{6.4}$$

$$M = \begin{cases} 00, & \lambda_2 < X \le A \\ 01, & A < X \le B \\ 10, & B < X \le C \\ 11, & C < X < \lambda_1 \end{cases} \tag{6.5}$$

λ_1 is selected according to the maximum noise variance and λ_2 is selected according to the minimum noise variance. If detected signals fall inside any one of the quantized intervals, it will generate its respective decimal values (DVs) as follows:

$$DV = \begin{cases} \text{If } M = 00, \text{ respective decimal value} - 0 \\ \text{If } M = 01, \text{ respective decimal value} - 1 \\ \text{If } M = 10, \text{ respective decimal value} - 2 \\ \text{If } M = 11, \text{ respective decimal value} - 3 \end{cases} \qquad (6.6)$$

This equation shows the DVs, which are compared with the threshold (λ) to make local decision at a fixed P_f, that is, 0.1. Outside the confused region, it will generate 0 or 1 depending upon signal existence.

6.6.2 ED Implies Multiple Antennas with ADTs

Conventional ED [19] in a single antenna-based CRN for improving reliability in detecting a spectrum hole has been studied considerably in recent times [21–23]. In [24,25], it is shown that the reliability of spectrum sensing can be improved by the CR by using multiple antennas. Figure 6.11 shows the system model of the proposed ED comprising multiple antennas with ADTs (ED_MA_ADT). In the figure, assume that M numbers of antennas are composed of a single ED at each CR user. N is the number of samples transmitted by each M number of antennas. Maximal-ratio combining scheme is not considered since it has spectrum sensing overhead due to channel estimation. Moreover, a combining scheme based on the sum of the decision statistics of all antennas in the CR is not analytically tractable. Therefore, we assume that each CR contains a selection combiner (SC) that outputs the maximum value out of M decision statistics calculated for different diversity branches as $x = \max(x_1, x_2, x_3,..., x_M)$. The output of the SC is applied to adaptive threshold-based ED that takes decision of a spectrum hole. If the detected signal energy is greater than or equal to the desired threshold (λ), it shows that the channel is busy; otherwise, the channel is empty.

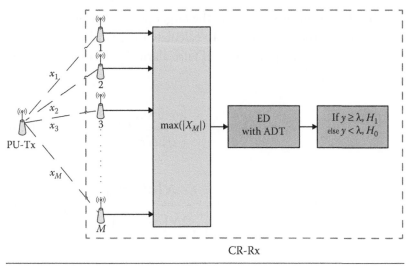

Figure 6.11 Proposed model: ED_MA_ADT. Tx, Transmitter.

6.6.3 Energy Detector

Energy detection comes under the category of blind spectrum sensing technique and is used to detect the PU licensed signal in CRN [19]. This detection method calculates the energy of the received signal and compares it to a threshold (γ) to take the local decision that the PU signal is present or absent. There is a mathematical expression to calculate the energy of any received PU signal given as [26]

$$X = \frac{1}{N}\sum_{n=1}^{N} |x(n)|^2 \qquad (6.7)$$

where:
$x(n)$ is the received input signal
N is the number of samples
X denotes the energy of the received input signal, which is compared with threshold to make the final decision

The threshold value is set to meet the target probability of false alarm P_f according to the noise power. To calculate the probability of detection alarm P_d and the probability of false alarm P_f, the expression can be defined as [20]

$$P_f = P_r(X < \lambda) = Q\left(\frac{\lambda - N\sigma_\omega^2}{\sqrt{2N\sigma_\omega^4}}\right) \tag{6.8}$$

$$P_d = P_r(X \geq \lambda) = Q\left(\frac{\lambda - N\left(\sigma_S^2 + \sigma_\omega^2\right)}{\sqrt{2N\left(\sigma_S^2 + \sigma_\omega^2\right)^2}}\right) \tag{6.9}$$

where:

$\sigma\omega^2$ and σ_S^2 are the noise variance and the signal variance, respectively
$Q()$ denotes the Gaussian tail probability Q-function

Now, the total error rate is the sum of the probability of false alarm P_f and the probability of missed detection P_m or $(1 - P_d)$. Thus, the total error rate is given by

$$P_e = P_f + (1 - P_d) \tag{6.10}$$

where:

$(1 - P_d)$ shows the probability of missed detection (P_m)

Figure 6.12 shows the internal architecture of ED with adaptive thresholds (ED_ADT). Here, the square-law device detects the signal and shows the signal energy X.

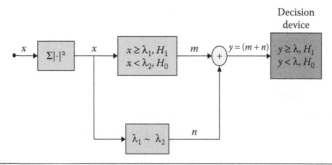

Figure 6.12　Internal architecture of ED with ADT scheme.

The threshold value can be calculated using the mathematical expression given as

$$\lambda = Q^{-1}\left(\overline{P_f}\right) \times \sqrt{2N\sigma_\omega^4} + N\sigma_\omega^2 \qquad (6.11)$$

where:

$Q^{-1}()$ denotes the inverse Gaussian tail probability Q-function

In the proposed double-threshold decision, the value of maximum noise variance shows the value of upper threshold λ_1 and the value of minimum noise variance shows the value of lower threshold λ_2. Hence,

$$\lambda_1 = Q^{-1}\left(\overline{P_f}\right) \times \sqrt{2N\rho\sigma_\omega^4} + N\rho\sigma_\omega^2 \qquad (6.12)$$

$$\lambda_2 = Q^{-1}\left(\overline{P_f}\right) \times \sqrt{\frac{2N}{\rho\sigma_\omega^4} + \frac{N}{\rho\sigma_\omega^2}} \qquad (6.13)$$

If the detected energy values (X) fall outside/between λ_1 and λ_2, using Equations 6.2, 6.5, and 6.6, it generates the values as

$$m = \begin{cases} 0, & X \le \lambda_2 \\ 1, & \lambda_1 \le X \end{cases} \qquad (6.14)$$

$$n = \{DV, \qquad \lambda_2 < X < \lambda_1 \qquad (6.15)$$

$$y = (m+n) \qquad (6.16)$$

Finally, the local decision L is expressed using Equations 6.14 through 6.16, which is the final output of ED_MA_ADT as follows:

$$L = \begin{cases} 1, & \lambda \le y \\ 0, & y < \lambda \end{cases} \qquad (6.17)$$

Algorithm 1: ED implies multiple antennas with adaptive threshold scheme (ED_MA_ADT)

1: Given $\{x_1, x_2, x_3,..., x_M\}$

2: Given $\{x_1, x_2, x_3,..., x_N\}$

3: Given $\{\lambda_1, \lambda_2; \lambda\}$

4: Distribute uniformly $\{\lambda_2, \lambda_1\}$
   as $\{\lambda_2 < A < B < C < \lambda_1\}$

5: Define range $R_0 = \{\lambda_2, A\}$, $R_1 = \{A, B\}$,
   $R_2 = \{B, C\}$, $R_3 = \{C, \lambda_1\}$

6: Values for ranges $n = \{0, 1, 2, 3\}$ for
   $\{R_0, R_1, R_2, R_3\}$

7: $X = 0$;

8: **for** $i = 1, 2,.. , N$

$x_i = \max(x_1, x_2, x_3,..., x_M)$;

$X = X + x_i^2$;

endfor

9: **if** $X \geq \lambda_1$

   $m = H_1$;

   else if $X \leq \lambda_2$

   $m = H_0$;

   else

   for $j = 0, 1, 2, 3$

   if $X \in R_j$

      $n = j$;

   endif

   endfor

   endif

10: $y = m + n$;

11: **if** $y \geq \lambda$

   $L = H_1$;

   else

   $L = H_0$;

   endif

6.7 Simulation Results and Analysis for ED_MA_ADT

In this system model, we have assumed that the total number of samples (N) is 1000, the SNR ranges from –20 to –6 dB, $P_f = 0.1$, M is the number of antennas, and Quadrature phase shift keying modulation (QPSK) modulation is considered in Rayleigh fading channel.

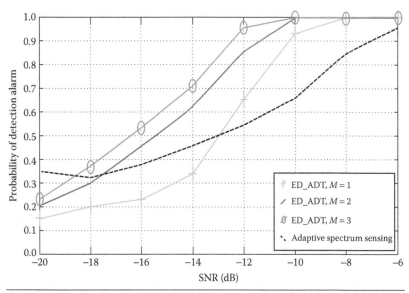

Figure 6.13   Probability of detection versus SNR at $P_f = 0.1$ with $N = 1000$, number of antennas $M = 1, 2, 3$, QPSK modulation scheme, and Rayleigh fading channel.

It can be seen from Figure 6.13 that when the number of antennas increases, the probability of detection increases. The ED_ADT with $M = 3$ outperforms ED_ADT with $M = 2$ and 1, and the adaptive spectrum sensing scheme by 10.5%, 30.6%, and 41.8% at −12 dB SNR with $P_f = 0.1$, respectively.

Figure 6.14 shows that the ED_ADT scheme with $M = 3$ has the minimum error rate compared to ED_ADT with $M = 2$, 1, and the adaptive spectrum sensing scheme. Hence, ED_ADT scheme with $M = 3$ has minimum error rate that is, 0.1 at −10 dB SNR.

Figure 6.15 shows the receiver operating characteristic (ROC) curves. The ROC curves exhibit the relationship between sensitivity (probability of detection alarm) and specificity (probability of false alarm) [27] of a spectrum sensing method under different SNR values for ED_MA_ADT when $M = 2$. This implies that when $P_f = 0.1$ and SNR = −10 dB, the probability of detection is close to 0.9, that is, 0.9950, which is the spectrum sensing requirement of IEEE 802.22 [28].

It can be seen from Figure 6.16 that there is an inverse relation between the probability of detection and the threshold for a fixed value of SNR. The figure shows that as the value of SNR increases, the probability of detection increases up to a level with respect to

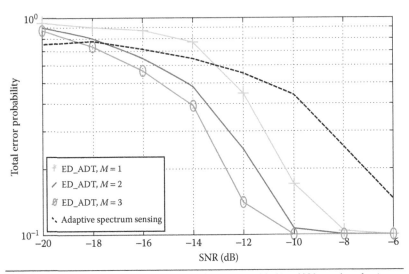

Figure 6.14   Probability of error versus SNR at $P_f = 0.1$ with $N = 1000$, number of antennas $M = 1, 2, 3$, QPSK modulation scheme, and Rayleigh fading channel.

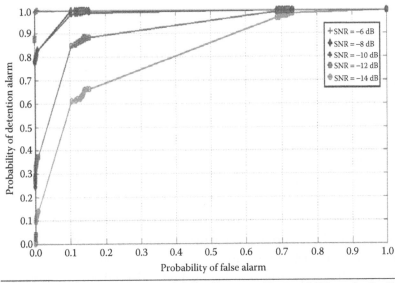

Figure 6.15   ROC curves at $N = 1000$, number of antennas $M = 2$, and SNR $= -6, -8, -10,$ $-12, -14$ dB.

the threshold. For SNR = -6 dB, the probability of detection is approximately 1.0 throughout the range of threshold (λ), which is better compared to other SNR values. It implies that the proposed ED_MA_ADT scheme when $M = 2$ can detect the PU signal at -6 dB SNR for $N = 1000$ and $\lambda = 3.0$.

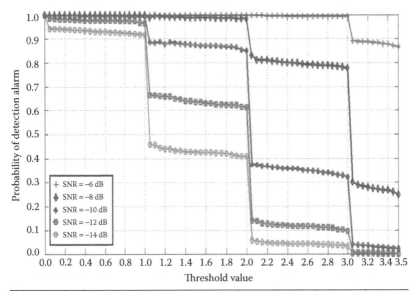

Figure 6.16   Probability of detection versus threshold values at SNR = −6, −8, −10, −12, −14 dB with $N = 1000$, QPSK modulation scheme, and Rayleigh fading channel.

6.8 Collaborative Spectrum Sensing Technique Based on ED Implies Multiple Antennas with ADTs in CRNs (CSS_ED_MA_ADT)

CSS is used to mitigate shadowing and fading in order to improve sensing performance of both local sensing performance and global sensing performance in a CRN. Here, all CRs are using the ED_ MA_ADT scheme to detect a signal. Once all CRs have taken the local decision, they transmit decisions in the form of 0 or 1 to the FC over error-free orthogonal channels to take a final decision. In Figure 6.17, assume that there are k number of CR users and all of them send the local decision L_i to a single FC.

Figure 6.17 shows the CSS technique with adaptive threshold scheme. Finally, the FC combines these binary decisions to find the presence or absence of the PU as follows:

$$Y = \sum_{i=1}^{k} L_i \tag{6.18}$$

$$L_i = \begin{cases} 1, & \lambda \le y_i \\ 0, & y_i < \lambda \end{cases} \tag{6.19}$$

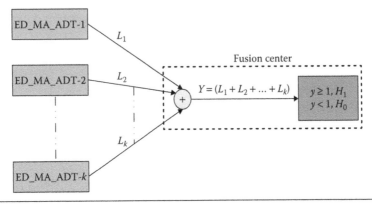

Figure 6.17 CSS technique based on ED_MA_ADT in CRNs.

In Equation 6.18, Y is the sum of all local decisions L_i from the CRs. The FC uses a hard decision OR rule for deciding the presence or absence of the PU. As per the hard decision OR rule, if Y is greater than or equal to 1, then the signal is detected; if Y is smaller than 1, then the signal is not detected. The mathematical expression can be written as

$$FC = \begin{cases} 0, & \sum_{i=1}^{k} L_i < 1 \\ 1, & \sum_{i=1}^{k} L_i \geq 1 \end{cases} \tag{6.20}$$

$$FC = \begin{cases} Y < 1, & H_0 \\ Y \geq 1, & H_1 \end{cases} \tag{6.21}$$

Finally, Equation 6.21 shows the global/final decision of FC. Now, the performance of the overall proposed system can be analyzed via P_F and P_D. Hence, the probability of false alarm (P_F) of the FC for CSS using OR rule can be expressed as follows:

$$P_F = P_r\{Y \geq 1 \mid H_0\} = P_r\left\{\sum_{i=1}^{k} y_i \geq 1 \mid H_0\right\} \tag{6.22}$$

$$P_F = 1 - \prod_{i=1}^{k}\left(1 - P_{f,i}\right) \tag{6.23}$$

The probability of detection alarm (P_D) of the FC for CSS using OR rule can be expressed as follows:

$$P_D = P_r\{Y \geq 1 \mid H_1\} = P_r\left\{\sum_{i=1}^{k} y_i \geq 1 \mid H_1\right\} \qquad (6.24)$$

$$P_D = 1 - \prod_{i=1}^{k}(1 - P_{d,i}) \qquad (6.25)$$

where:

P_f and P_d are the probability of false alarm and the probability of detection alarm of individual CR users, respectively

---

Algorithm 2: CSS with ED_MA_ADT scheme
1: Given $\{L_1, L_2, L_3,..., L_k\}$
2: Given $\{H_1, H_0\}$ as $\{1, 0\}$
3: $Y = L_1 + L_2 + L_3 + ... + L_k$
4: **if** $Y \geq 1$
      FC = H_1;
  else
      FC = H_0;
  endif

6.9 Simulation Results and Analysis for CSS_ED_MA_ADT

Figure 6.18 shows the graph of the probability of detection (P_d) versus SNR. In CSS, we consider only three CR users. Numerical results show that CSS with ED_MA_ADT outperforms the hierarchical with quantization. ED_MA_ADT improves the detection performance around 26.8% compared to the hierarchical with quantization at -12 dB SNR.

In Figure 6.19, the probability of detection (P_d) versus SNR is plotted for different number of cooperative CR users ($k = 1, 2,...,10$; $P_f = 0.1$; $M = 2$; and $N = 1000$). It can be seen from the figure that the probability of detection increases with increasing the value of SNR for different number of CR users. The probability of detection is maximum

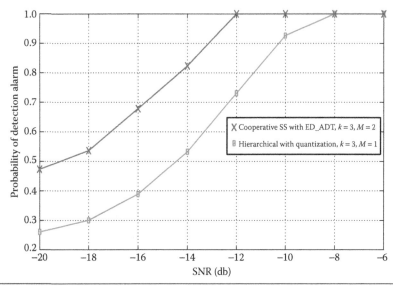

Figure 6.18 Probability of detection versus SNR at $P_f = 0.1$ with $N = 1000$, $M = 2$, total number of cooperative CR users $k = 3$, QPSK modulation scheme, and Rayleigh fading channel.

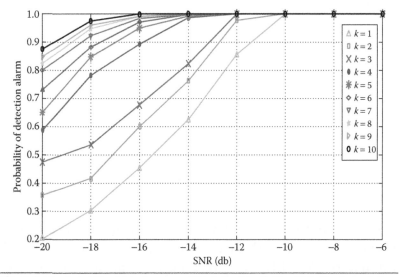

Figure 6.19 Probability of detection versus SNR at $P_f = 0.1$ with $N = 1000$, $M = 2$, $k = 1, 2, \ldots$, 10, QPSK modulation scheme, and Rayleigh fading channel.

for $k = 10$; it implies that for $N = 1000$, $M = 2$, and $P_f = 0.1$, only 10 CR users are required for deciding the presence of the PU by using CSS_ED_MA_ADT. When $k = 10$, $P_f = 0.1$, and SNR = −19.5 dB, the probability of detection value of 0.9 is achieved, which is the spectrum sensing requirement of IEEE 802.22 [28].

6.10 Conclusion and Future Work

In this chapter, a new spectrum sensing technique based on ED with adaptive threshold scheme has been proposed. The proposed sensing technique is capable of overcoming the sensing failure problem without requiring multiple CRs and FC. Numerical results shown that the proposed ED_MA_ADT scheme improves sensing performance as the number of antennas (M) increases, that is as $M = 3$ the detector outperforms $M = 2$, 1, detector and adaptive spectrum sensing detector by 10.5%, 30.6%, and 41.8% at –12 dB SNR, respectively. Therefore, we conclude that the proposed ED_MA_ADT scheme optimizes detection performance and reduces the total error rate at very low SNR. Moreover, the CSS scheme utilizes ED_MA_ADT-based CRs over error-free orthogonal reporting channels. When $k = 3$, $M = 2$, and $P_f = 0.1$, this scheme outperforms hierarchical with quantization scheme by 26.8% at –12 dB SNR. Simulation results indicate that the proposed scheme improves detection performance at low SNR. In future, we will try to make a robust spectrum sensing technique that will be able to sense the PU signal at very low SNR and will improve detector sensing performance as well.

Acknowledgments

We thank our parents for their support and motivation, for without their blessings and God's grace this paper would not be possible. This work is dedicated to my loving niece Anushka Gaur, loving nephew Anshul Semwal, and all Women's Institute of Technology (WIT) family members including technical and nontechnical staff for their valuable and fruitful support.

References

1. A. Bagwari and B. Singh, "Comparative performance evaluation of spectrum sensing techniques for cognitive radio networks," *The 4th IEEE International Conference on Computational Intelligence and Communication Networks*, pp. 98–105, November 3–5, 2012, Mathura, India.
2. D. Cabric, S. M. Mishra, and R. W. Brodersen, "Implementation issues in spectrum sensing for cognitive radios," *Proceedings of the Asilomar Conference on Signals, Systems & Computers*, vol. 1, pp. 772–776, 2004.

3. Y. Zeng, Y.-C. Liang, A. Tuan Hoang, and R. Zhang, "A review on spectrum sensing for cognitive radio: Challenges and solutions," *EURASIP Journal on Advances in Signal Processing*, vol. 2010, pp. 1–15, 2010.

4. C. Song, Y. D. Alemseged, H. Nguyen Tran, G. Villardi, C. Sun, S. Filin, and H. Harada, "Adaptive two thresholds based energy detection for cooperative spectrum sensing," *Proceedings of the IEEE on CCNC*, pp. 1–6, January 9–12, 2010 Las Vegas, NV.

5. R. Tandra and A. Sahai, "SNR walls for signal detection," *IEEE Journal of Selected Topic in Signal Processing*, vol. 2, no. 1, pp. 4–16, 2008.

6. W. Ejaz, N. U. Hasan, and H. S. Kim, "SNR-based adaptive spectrum sensing for cognitive radio networks," *International Journal of Innovative Computing, Information and Control*, vol. 8, no. 9, pp. 6095–6105, 2012.

7. I. F. Akildiz, B. F. Lo, and R. Balakrishnan, "Cooperative spectrum sensing in cognitive radio networks: A survey," *Physical Communication*, vol. 4, pp. 40–62, 2011.

8. Y. Hur, J. Park, W. Woo, K. Lim, C.-H. Lee, H. S. Kim, and J. Laskar, "A wideband analog multi-resolution spectrum sensing (MRSS) technique for cognitive radio (CR) systems," *IEEE Proceedings of the ISCAS*, pp. 4090–4093, May 21–24, 2006, Island of Kos, Greece.

9. L. Luo, N. M. Neihart, S. Roy, and D. J. Allstot, "A two-stage sensing technique for dynamic spectrum access," *IEEE Transactions on Wireless Communications*, vol. 8, no. 6, pp. 3028–3037, 2009.

10. S. Maleki, A. Pandharipande, and G. Leus, "Two-stage spectrum sensing for cognitive radios," *IEEE International Conference on Acoustic Speech and Signal Processing*, pp. 2946–2949, March 14–19, 2010, Dallas, TX.

11. W. Yue, B. Zheng, Q. Meng, and W. Yue, "Combined energy detection and one-order cyclostationary feature detection techniques in cognitive radio systems," *The Journal of China Universities of Posts and Telecommunications*, vol. 17, no. 4, pp. 18–25, 2010.

12. A. Fehske, J. D. Gaeddert, and J. H. Reed, "A new approach to signal classification using spectral correlation and neural networks," *Proceedings of the IEEE DySPAN*, pp. 144–150, November 8–11, 2005, Baltimore, MD.

13. S.-Q. Liu, B.-J. Hu, and X.-Y. Wang, "Hierarchical cooperative spectrum sensing based on double thresholds energy detection," *IEEE Communications Letters*, vol. 16, no. 7, pp. 1096–1099, 2012.

14. C. H. Sun, W. Zhang, and K. Ben Letaief, "Cooperative spectrum sensing for cognitive radios under bandwidth constraints," *Proceedings of the IEEE Wireless Communications and Networking Conference*, pp. 1–5, March 11–15, 2007, Kowloon, Hong Kong.

15. S. K. Srivastava and A. Banerjee, "'n-Ratio' logic based cooperative spectrum sensing using double threshold energy detection," *Proceedings of the International Conference on Cognitive Radio Oriented Wireless Networks and Communications*, pp. 312–317, June 22–24, 2009, Hannover, Germany.

16. S. Y. Xu, Y. L. Shang, and H. M. Wang, "Double thresholds based cooperative spectrum sensing against untrusted secondary users in cognitive radio networks," *Proceedings of the IEEE Vehicular Technology Conference*, pp. 1–5, April 26–29, 2009, Barcelona, Spain.

17. J. B. Wu, T. Luo, J. F. Li, and G. X. Yue, "A cooperative double-threshold energy detection algorithm in cognitive radio systems," *Proceedings of the International Conference on Wireless Communications, Networking and Mobile Computing*, pp. 1–4, September 24–26, 2009, Beijing, People's Republic of China.

18. J. Zhu, Z. G. Xu, F. R. Wang, B. X. Huang, and B. Zhang, "Double threshold energy detection of cooperative spectrum sensing in cognitive radio," *Proceedings of the International Conference on Cognitive Radio Oriented Wireless Networks and Communications*, pp. 1–5, May 15–17, 2008, Singapore.

19. H. Urkowitz, "Energy detection of unknown deterministic signals," *Proceedings of the IEEE*, vol. 55, no. 4, pp. 523–531, 1967.

20. M. López-Benítez and F. Casadevall, "Improved energy detection spectrum sensing for cognitive radio," *IET Communications*, vol. 6, no. 8, pp. 785–796, 2012.

21. K. B. Letaief and W. Zhang, "Cooperative communications for cognitive radio," *Proceedings of the IEEE*, vol. 97, no. 5, pp. 878–893, 2009.

22. W. Zhang, R. K. Mallik, and K. B. Letaief, "Optimization of cooperative spectrum sensing with energy detection in cognitive radio networks," *IEEE Transactions on Wireless Communications*, vol. 8, no. 12, pp. 5761–5766, 2009.

23. W. Zhang and K. B. Letaief, "Cooperative spectrum sensing with transmit and relay diversity in cognitive radio networks," *IEEE Transactions on Wireless Communications*, vol. 7, no. 12, pp. 4761–4766, 2008.

24. A. Pandharipande and J.-P. M. G. Linnartz, "Performance analysis of primary user detection in a multiple antenna cognitive radio," *Proceedings of the IEEE International Conference on Communications*, pp. 6482–6486, June 24–28, 2007, Glasgow, UK.

25. A. Taherpour, M. Nasiri-Kenari, and S. Gazor, "Multiple antenna spectrum sensing in cognitive radios," *IEEE Transactions on Wireless Communications*, vol. 9, no. 2, pp. 814–823, 2010.

26. D. Chen, J. Li, and J. Ma, "Cooperative spectrum sensing under noise uncertainty in cognitive radio," *Wireless Communications, Networking and Mobile Computing*, pp. 1–4, October 12–14, 2008, IEEE, Dalian, People's Republic of China.

27. T. Yucek and H. Arslan, "A survey of spectrum sensing algorithms for cognitive radio applications," *IEEE Communication Serveys and Tutorials*, vol. 11, no. 1, pp. 116–130, 2009.

28. T. Do and B. L. Mark, "Improving spectrum sensing performance by exploiting multiuser diversity," *Foundation of Cognitive Radio Systems*, S. Cheng (Ed.), pp. 119–140, March 16, 2012, InTech.

Biographical Sketches

Ashish Bagwari received his BTech (Hons) and MTech (Hons and Gold Medalist) degrees in electronics and communication engineering in 2007 and 2011, respectively. He is currently pursuing his

PhD degree in wireless communication from Uttarakhand Technical University (UTU), Dehradun, India, under the guidance of Professor G. S. Tomar. He is currently an assistant professor of the electronics and communication engineering department, Women's Institute of Technology, Constituent College of Uttarakhand Technical University (State Government Technical University), Dehradun, India. He has more than 5 years of industrial, academic, and research experience. He has published more than 45 research papers in various international journals [including Science Citation Index/Institute for Scientific Information (ISI/SCI) indexed] and Institute of Electrical and Electronics Engineers (IEEE) international conferences. His current research interests include robust spectrum sensing techniques in cognitive radio networks. Mr. Bagwari is a member of IEEE-USA (Washington, DC) since 2010; Machine Intelligence Research Labs, Gwalior, India; Institution of Electronics and Telecommunication Engineers (IETE); Association for Computer Machinery (ACM); and International Association of Engineers (IAENG). He has also been the editor, advisor, and reviewer of several well-known international journals from IEEE, Taylor & Francis, Springer, and Elsevier; *IJCCN, ISTP, IJATER, JREEE, JCSR,* and *JNMS*; and CICN (2011, 2013, 2014), CSNT (2014), ICMWOC (2014), I4CT (2014), and ICEPIT (2014).

Geetam Singh Tomar, born in 1964, received his undergraduate, postgraduate, and PhD degrees in electronics engineering from the universities of India and PDF from the University of Kent, Canterbury, Kent. He worked as a visiting professor at the University of Kent and a reputed institute in India. He is currently working at the department of electrical computer engineering, University of West Indies, St. Augustine, Trinidad and Tobago. He is the director of Machine Intelligence Research Labs, Gwalior, India. He is a visiting professor in Hannam University, Daejeon, South Korea; Thapar University Patiala, Patiala, Punjab, India; and many other reputed institutes. He is associated with professional societies such as the Institute of Electrical and Electronics Engineers (IEEE) as senior member and the IETE and IE (I) as fellow. His research areas of interest are air interface and advanced communication networks, wireless communication and digital systems, sensors and sensor networks, and digital design. He is actively involved in IEEE activities and has organized more than 15 IEEE international conferences in India and other countries. He delivered keynote in many conferences abroad. He is a chief editor of five international journals, has published more than 100 research papers in international journals/conferences, and has written five books and book chapters in IGI Global publication.

PART III

NETWORK CODING AND DESIGN

7

RADIOELECTRIC CHANNEL MODELING AND IMPACT ON GLOBAL WIRELESS SYSTEM DESIGN

FRANCISCO FALCONE, LEIRE AZPILICUETA, JOSÉ JAVIER ASTRÁIN, AND JESÚS VILLADANGOS

Contents

7.1 Introduction

Wireless systems are being widely adopted in a broad range of applications, spanning from conventional mobile services, remote telemetry and control, m-commerce, e-health, and social networking, just to name a few examples. This leads to an intensive use of the wireless spectrum and, in parallel, to more complex network architectures. The result is a heterogeneous wireless environment, with multisystem capable transceivers located in diverse topologies. In order to optimize the capacity as well as the grade of service while minimizing energy consumption, it is compulsory to adequately analyze and model the wireless channel as a function of the applied technologies as well as the topology and morphology of the operational scenarios. Further considerations, such as the need to reduce energy consumption, size, and cost of transceivers, pose further restrictions that impact on the overall system performance. Taking into account the impact of mobile users and their increased presence will derive in the need of adaptive radios and agile transceivers. In this scenario, the analysis of the overall system performance is a relevant parameter in order to assess the user experience and eventual adoption of this type of services.

Moreover, ubiquitous and pervasive computing and Internet of things are some of the causes of the emergence and success of wireless sensor networks (WSNs). While localization is a key aspect of such networks, since sensor's location is critical in order to provide location-based services or even to interact with the environment, another key aspect is the correct choice of the communication topology and the optimal placement of sensors. As indicated in [1], topology control is an attractive mechanism because it can simultaneously improve energy efficiency and network performance. Topology control can reduce power consumption and channel contention in WSNs by adjusting the transmission power. Once the designer of a WSN identifies the workspace, and the nature and type of services to be provided, he must choose between a cluster and a chain-based topology.

The WSN's characteristics are particularly dynamic and complex indoor environments, where the communication link quality varies significantly over time due to human activity and multipath effects. Topology control is an attractive mechanism because it can simultaneously improve energy efficiency and network performance. In such context, the Institute of Electrical and Electronics Engineers (IEEE) 802.15.4

standard [2] is probably the most widely used standard for WSNs, which provides low-rate and energy-efficient data transmissions by means of a hierarchical network architecture with low-power consumption. One of the frequency bands that IEEE 802.15.4 uses is the 2.4 GHz band, where many technologies and systems coexist, making difficult the optimization of both operation and network design issues.

Different routing strategies have been proposed in the literature; some of them follow a chain-based strategy, whereas others follow a cluster-based strategy [3–5]. The reference cluster-based proposal is low-energy adaptive clustering hierarchy (LEACH) [3], which organizes a WSN into a set of clusters so that the energy consumption can be evenly distributed among all the sensor nodes. Cluster approaches provide good fault tolerance at the cost of increased energy consumption, whereas chain-based algorithms such as power-efficient gathering in sensor information systems (PEGASIS) [6] reduce such energy consumption, causing longer latency values. Even cluster–chain-based routing protocols have also been proposed [7–11] since both LEACH and PEGASIS cannot directly be applied to large-scale WSNs due to the energy consumption and latency associated, respectively. Hybrid proposals try to make full use of the advantages of LEACH and PEGASIS and provide improved performance by dividing the WSN into chains and running in two stages. A wide survey on clustering routing protocols in WSNs can be obtained in [12].

The existing research results on WSNs generally consider a flat or a hierarchical architecture, where randomly distributed sensor nodes constitute a self-organizing network with a sink (or base station) connecting to outdoor wired or wireless networks. Each of the nodes (also called sensors) of the network has the capability to collect and route data to the sink. Each node may transmit to and receive from its neighbor nodes or to the sink, having very different energy consumptions in both cases. Energy saving is always the common and the most important goal of the routing protocols already proposed. The operation states of a sensor node can be categorized as communicating (either transmitting or receiving), idle, and hibernating. The greatest energy consumption corresponds to the transmission activity, followed by reception, idle, and hibernation activities, respectively. In this regard, it is essential to adequately characterize the communication channel in order to design a communication mechanism that ensures a minimum consumption with an adequate guarantee of message delivering.

The WSNs in indoor scenarios are characterized in terms of received signal strength indicator (RSSI), latency, and throughput. Both radio channel simulations and real sensor measurements agree in pointing out the dependence of the network topology. In order to gain insight into the impact of the scenario on wireless system operation, this chapter reviews starting from the physical layer (i.e., radio propagation behavior), moving up to a real application within a WSN.

Cognitive radio (CR) provides high bandwidth to mobile users via heterogeneous wireless network architectures and dynamic spectrum access techniques. It deals with increasing the utilization of the scarce radio-frequency (RF) spectrum, obtaining performance improvement while avoiding interferences, and adapting to their operation environment and channel conditions. However, the high cost related to the management and configuration of large-scale networks, the great variability of the available spectrum, the distributed nature of RF and its mobility, the ever-increasing complexity of network architectures, and the different quality-of-service (QoS) requirements of RF-based applications motivate the need of any kind of self-organization mechanisms [13,14]. According to [15], a CRWSN is a distributed network of CR wireless sensor nodes, which collect collaborative information and dynamically communicate it in a multihop manner to a sink over the available spectrum bands. Such networks are indicated for bursty traffic.

Some techniques, such as catching, which are well known and widely used in WSNs, cannot be directly applied to CRWSNs since the link transmission delay cannot be known a priori. In the same way, CRWSNs require dynamic time-spectrum block allocation in response to demand and a periodic readjustment of the bandwidth. This makes it necessary to design and develop a framework for testing CRWSN routing protocols with cost-efficient and large-scale deployability as described in [16], a depth study of the propagation model on the working scenario on which our paper focuses, and probably the use of machine-learning techniques in CR [17].

7.2 Interaction of Radio Waves with Surrounding Media and Propagation Loss Estimation

In order to guarantee that a communication link can be established between different nodes, verification of the received power level versus the receiver sensitivity is performed. The value of power available at the receiver end is dependent on the configuration of the transceivers as well on the total propagation losses within the particular link under analysis. However, the receiver sensitivity is given mainly by the election of modulation and coding schemes, channel capacity, and technological parameters, such as the overall noise factor. Since the value of the received power is strongly dependent on the distance between the transmitter and the receiver, coverage radius estimations can be obtained. Moreover, since sensitivity is strongly dependent on the desired bit rate, this gives rise to more complex coverage–capacity relations.

The validation of the coverage radius as well as the coverage–capacity relations is given in first approximation by a wireless link balance, which determines the expected value of the received RF power within the receiver end. In general terms, the link balance can be given by the expression in the following equation:

$$P_{\text{RX}} = P_{\text{TX}} - L_{\text{TX feeder}} + G_{\text{TX antenna}} - L_{\text{propagation}}$$
$$+ G_{\text{RX antenna}} - L_{\text{RX feeder}}$$

$$(7.1)$$

where:

P_{RX} is the value of the received power in the receiver end

P_{TX} is the nominal transmitter power level

$L_{\text{TX feeder}}$ and $L_{\text{RX feeder}}$ are the losses due to cables and connectors in transmitter (TX) and receiver (RX), respectively

$G_{\text{TX antenna}}$ and $G_{\text{RX antenna}}$ are the values of antenna gain in the main lobe for TX and RX, respectively

$L_{\text{propagation}}$ is the propagation loss

All of the values in the previous link balance are dependent on the configuration and hardware of the transceivers, except for the case of the contribution of propagation losses. This last term is basically dependent on the frequency of operation, the distance between the transmitter and the receiver, and the topology and morphology of the scenario under analysis. Therefore, by estimating the value of radio propagation losses, a correlation with the maximum TX–RX distance can be inferred, leading to a coverage radius.

The basic mechanism of radio propagation losses is by increasing the distance between transceivers, given by an ideal dependence (in the absence of any type of material element within the propagation scenario, i.e., free space propagation) of the received power with the square of the radius. If additional elements such as ground surface, vegetation, and buildings are present in the scenario, further contributions such as multipath propagation, material absorption, diffraction, and diffuse scattering take place. Material absorption is given by the fact that the electric properties of materials are given by the dielectric constant, of dispersive nature and with real and imaginary parts. The imaginary part is responsible for losses, given by the conversion of electromagnetic energy into heat when radio waves interact with the material.

Diffraction losses are given when an object is within the propagation path of the radiowave. Depending on the physical dimensions of the object in relation with operational wavelength and the shape of the object, new wavefronts are generated by the impinging wave, as a consequence of Huygens' principle and the existence of source currents derived from Fresnel–Kirchhoff diffraction integral [18]. The consequence of the existence of diffraction is added radio-wave propagation losses.

In order to obtain the complete estimation of losses within a wireless link, a profound knowledge of the scenario in which the wireless system will operate is necessary. The application of Maxwell's laws with the corresponding boundary conditions in principle can give a precise result. However, due to the great amount of detail in the application of such boundary conditions, this method cannot be applied in general. Different approximations can be used from empirical-based formulation to deterministic methods. Empirical-based methods are based on propagation loss determination by extraction of a modeling equation, obtained by regression from measurement campaigns as well as by semianalytical approximations. These methods (i.e., one-slope, dual-slope, Cost-231, etc.) offer rapid results with the drawback of lower accuracy as a consequence of not taking into account the surrounding environment. However, deterministic methods are based on the application of computationally demand calculations (e.g., full-wave electromagnetic simulation such as finite-element methods, finite-difference time domain, or finite integration time domain, or geometrical optics [GO], such as ray tracing or ray launching techniques), which can fully account for the topology and the morphology of the surrounding environment. In general terms, deterministic methods are capable of offering estimations with mean error values close to zero and smaller standard deviation than in the case of empirical-based methods. From the possibilities available in deterministic simulation tools, those based on GO, such as ray launching and ray tracing, are computationally more efficient for scenarios that are many times larger than the wavelength of the carrier frequency of the wireless system. Therefore, in this chapter, we describe wireless channel characterization of complex indoor scenarios with the aid of a three-dimensional (3D) ray launching simulation code implemented ad hoc for such purpose. A summary of various propagation models dealing with path loss is shown in Table 7.1, showing that ray tracing methods lead to a trade-off between accuracy and computational time of simulations.

Ray theory is an accurate site-specific method that is emerging as highly promising to obtain useful simulation results. Ray tracing is a technique based on GO and geometrical theory of diffraction (GTD) that can easily be applied as an approximate method for estimating the levels of high-frequency electromagnetic fields. GO principle is that energy can be considered to be radiated through infinitesimally small tubes, often called rays. These rays are normal to the surface of equal signal power. They lie along the direction of propagation and travel in straight lines. Therefore, signal propagation can be modeled via ray propagation. By using the concept of ray tracing, rays can be launched from a transmitter location, and the interaction of the rays can be described using the well-known theory of refraction and reflection and

Table 7.1 Comparison of Different Path Loss Propagation Models

MODEL NAME	SUITABLE ENVIRONMENT	COMPLEXITY	DETAILS OF ENVIRONMENT	ACCURACY	TIME
Free space model	Macrocell	Simple	No	Good	Little
Diffracting screens model	Macrocell	Simple	No	Good	Little
Okumura model	Macrocell	Simple	No	Good	Little
Hata model	Macrocell (early cellular)	Simple	No	Good	Little
COST-231	Microcell (outdoor)	Simple	No	Good	Little
Ray tracing	Outdoor and indoor	Complex	Yes	Very good	Very much
FDTD	Indoor (small)	Complex	Every detail	Best	Very much
MoM	Indoor (small)	Complex	Every detail	Best	Very much

FDTD, Finite difference time domain; MoM, Method of moments.

interactions with the neighboring environment. The rays considered in GO are only direct, reflected, and refracted rays. For this reason, abrupt transition areas may occur, corresponding to the boundaries of the regions where these rays exist. To complement the GO theory, the diffracted rays are introduced with the GTD and its uniform extension, the uniform GTD (UTD). The purpose of these rays is to remove the field discontinuities and to introduce proper field corrections, especially in the zero-field regions predicted by GO.

The advantages of this model are very accurate, and it can be used in a fixed location, which also can predict broadband parameters. Its drawback is relatively slow in operation so that preprocess and simplification must be applied. Moreover, the details of the geometrical location of obstacles and the electromagnetic parameters of all the materials are required accurately. In the 3D coordinate system, the deterministic ray modeling approach is generally defined by establishing the transmitter and receiver as reference points in the 3D coordinate system. Generally, the indoor walls, ceilings, and floors are defined as a plane with dielectric constant and thickness, whose surfaces are processed as plane sections. The simulation path is composed of those rays emitted from the transmitter through walls, ceilings, floors, tables, and other objects, reflected, and reach the receiver through these objects. In Section 7.2.2, a description of the algorithm employed to perform these calculations is given.

7.2.1 Ray Tracing Methods

The image method is a simple and accurate method for determining the ray trajectory between the transmitter and the receiver. Figure 7.1 represents the basic idea of the image method [19]. For this simple case, first the image of the transmitter 1 (Tx1) due to the wall 1 (W1) is determined. Then, it is calculated from the image of Tx1 due to wall 2 (W2), (Tx2). A reflection point (P2) can be found connecting Rx and Tx2. another reflection point (P1) can be found connecting P2 and Tx1. The image method is efficient, but it can only handle simple environments. Many environments with which we are concerned in our daily life are complicated, and the conventional image method is not adequate. For realistic applications, special techniques such as the hybrid and acceleration methods have to be used to reduce the computation time.

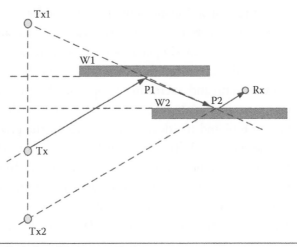

Figure 7.1   Illustration of the image method.

The ray tracing method is simple and is most widely used in the area of site-specific propagation prediction. However, it can be very computationally inefficient, leading to efforts focused on the acceleration of ray tracing algorithms. In [20], ray tracing is accelerated considerably by using the angular Z-buffer (AZB) technique. The philosophy behind this method consists of reducing the number of rigorous tests that have to be made by reducing the number of facets that each ray has to treat. It also consists of dividing the space seen from the source in angular regions and storing the facets of the model in the regions where they belong. In this way, for each ray, only the facets stored in its region need to be analyzed. This method can accelerate the ray tracing algorithm, but, when multiple reflections are needed, the preprocessing is not easy. Liang and Bertoni [21] proposed the vertical plane launch (VPL) technique for approximating a full 3D site-specific ray trace. The VPL technique employs the standard shooting and bouncing ray (SBR) method [22] in the horizontal plane while using a deterministic approach to find the vertical displacement of the unfolded ray paths. This approximation is valid since building walls are almost always vertical. In this method, the usual two-dimensional (2D) ray tracing is used in the horizontal plane and each ray in the 2D case represents a vertical propagation plane. When a ray hits a vertical wall, reflection from the vertical wall and diffraction from the rooftop horizontal edge can occur. When the ray hits a vertical edge, diffraction also occurs. The over-rooftop diffraction creates two vertical propagation planes:

one in the same direction as the incident ray and the other in the direction of reflection. Diffraction from the vertical edges creates a new source and many new rays in 2D planes should be launched.

Degli-Eposti et al. [23] introduced the scale-level concept to assess some criteria for a useful and interesting comparison between measurement and ray tracing simulation. They analyzed the validity domain of the ray tracing tool and the degree of accuracy of the prediction as a function of the scale level. The scale levels are related to the extent of the spatial regions representing the basic elements from which propagation parameters are extracted. The basic spatial regions can be grouped according to their characteristics at any scale level, from the smaller region (a point at space) to a whole class of buildings. In [24], an enhanced ray launching model is proposed to improve the efficiency and the speed of computation of propagation characteristics. Two methods are used to obtain the improvements: the effective propagation area method, which restricts the propagation region to be considered when the wave propagation characteristics are predicted. This method considers only the buildings included in the effective propagation area, which is defined as the region where the buildings that have dominant effects on the propagation characteristics are distributed. In the suggested model, the boundary of the effective propagation area (EPA) is defined in terms of the average delay time and the root-mean-square (RMS) delay spread of the received signal. The second method is the dominant corner extraction method, which is used to diminish the amount of the diffraction ray computation by considering only the particular building corners that have a significant influence on the received power. In the ray tracing process of the ray launching model, each building corner operates as a diffraction wedge and the corresponding diffracted ray paths are obtained by applying the ray launching method in which the corners are regarded repeatedly as new transmitters and receivers. As a result, the actual number of base stations to be considered increases with the number of the building corners. However, there exist dominant rays that provide most of the power and have a significant influence on the delay spread. In the dominant corner extraction method, the corners on which the dominant rays are diffracted are picked out and taken into calculation. Rappaport and colleagues [25] explained a new 3D ray tracing technique that reduces kinematic errors associated with ray launching algorithms based on the reception sphere and distributed wavefront methods. In their method, a sphere, centered on a candidate receiver location and sized according to the angular separation of the incoming rays, collects the rays that contribute to the overall electric field. The radius of a reception sphere sweeps out a circular area across the adjacent rays, but sometimes two rays fall within the sphere, which lead to double count errors. Distributed wavefront methods increase the accuracy of 3D ray tracing by eliminating the kinematic double-count errors while maintaining simplicity and speed.

The main drawback of ray propagation modeling is lengthy computation time due to 3D space analysis. This has given rise to the proposal of hybrid methods that are a combination of different methods to accelerate and improve efficiency of ray tracing models. Rossi and Gabillet [26] presented the geometrical ray implementation for mobile propagation modeling (GRIMM). In their approach, the ray construction mixes two techniques by splitting the 3D problem into two successive 2D stages, without loss of generality compared with the full 3D techniques. The first stage uses the basis of ray launching, whereas the second stage uses ray tracing. An SBR/image method was proposed in [27]. A deterministic approach using a modified SBR technique denoted as the SBR/image method is developed in the work to deal with the radio-wave propagation in furnished rooms that are composed of triangular facets. Conceptually, they use the SBR method to trace triangular ray tubes (not rays) bouncing in the room. If the Rx is within a ray tube, the ray tube will have a contribution to the received field at Rx and the corresponding equivalent source (image) can be determined. Besides, the first-order wedge diffraction from furniture is included, and the diffracted rays also can be attributed to the corresponding images. By summing all contributions of these images, the total received field at Rx can be obtained.

7.2.2 Radio Planning Process by Deterministic Channel Modeling

A 3D ray launching algorithm has been implemented in-house based on GO and GTD. Different applications of this algorithm can be found in the literature, such as analysis of wireless propagation in closed environments [28–32], interference analysis [33], or electromagnetic dosimetry evaluation in wireless systems [34]. The ray launching method considers a bundle of transmitted rays that may or may not reach the receiver. The available spatial resolution and the accuracy of the model are determined by the number of rays considered and the distance from the transmitter to the receiver location. The philosophy is that a finite sample of the possible directions of the propagation from the transmitter is chosen and a ray is launched for each such direction. If a ray hits an object, then a reflecting ray and a refracting ray are generated.

The algorithm has been designed for evaluating the communication in an indoor environment. Hence, it has the following features:

- Recreation of a realistic multipath
- Three-dimensional modeling of all types of rooms with different shapes and sizes
- Creation of complex environments as large as desired, with multiple plants and multiple buildings
- Characterization and modeling of any objects (windows, tables, chairs, walls, etc.) through its 3D shape and dielectric constant
- Modeling of reflection, refraction, and diffraction
- Calculation of the interaction between rays and objects taking into account the polarization wave
- Modeling of any transceiver
- Duration of the simulation of a ray determined by the number of rebounds and the maximum delay
- Analysis of the scenario by extracting parameters such as electric field strength, signal/interference, power delay, and dispersion

The use of advanced radio planning tools, such as the 3D ray launching algorithm, is usually divided into three phases:

1. *Phase 1: Creation of the scenario.* This phase sets the scenario, consisting of rooms with objects, transmitters, and receivers.
2. *Phase 2: Simulation of ray tracing in three dimensions.* In this phase, the rays are launched from each transmitter, keeping the parameters in each position at the space.
3. *Phase 3: Analysis of the results.* In this phase, the values are obtained from the simulation to calculate the desired parameters.

7.2.2.1 Creation of the Scenario The scenario is created with the information that characterizes the room and the objects in it, the interconnections of the different rooms, and the transmitters and receivers.

The room's characterization is given by its size. In this algorithm, the rooms are defined as different hexahedra with different dimensions in the x-, y-, and z-axes. Besides, the material that is made for the room can be defined. Normally, this material is air, but it is possible to consider any material for the room.

Complex forms can be created by interconnecting several hexahedra. Moreover, the fact of dividing an environment into multiple hexahedra gives us the ability to analyze each room with different resolution, providing us more flexibility. Each room is divided into multiple hexahedra and the parameters of the ray that reach that hexahedron are stored. Thus, the resolution in each axis of each room is defined as the number of hexahedra of that axis. Therefore, each room is divided into multiple equidimensional hexahedra and the precision of each axis can be chosen. Figure 7.2 represents the typical office with the rays launched from the transmitter.

Objects are defined as different hexahedra in the algorithm. By this basic geometric shape, it is very easy to form other objects much more complex, such as tables, chairs, and shelves, and place them into the room. In a generic room, walls can be formed by windows, doors, frames, and so on. Therefore, to characterize the walls of a room, each discontinuity on the wall must be characterized. This will define each part of the wall as an object by its central position on the wall, the width in each dimension, and the material that is made. Rooms are created individually and interconnected with the interconnection matrix, which defines the portion of a wall of a room that is connected with the wall portion of another room.

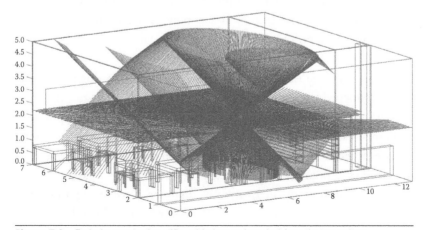

Figure 7.2 Typical scenario of an office with the rays launched from the transmitter.

In this phase, transmitters are defined in a generic way. Thus, performing a single simulation of phases 1 and 2, the parameters of the transceiver, such as the radiated power and the directivity of the antennas, can vary in phase 3, saving a lot of calculation time. Therefore, each transceiver can be defined with the following parameters:

- Location (room and coordinates)
- Number of launching rays and directions
- Maximum number of bounces of the rays
- Maximum delay of the rays
- Frequency of emission f_c
- Radiated power (P_r)
- Directivity $D(\theta, \Phi)$
- Polarization ($X^\perp, X^\parallel$)

A receiver is modeled taking into account that it absorbs the rays that are received in certain directions. Therefore, to define a receiver, the following parameters are characterized:

- Location (room and coordinates)
- Angular range in which the receiver absorbs the rays

Besides, in this phase of simulation, there are several options to be chosen, such as the calculation of parameters inside the obstacles. This option is useful to analyze the radiated power inside a certain object, for example, a human body. Nevertheless, the different materials of the objects are always taken into account. Consideration of diffraction or not can also be chosen in this phase of the algorithm. Commitment between computational time and accuracy of results must be taken into account to obtain best results of the simulation.

7.2.2.2 Simulation of Ray Tracing in Three Dimensions   In this phase, the rays are launched from each of the transmitters. They propagate through the space interacting with the obstacles in their path, causing physical phenomena such as reflection and refraction. The parameters of these rays are stored as they enter each hexahedron until the rays have a certain number of bounces or exceed the pre-propagation time set.

The algorithm works in an iterative manner, considering a ray and its reflections, storing the created ray inorder to introduce diffraction at a later stage.

The ray launching algorithm in three dimensions has three recurrent steps. The first step is to take all the antennas of the same room and introduce them to simulate. The second step is to simulate the room, and the third step is to transform the rays coming out of the room through new antennas in other rooms.

The most important steps of the algorithm are as follows:

- Calculate the impact point.
- Travel from the initial point to the point of impact storing the parameters.
- Calculate the reflected ray.
- Calculate the transmitted ray.

Figure 7.3 describes the different steps of the algorithm.

The calculation of the impact point is performed decomposing the 3D problem in a dimensional problem, calculating the propagation velocities in each axis. Thus, the impact happens in the first instant of time that the three projections are in the range of the projection of the obstacle. If it does not impact with any obstacles, the algorithm calculates in the same way the point of impact with the wall.

Once the point of impact is calculated, it is necessary to follow the straight line joining the initial point (x_i, y_i, z_i) and the final impact point (x_f, y_f, z_f) saving the parameters of each ray. The information stored in each hexahedron that the ray passes in its trajectory is as follows:

- Time taken from the ray to arrive (τ)
- Distance traveled by the ray (d)
- Loss coefficient in each polarization $(L^{\perp}, L^{\parallel})$
- Ray direction in the transmitter (θ_t, ϕ_t)
- Ray direction in the receiver (θ_r, ϕ_r)
- Transmitting antenna (n)
- Diffraction

Figure 7.4 represents the phenomena of reflection and transmission when the ray impacts with an obstacle and the diffraction phenomena when the ray impacts with an edge.

A plane electromagnetic wave falling to the planar interface between two regular semi-infinite media 1 and 2 gives rise to two plane waves:

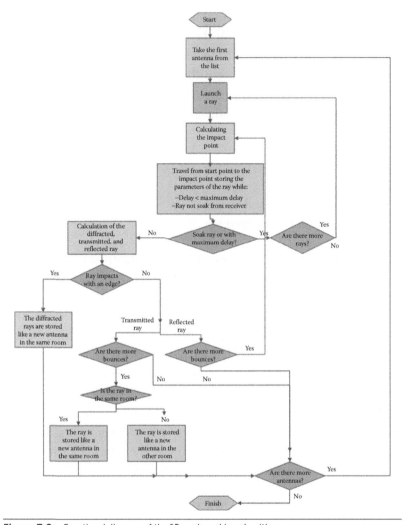

Figure 7.3 Functional diagram of the 3D ray launching algorithm.

reflected and transmitted (or refracted). According to Snell's law [18], the reflection coefficient $R^\perp$ and the transmission coefficient $T^\perp$ are calculated as follows:

$$T^\perp = \frac{E_t^\perp}{E_i^\perp} = \frac{2\eta_2 \cos(\Psi_i)}{\eta_2 \cos(\Psi_i) + \eta_1 \cos(\Psi_t)} \tag{7.2}$$

$$R^\perp = \frac{E_r^\perp}{E_i^\perp} = \frac{\eta_2 \cos(\Psi_i) - \eta_1 \cos(\Psi_t)}{\eta_2 \cos(\Psi_i) + \eta_1 \cos(\Psi_t)} \tag{7.3}$$

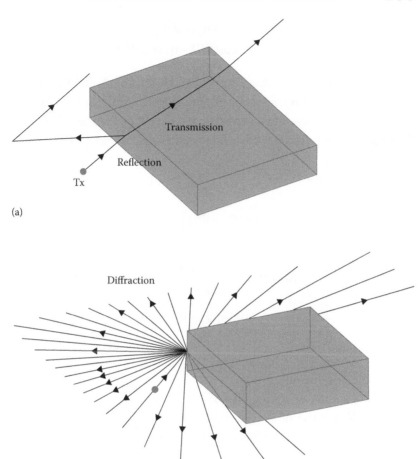

(a)

(b)

Figure 7.4 Principle of ray launching method: (a) Reflection and transmission; (b) diffraction.

where:

$$\eta_1 = 120\pi / \sqrt{\varepsilon_{r1}}$$
$$\eta_2 = 120\pi / \sqrt{\varepsilon_{r2}}$$

Ψ_i, Ψ_r, and Ψ_t are the incident, reflected, and transmitted angles, respectively

For the parallel (or magnetic) polarization, the magnetic field vector of the incident wave is perpendicular to the plane of incidence. Then, the reflection and transmission coefficients $R^{\parallel}$ and $T^{\parallel}$ can be calculated as follows:

$$R^{\parallel} = \frac{E_r^{\parallel}}{E_i^{\parallel}} = \frac{\eta_1 \cos(\Psi_i) - \eta_2 \cos(\Psi_t)}{\eta_1 \cos(\Psi_i) + \eta_2 \cos(\Psi_t)} \tag{7.4}$$

$$T^{\parallel} = \frac{E_t^{\parallel}}{E_i^{\parallel}} = \frac{2\eta_2 \cos(\Psi_i)}{\eta_1 \cos(\Psi_i) + \eta_2 \cos(\Psi_t)} \tag{7.5}$$

Once the parameters of transmission T and reflection R and the angles of incidence Ψ_i and transmission Ψ_t are calculated, the new angles of the reflected and transmitted waves (θ_r, ϕ_r) and (θ_t, ϕ_t) can be calculated. The finite conductivity 2D diffraction coefficients are given by [35,36]

$$D^{\parallel\perp} = \frac{-e^{(-j\pi/4)}}{2n\sqrt{2\pi k}} \left\{ \begin{aligned} &\cot g\left[\frac{\pi + (\Phi_2 - \Phi_1)}{2n}\right] F[kLa^+(\Phi_2 - \Phi_1)] \\ &+ \cot g\left[\frac{\pi - (\Phi_2 - \Phi_1)}{2n}\right] F[kLa^-(\Phi_2 - \Phi_1)] \\ &+ R_0^{\parallel\perp} \cot g\left[\frac{\pi - (\Phi_2 + \Phi_1)}{2n}\right] F[kLa^-(\Phi_2 + \Phi_1)] \\ &+ R_n^{\parallel\perp} \cot g\left[\frac{\pi + (\Phi_2 + \Phi_1)}{2n}\right] F[kLa^+(\Phi_2 + \Phi_1)] \end{aligned} \right\} \tag{7.6}$$

where:

$n\pi$ is the wedge angle

F, L, a^+, and a^- are defined in [35]

R_0 and R_n are the reflection coefficients for the appropriate polarization for 0 and n faces, respectively

Φ_2 and Φ_1 refer to the angles in Figure 7.5.

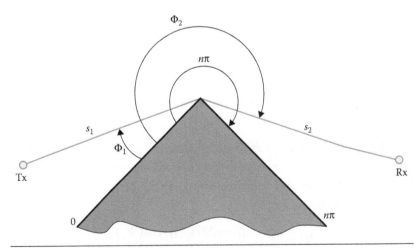

Figure 7.5 Geometry for wedge diffraction coefficients.

7.2.2.3 Analysis of the Results In the third phase of the simulation procedure, the results are analyzed using the stored parameters of each ray. The electric field $\vec{E}$ created by an antenna with a radiated power P_{rad} in (θ,ϕ) directions with a directivity $D(\theta,\phi)$ and a polarization ratio $(X^{\perp}, X^{\parallel})$ at distance d in the free space is calculated by [28]

$$E_i^{\perp} = \sqrt{\frac{P_{rad} D_t(\theta_t, \phi_t) \eta_0}{2\pi}} \frac{e^{-j\beta_0 r}}{r} X^{\perp} L^{\perp} \qquad (7.7)$$

$$E_i^{\parallel} = \sqrt{\frac{P_{rad} D_t(\theta_t, \phi_t) \eta_0}{2\pi}} \frac{e^{-j\beta_0 r}}{r} X^{\parallel} L^{\parallel} \qquad (7.8)$$

where:
$\beta_0 = 2\pi f_c \sqrt{\varepsilon_0 \mu_0}$
$\varepsilon_0 = 8,854 \times 10^{-12}$
$\mu_0 = 4\pi \times 10^{-7}$
$\eta_0 = 120\pi$

The diffracted field is calculated by [18]

$$E_{UTD} = e_0 \frac{e^{-jks_1}}{s_1} D^{\perp \parallel} \sqrt{\frac{s_1}{s_2(s_1 + s_2)}} e^{-jks_2} \qquad (7.9)$$

where:
$D^{\perp \parallel}$ are the diffraction coefficients in Equation 1.5
s_1 and s_2 are the distances represented in Figure 7.6, from the source to the edge and from the edge to the receiver point

The received power is calculated at each point taking into account the losses of propagation through a medium $(\varepsilon, \mu, \sigma)$ at distance d, with the attenuation constant α (Np/m) and the phase constant β (rad/m).

In order to gain insight into topomorphological impact on wireless channel behavior, different systems have been analyzed in a complex indoor office environment. The employed scenario is depicted in Figure 7.6, in which different office elements, such as furniture and workstations, have been considered. The characteristics of the simulated systems are given in Table 7.2.

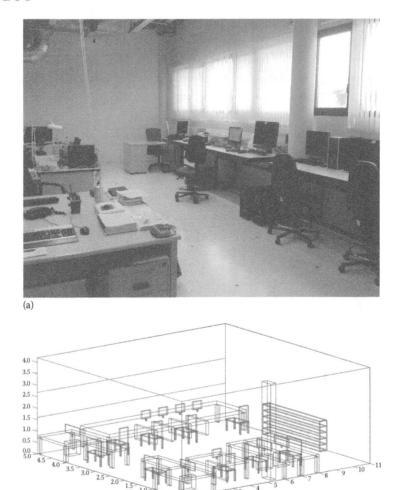

Figure 7.6 Scenario employed in order to analyze topomorphological impact on different wireless systems: (a) Image of the test scenario; (b) implemented simulation model.

7.2.3 Universal Mobile Telecommunications System

The use of third-generation (3G) networks within the considered indoor environment is mainly for mobile/smartphone connection, as well as for future femtocell deployment. The bidimensional power distribution reveals that the received power levels strongly depend on the location even in the case of a small scenario. The impact of multipath propagation can clearly be observed from the power delay profile estimation, as presented in Figure 7.7.

Table 7.2 Parameters for Different Systems Considered in the Radio Planning Analysis

SYSTEM/ PARAMETERS	FREQUENCY (GHz)	TRANSMITTED POWER (dBm)	BIT RATE (Mbps)	ANTENNA GAIN (dBi)	Rx SENSITIVITY (dBm)
UMTS	2.1	23	5.5	3	
WiFi (b/g)	2.4	20	54	5	
WiFi (a)	5	18	54	8	
ZigBee	2.4	20		5	−100
ZigBee	0.868	25		5	−112
Bluetooth (v2.1)	2.4	3	3	5	
Bluetooth (v4.0)	2.4	0	24	5	
LTE	0.7/0.8/1.9/2.1/2.6	0	100		

UMTS, Universal Mobile Telecommunications System.

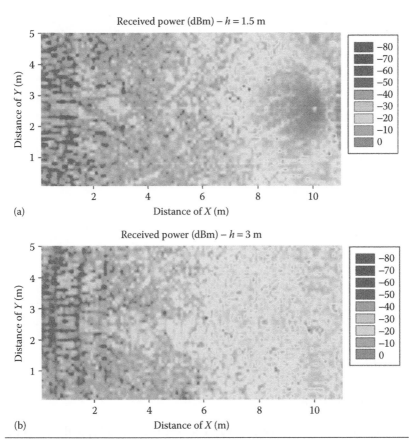

Figure 7.7 Estimations of received power levels in 3G UMTS systems. The impact of multipath propagation can be seen in the power delay profile estimations, which derive in an overall delay spread of 74.3 ns.: (a) Estimation of received power for $h = 1.5$ m; (b) Estimation of received power for $h = 3$ m; (c) received power profile for both heights; (d) Power Delay Profile estimation.

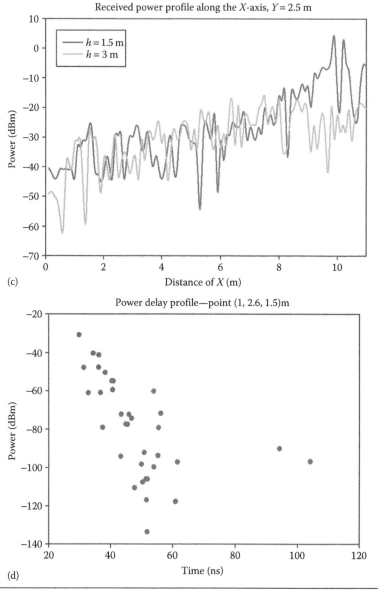

Figure 7.7   (Continued)

7.2.4 IEEE 802.11b/g

Wireless local area networks (WLAN WiFi) are also typically present within indoor scenarios. Again in this case, there is a clear dependence on the topology and morphology of the scenario. Typical links in this case will be formed between the WLAN router/access point and the portable devices (laptops, netbooks, ultrabooks, tablets, and smartphones), with high connection speeds and quasistatic mobility. Higher propagation losses are present in the case of 802.11a, given by the use of the 5 GHz band instead of the 2.4 GHz band. However, when performing a coverage–capacity analysis, signal-to-noise ratios in the 5 GHz band are in principle subject to lower values of interference, due to lower intensity in the use of this frequency band. Multipath propagation again plays a fundamental role in the loss mechanism within the simulation scenario, as depicted in Figure 7.8.

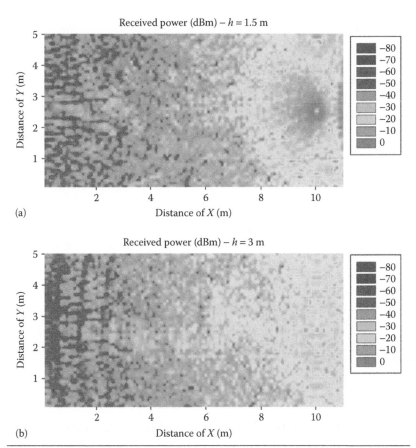

Figure 7.8 Estimations of received power levels in WLAN systems. The impact of multipath propagation can be seen in the power delay profile estimations, which derive in an overall delay spread of 74.25047 ns.

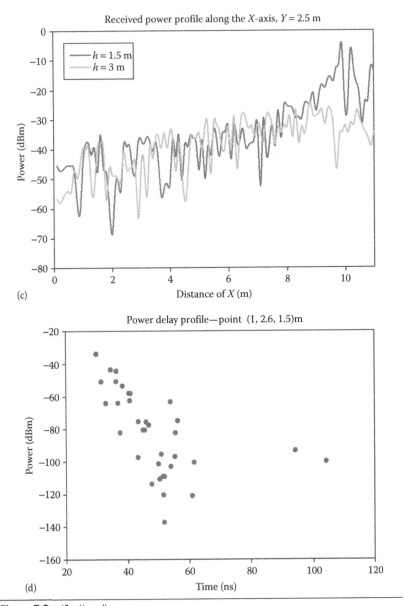

Figure 7.8   (Continued)

7.2.5 IEEE 802.11a

In this case, the operational frequency band is within 5 GHz, leading in principle to increased propagation losses, as can be seen from the bidimensional coverage plots presented. Coverage radius is therefore smaller in this case. In terms of link balance analysis, the effect may not be so negative, due to the fact that antenna gains can be larger (if maintaining the same antenna dimensions as in the previous case). Moreover, interference levels are in principle smaller [due to less intensive use than the 2.4 GHz industrial, scientific, and medical (ISM) band], enhancing receiver sensitivity, as shown in Figure 7.9.

7.2.6 ZigBee (868 MHz)

Personal area networks are being widely adopted within indoor scenarios in the implementation of WSNs, which can be applied for home automation, ad hoc networks for low-speed data communications,

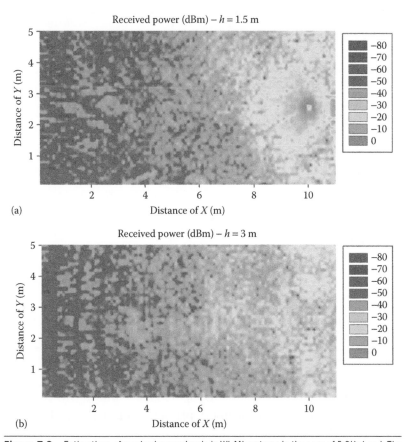

Figure 7.9 Estimations of received power levels in WLAN systems in the case of 5 GHz band. The impact of multipath propagation can be seen in the power delay profile estimations, which derive in an overall delay spread of 74.25047 ns.

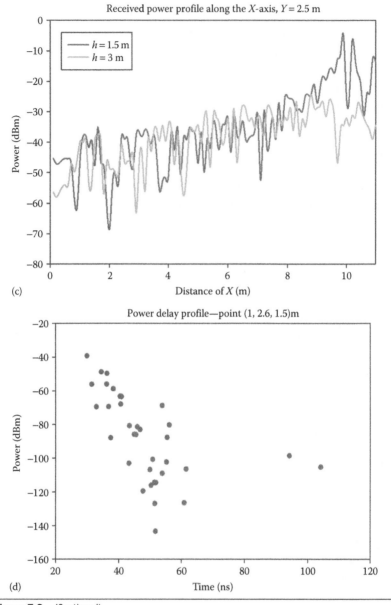

Figure 7.9   (Continued)

or monitoring and control of patients, to name a few. In this context, systems such as ZigBee or Bluetooth play a key role in the deployment of these WSNs, given by their moderate cost, reduced size, and optimized power consumption. Simulation results have been presented for several cases in several frequency bands, following the parameters previously stated in Table 7.1. As can be seen from the bidimensional received power planes, there is a considerable difference in the power levels as a function mainly of frequency. In practice, transceivers operating at 868 MHz are successfully applied in outdoor applications due to their increased coverage radius. The results are depicted in Figure 7.10 and the results for the 2.4 GHz band are shown in Figure 7.11.

7.2.7 Bluetooth

An interesting alternative for the deployment of WSNs is the use of Bluetooth version 4.0, characterized by low-power consumption.

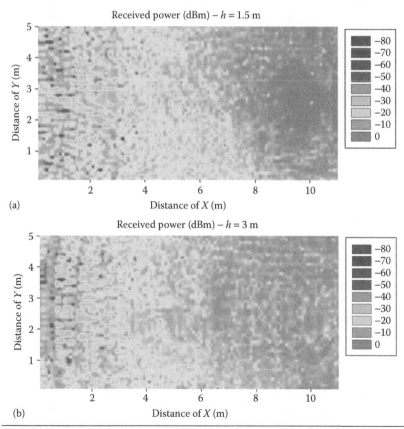

Figure 7.10 Estimations of received power levels in ZigBee systems in the case of 868 MHz band. The impact of multipath propagation can be seen in the power delay profile estimations, which derive in an overall delay spread of 32.8761 ns.

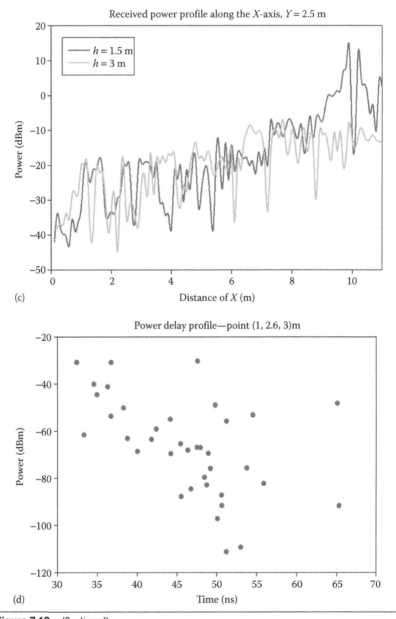

(c)

(d)

Figure 7.10 (Continued)

In this case, coverage levels are smaller than in other systems previously presented. This initial drawback can be surpassed by the use of multiple Bluetooth devices interconnected, leading to an ad hoc network configuration. It is interesting to note that several smartphone vendors have started to introduce Bluetooth v4.0 chipsets, which will

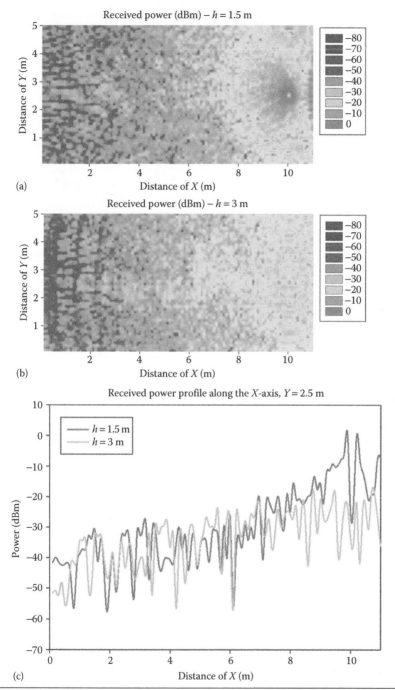

Figure 7.11   Estimations of received power levels in ZigBee systems in the case of 2.4 GHz band. The impact of multipath propagation can be seen in the power delay profile estimations, which derive in an overall delay spread of 32.8761 ns.

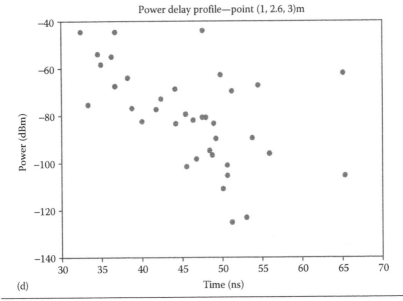

Figure 7.11   (Continued)

increase the adoption by users and will impact on application performance. Both cases (earlier versions as well as v4.0) are depicted,
respectively, in Figures 7.12 and 7.13.

7.2.8 Long-Term Evolution

The next step in the evolution of wireless networks is the adoption of
fourth-generation (4G) systems, of which long-term evolution (LTE)
is the main driver. LTE is the evolution of the systems developed
within the framework of third generation partnership project (3GPP),
based on more advanced radio interface (larger spectrum blocks in
different frequency bands, higher level modulation schemes, or multiple antenna systems, to name a few) as well as higher level network
operation and scheduling. One of the most interesting scenarios for
LTE is the deployment of femtocells within indoor scenarios, in
which a compact size transceiver (similar to domestic WiFi routers)
will provide multisystem connectivity within a household or a con-
fined workspace. In this context, the result is a heterogeneous wireless

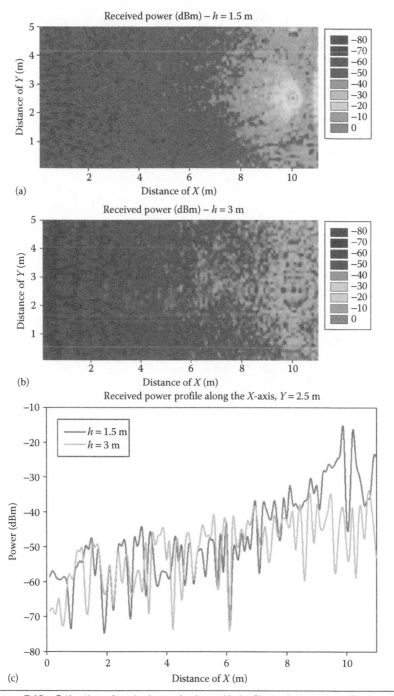

Figure 7.12 Estimations of received power levels considering Bluetooth transceivers. The impact of multipath propagation can be seen in the power delay profile estimations, which derive in an overall delay spread of 76.4549 ns.

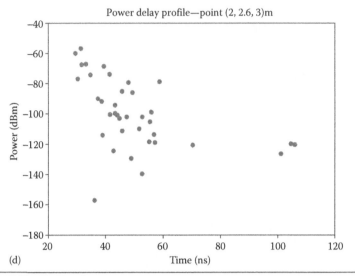

(d)

Figure 7.12 (Continued)

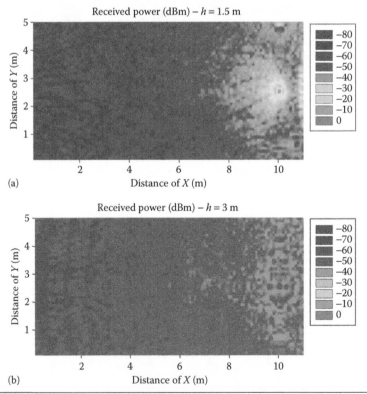

Figure 7.13 Estimations of received power levels considering Bluetooth v4.0 transceivers. The impact of multipath propagation can be seen in the power delay profile estimations, which derive in an overall delay spread of 84.0497 ns.

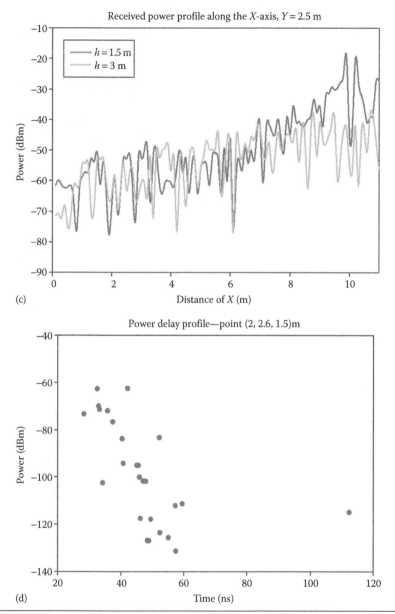

Figure 7.13   (Continued)

scenario, in which coverage–capacity relations will be dynamically changing as a function of traffic demand, user location, and allocated frequency of operations. Results for several frequencies of operation of LTE for the simulation scenario are presented in Figure 7.14 and a clear dependence on such frequency of operation can be observed.

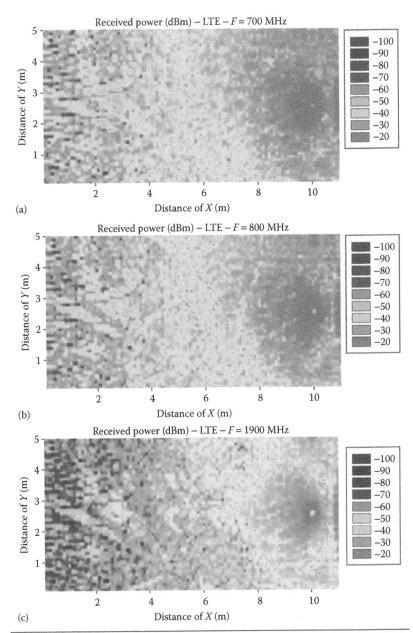

Figure 7.14 Estimation of received power levels for different frequency bands of operation in LTE systems. A comparison of all of the systems within the scenario is presented in (d), in which the frequency dependence for radio propagation losses can be observed, even in the case of a small scenario: (a) received power, frequency of 700 MHz; (b) received power, frequency of 800 MHz; (c) received power, frequency of 1900 MHz; (d) received power, frequency of 2100 MHz; (e) received power, frequency of 2600 MHz; (f) comparison of received power distributions for the previous frequencies under analysis.

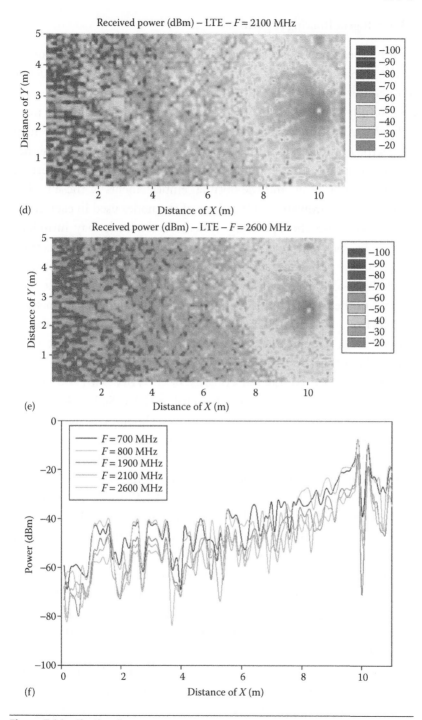

Figure 7.14 (Continued)

7.3 Intelligent Routing Algorithms

As it has been previously shown, a clear dependence on the topology of the scenario as well as on the system parameters (i.e., frequency of operation and overall available RF power) is observed and must be carefully considered in order to assess the performance of multiple systems within a particular scenario. The analysis of the RF map allows the optimization of the design of the WSN since the designer can select the minimum number of nodes required to grant a certain communication level, and the optimal emplacements of these nodes. However, the hardware specifications of the nodes used in each radio system, and also the operating systems embedded, may introduce some behaviors that must be conveniently analyzed. In this section, field trials are performed and the results in relation with their location are derived, for the case of WSNs. In such a way, we perform a wide analysis of the frequency between messages (considering transmission periods from 45 to 10,000 ms), concluding that some communication periods are inadequate due to the great number of messages lost without reaching their destination.

The workspace considered consists of a WSN with Waspmote nodes that communicate among them using IEEE 802.15.4 technology. The election of this communication technology is due to low-power consumption and also to low-cost and low-rate communication criteria. Waspmote nodes include an 8 MHz ATmega1281 microcontroller with an 8 KB static random-access memory (SRAM), an erasable programmable read-only memory (EEPROM) of 4 KB, and a flash memory of 128 KB. The clock frequency is 32 KHz, and the power consumptions are 8 mA in ON mode, 62 μA in sleep mode, and 0.7 μA in hibernate mode. The IEEE 802.15.4-compliant RF device works on the 2.40–2.48 GHz band, with a transmission power of 1 mW, a sensitivity of –92 dBm, a coverage distance of ~500 m, and 16 channels. Figure 7.15 illustrates the experimental scenario, where seven motes are considered and a gateway acts as a sink gathering all the data aggregated through the WSN.

First of all, a model of the workspace is built and validated with the aid of a ray tracing simulator and a spectrum analyzer, respectively. A chain of eight Waspmote nodes is then built in charge of data collecting and aggregating. Data collected at each node include

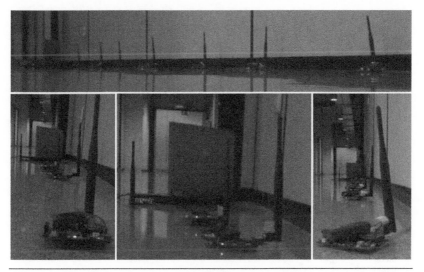

Figure 7.15   Experimental scenario employed for WSN traffic analysis.

the temperature sensed, the battery level of the node, and the RSSI of the last message received. The initiator node (1) starts the process sensing the temperature value and including it with the battery level in a message, which is sent to the following node (2). Node 2 receives the message sent by the initiator and aggregates its own information (concerning battery and temperature) to this message and resends it to the following node (3). This process is repeated until it reaches the last node of the network (7), and then the gateway node (GW) obtains the information aggregated and supplies it to the corresponding application.

We analyze the proposed chain sending 12,540,000 messages considering different transmission periods from 50 to 10,000 ms. Results obtained show a reception rate of 83.19%, whereas 16.81% of the messages are lost due to communication problems. This high rate of loss is due to the saturation of the reception buffer of the radio interface. The lower the transmission period, the higher the number of losses occurs. Figure 7.16 shows the number of messages received and lost for different transmission periods (in logarithmic scale in order to facilitate its understanding) and the variation on the number of messages lost. One can observe a special behavior in the range of 55–75 ms, where the highest number of losses occurs. Due to internal features of the nodes, the operating system of the nodes and the size of the

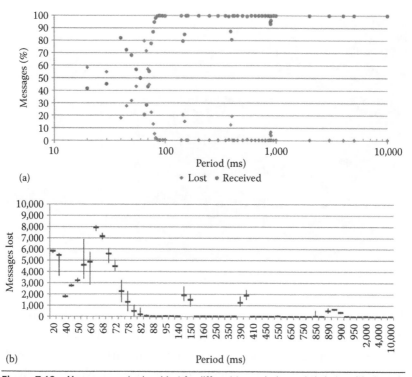

Figure 7.16   Messages received and lost for different transmission periods in logarithmic scale (a) and variation on the number of messages lost (b).

communication buffers interact causing a relevant loss of messages. Higher deviations occur at the transient period (20–85 ms), but also occur at 145–150, 390–400, and 850–900 ms. Message losses that occur during these periods are mainly due to the saturation of the communication buffer following the completion of other tasks by the microprocessor of the mote. The number of messages lost is irrelevant for higher transmission periods.

Figure 7.17 illustrates the different transmission periods where communication is most suitable. One can observe four different communication windows: 90–140, 160–350, 410–880, and up to 905 ms. Figure 7.17a shows the rate of messages received according to the transmission period, whereas Figure 7.17b shows the rate of messages lost for the same period.

Figure 7.18 shows the distribution of messages lost by node. The initiator node (1) sends periodically the information collected to the second node following a constant transmission rate of 20, 45, 50, 55,

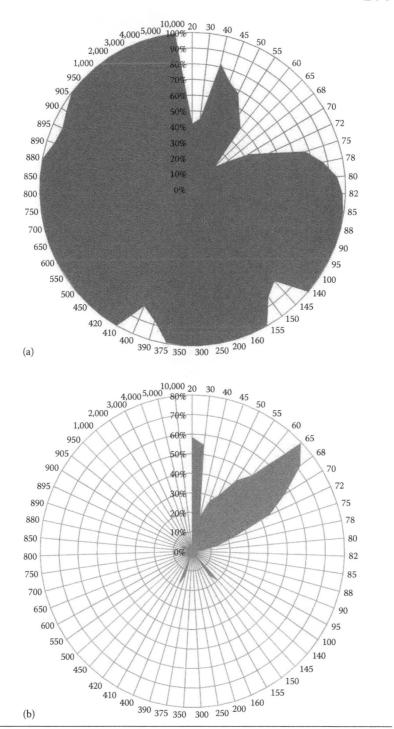

(a)

(b)

Figure 7.17   Messages received (a) and lost (b) for different transmission periods.

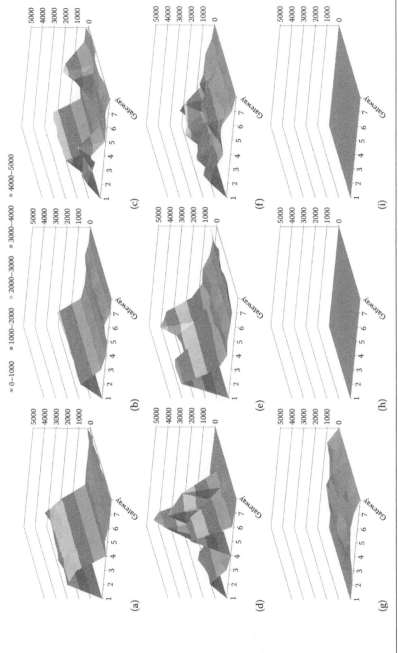

Figure 7.18 Distribution of messages lost by node. Transmission periods of 20 (a), 45 (b), 50 (c), 55 (d), 60 (e), 65 (f), 75 (g), 100 (h), and 200 ms (i).

60, 65, 75, 100, and 200 ms. Shorter transmission periods cause a high number of losses on node 2; its communication buffer collapses and then node 2 acts as a stop filter that ensures a transmission rate that can be easily accomplished by the rest of nodes. When considering transmission periods from 20 to 65 ms, the number of messages lost is notably high reaching up to 5000 messages, equivalent to a message loss of approximately 50 %. Graphics corresponding to 55 and 60 ms show dual behavior. When node 2 suffers high losses, the rest of the nodes do not suffer any congestion and the number of messages lost is low. Conversely, when node 2 suffers from moderate losses, node 6 in the case of 55 ms and 5 and 7 in the case of 60 ms suffer heavy losses due to congestion. When considering 75, 100, or 200 ms transmission periods, the number of messages lost is very low, below 50 messages in the case of 100 ms and below 10 in the case of 200 ms. Furthermore, the distribution of losses is fairly uniform.

As the period increases, the loss distribution becomes uniform among the different nodes involved in the communication. As Figure 7.19 shows, initially node 2 suffers a greater number of losses, along with node 3, but as the transmission period increases, the losses decrease and the first node losses are matching those of the other nodes. We observe a marked reduction in the total number of messages lost, along with a more homogeneous distribution of losses. This is because the communication buffer used by the radio interface is not saturated, and the losses are due to multipath propagation and not to buffer overrun problems. The number of losses of node 1 is always 0, since it is the initiator node and never receives any message. The gateway, which is the node in charge of data collection, is the sink node of the chain and has the same features as the other nodes of the network; then its loss distribution is analogous to those of other nodes.

Simulation aids to minimize the time and resources needed to characterize the RF scenario in order to design and develop intelligent routing algorithms for CRWSN. Experimentation shows the validity of the results obtained from simulation. An adequate characterization of both the RF spectrum and the working scenario, as well as the data flow of the WSN, allows an optimal use of the RF channel, maximizing the throughput of the network and its effectiveness. Some recommendations may be achieved as a result of this analysis. The transmission periods can significantly affect the efficiency of

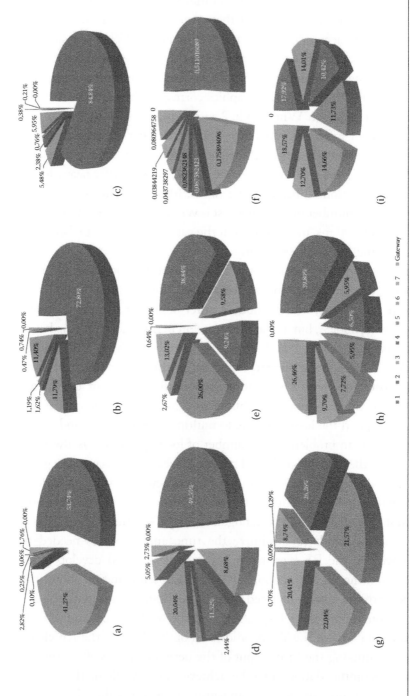

Figure 7.19   Distribution of losses by node. Transmission periods of 20 (a), 45 (b), 50 (c), 55 (d), 60 (e), 65 (f), 75 (g), 100 (h), and 200 ms (i).

the communication, so one should choose those that maximize the throughput rate (e.g., 100–200 ms). Furthermore, first nodes of the chain act as stoppers of the traffic flow when the transmission period is quite short, so intelligent routing algorithms must take into account this bottleneck and vary over time the transmission period or even alter the transmission order of the WSN chain.

7.4 Conclusions

The increased adoption of wireless systems, given by incumbent 4G systems, the popularity of machine-to-machine (M2M) communications, and the advent of Internet of Things and Smart Cities lead to a heterogeneous wireless ecosystem. In such a scenario, great challenges in terms of time-varying and location-dependent interference sources must be considered in order to guarantee QoS. The use of deterministic wireless channel modeling can aid in the correct definition of each one of the employed wireless systems, and hence in the adoption of the adequate solution within multisystem wireless transceivers, supported by CR strategies in order to maximize overall spectral efficiency and user experience.

Acknowledgments

This work has been supported by the Spanish Government under research grants: TIN2010-17170, TIN2011-28347-CO1-02, and TIN2011-28347-CO2-02, project IPT-370000-2010-36-RailTrace and project IIM13185.RI1-Faster.

References

1. G. Hackmann, O. Chipara, and C. Lu, "Robust topology control for indoor wireless sensor networks," *Proceedings of the 6th ACM Conference on Embedded Network Sensor Systems*, pp. 57–70, ACM, New York, November 4–7, 2008.
2. IEEE, *IEEE 802.15.4 Std: Wireless Medium Access Control (MAC) and Physical Layer (PHY) Specifications for Low-Rate Wireless Personal Area Networks (LR-WPANs)*, IEEE Computer Society Press, New York, pp. 1–679, October 2003.

3. W. R. Heinzelman, J. Kulik, and H. Balakrishnan, "Adaptive protocols for information dissemination in wireless sensor networks," *Proceedings of the 5th ACM International Conference on Mobile Computing and Networking*, pp. 174–185, Seattle, WA, August 15–20, 1999.

4. P. Ding, J. Holliday, and A. Celik, "Distributed energy efficient hierarchical clustering for wireless sensor networks," *Proceedings of the 8th IEEE International Conference on Distributed Computing in Sensor Systems*, pp. 322–339, Marina Del Rey, CA, June 8–10, 2005.

5. L. Buttyan and P. Schaffer, "PANEL: Position-based aggregator node election in wireless sensor networks," *International Journal of Distributed Sensor Networks*, 1–16, 2010. http://www.hindawi.com/journals/ijdsn/2010/679205/cta/.

6. S. Lindsey, C. Raghavendra, and K. M. Sivalingam, "Data gathering algorithms in sensor networks using energy metrics," *IEEE Transactions on Parallel and Distributed Systems*, 13: 924–935, 2002.

7. R. Sheikhpour and S. Jabbehdari, "An energy efficient cluster-chain based routing protocol for time critical applications in wireless sensor networks," *Indian Journal of Science and Technology*, 5(5): 2741–2749, 2012.

8. J. D. Yu, K. T. Kim, B. Y. Jung, and H. Y. Youn, "An energy efficient chain-based clustering routing protocol for wireless sensor networks," *International Conference on Advanced Information Networking and Applications Workshops*, pp. 383–388, Bradford, UK, May 26–29, 2009.

9. O. Younis and S. Fahmy, "HEED: A hybrid, energy-efficient, distributed clustering approach for ad hoc sensor networks," *IEEE Transactions on Mobile Computing*, 3(4): 366–379, 2004.

10. D. E. Boubiche and A. Bilami, "HEEP (Hybrid Energy Efficiency Protocol) based on chain clustering," *International Journal of Sensor Networks*, 10(1/2): 25–35, 2011.

11. F. Tang, I. You, S. Guo, M. Guo, and Y. Ma, "A chain-cluster based routing algorithm for wireless sensor networks," *Journal of Intelligent Manufacturing*, 23(4): 1305–1313, 2010.

12. X. Liu, "A survey on clustering routing protocols in wireless sensor networks," *Sensors*, 12: 11113–11153, 2012.

13. Z. Zhang and K. Long, "Self-organization paradigms and optimization approaches for cognitive radio technologies: A survey," *IEEE Wireless Communications*, 20(2): 36–42, 2013.

14. G. P. Joshi, S. Y. Nam, and S. W. Kim, "Cognitive radio wireless sensor networks: Applications, challenges and research trends," *Sensors*, 13: 11196–11228, 2013.

15. O. B. Akan, O. B. Karli, and O. Ergul, "Cognitive radio sensor networks," *IEEE Networks*, 23: 34–40, 2009.

16. A. Asal, A. Mamdouh, A. Salama, M. Elgendy, M. Mokhtar, M. Elsayed, and M. Youssef, "CRESCENT: A modular cost-efficient open-access testbed for cognitive radio networks routing protocols," *Proceedings of the 19th Annual International Conference on Mobile Computing & Networking*, pp. 179–182, ACM, New York, 2013.

17. M. Bkassiny, Y. Li, and S. K. Jayaweera, "A survey on machine-learning techniques in cognitive radios," *IEEE Communications Surveys & Tutorials*, 15(3): 1136–1159, 2013.
18. H. D. Hristov, *Fresnel Zones in Wireless Links, Zone Plate Lenses and Antennas*," Artech House, Boston, MA, 2000.
19. M. F. Iskander and Z. Yun, "Propagation prediction models for wireless communication systems," *IEEE Transactions on Microwave Theory and Techniques*, 50: 662–673, 2002
20. F. S. de Adana, O. Gutierrez Blanco, I. G. Diego, J. Perez Arriaga, and M. F. Catedra, "Propagation model based on ray tracing for the design of personal communication systems in indoor environments," *IEEE Transactions on Vehicular Technology*, 49: 2105–2112, 2000.
21. G. Liang and H. L. Bertoni, "A new approach to 3D ray tracing for propagation predictions in cities," *IEEE Transactions on Antennas and Propagation*, 46:853–863, 1998.
22. H. Ling, R. C. Chou, and S. W. Lee, "Shooting and bouncing rays: Calculating the RCS of an arbitrarily shaped cavity," *IEEE Transactions on Antennas and Propagation*, 37:194–205, 1989.
23. V. Degli-Eposti, G. Lombardi, C. Passerini, and G. Riva, "Wide-band measurement and ray-tracing simulation of the 1900MHz indoor propagation channel: Comparison criteria and results," *IEEE Transactions on Antennas and Propagation*, 49:1101–1110, 2001.
24. K. R. Chang and H. T. Kim, "Improvement of the computation efficiency for a ray launching model," *IEEE Proceedings—Microwaves, Antennas and Propagation*, 145:98–106, 1998.
25. G. Durgin, N. Patwari, and T. S. Rappaport, "Improved 3D ray launching method for wireless propagation prediction," *Electronic Letters*, 33: 1412–1413, 1997.
26. J. P. Rossi and Y. Gabillet, "A mixed ray launching/tracing method for full 3D UHF propagation modeling and comparison with wide-band measurements," *IEEE Transactions on Antennas and Propagation*, 50: 517–523, 2002.
27. S. H. Chen and S. K. Jeng, "An SBR/image approach for radio wave propagation in indoor environments with metallic furniture," *IEEE Transactions on Antennas and Propagation*, 45: 98–106, 1997.
28. L. Azpilicueta, F. Falcone, J. J. Astráin, J. Villadangos, I. J. García Zuazola, H. Landaluce, I. Angulo, and A. Perallos, "Measurement and modeling of a UHF-RFID system in a metallic closed vehicle," *Microwave and Optical Technology Letters*, 54(9): 2126–2130, 2012.
29. A. Moreno, I. Angulo, A. Perallos, H. Landaluce, I. J. G. Zuazola, L. Azpilicueta, J. J. Astráin, F. Falcone, and J. Villadangos, "IVAN: Intelligent van for the distribution of pharmaceutical drugs," *Sensors*, 12: 6587–6609, 2012.
30. J. A. Nazábal, P. Iturri López, L. Azpilicueta, F. Falcone, and C. Fernández-Valdivielso, "Performance analysis of IEEE 802.15.4 compliant wireless devices for heterogeneous indoor home automation environments," *International Journal of Antennas and Propagation*, 14pp. 2012. http://www.hindawi.com/journals/ijap/2012/176383/cta/.

31. I. Sesma, L. Azpilicueta, J. J. Astráin, J. Villadangos, and F. Falcone, "Analysis of challenges in the application of deterministic wireless cannel modelling in the implementation of WLAN-based indoor location system in large complex scenarios," *International Journal of Ad Hoc and Ubiquitous Computing*, 15(1–3): 171–184, 2014.

32. S. Led, L. Azpilicueta, E. Aguirre, M. Martínez de Espronceda, L. Serrano, and F. Falcone, "Analysis and description of HOLTIN service provision for AECG monitoring in complex indoor environments," *Sensors*, 13(4): 4947–4960, 2013.

33. P. L. Iturri, J. A. Nazábal, L. Azpilicueta, P. Rodriguez, M. Beruete, C. Fernández-Valdivielso, and F. Falcone, "Impact of high power interference sources in planning and deployment of wireless sensor networks and devices in the 2.4GHz frequency band in heterogeneous environments," *Sensors*, 12(11): 15689–15708, 2012.

34. E. Aguirre, J. Arpón, L. Azpilicueta, S. de Miguel, V. Ramos, and F. Falcone, "Evaluation of electromagnetic dosimetry of wireless systems in complex indoor scenarios within body human interaction," *Progress in Electromagnetics Research B*, 43: 189–209, 2012.

35. R. J. Luebbers, "A heuristic UTD slope diffraction coefficient for rough lossy wedges," *IEEE Transactions on Antennas and Propagation*, 37: 206–211, 1989.

36. R. J. Luebbers, "Comparison of lossy wedge diffraction coefficients with application to mixed path propagation loss prediction," *IEEE Transactions on Antennas and Propagation*, 36: 1031–1034, 1988.

Biographical Sketches

Francisco Falcone received his MSc in telecommunications engineering (1999) and PhD in communications engineering (2005), both at the Universidad Pública de Navarra (UPNA), Navarra, Spain. From 1999 to 2000, he worked as a microwave commissioning engineer at Siemens-Italtel Málaga, Spain. From 2000 to 2008, he worked as a radio network engineer in Telefónica Móviles Pamplona, Spain. In 2009, he cofounded Tafco Metawireless, a spin-off devoted to complex electromagnetic (EM) analysis. From 2003 to 2009, he was also an assistant lecturer at the UPNA and then became an associate professor in 2009. His research area includes artificial electromagnetic media, complex electromagnetic scenarios, and wireless system analysis. He has over 300 contributions in journal and conference publications. He has been a recipient of the CST Best Paper Award in 2003 and 2005; Best PhD in 2006 awarded by the Colegio Oficial de Ingenieros de Telecomunicación, Madrid, Spain; Doctorate award 2004–2006 awarded by the UPNA; Juan Lopez de Peñalver Young Researcher Award 2010 awarded by the Royal Academy of Engineering of Spain; and Premio Talgo 2012 for Technological Innovation.

Leire Azpilicueta received her integrated degree in telecommunications engineering from the Universidad Pública de Navarra (UPNA), Navarra, Spain, in 2009. In 2011, she obtained a master of communications from the same university. She is currently pursuing her PhD degree in telecommunications engineering. In 2010, she worked as a radio engineer in the R&D department of RFID Osés Peralta, Navarra, Spain. Her research interests include radio propagation, mobile radio systems, ray tracing, and channel modeling.

José Javier Astráin received his MSc in telecommunications engineering (1999) and PhD in computer science (2004), both at the Universidad Pública de Navarra, Navarra, Spain, where he works as lecturer. His current research interests concern wireless sensor networks, distributed systems, and ontology-driven algorithms. He was awarded with the Premio Talgo 2012 for Technological Innovation.

Jesús Villadangos received his MS in physics from the Universidad del País Vasco, Bilbao, Spain, in 1991 and his PhD in communications engineering (1999) from the Universidad Pública de Navarra, Navarra, Spain, where he became associate professor in 2000. He has worked in several R&D projects supported by national and international entities. He is the coauthor of 3 Spanish patents and more than 20 international papers. He was awarded with the Premio Talgo 2012 for Technological Innovation. His research interests are software engineering, distributed algorithms, vehicular networks, and ontology-driven algorithms.

8

Network Coding

An Optimized Solution for Cognitive Radio Networks

MUHAMMAD ZUBAIR FAROOQI, SALMA
MALIK TABASSUM, MUBASHIR HUSAIN
REHMANI, AND YASIR SALEEM

Contents

8.1 Introduction

Network coding (NC) is a technique of transmitting data in an encoded and a decoded form. It is applied on the nodes of a network. It increases the throughput by sending more information in less packet transmission and stabilizes the network. In cognitive radio network (CRN), one of the main advantages of NC is that the transmission time of secondary users (SUs) is reduced [1]. NC is an emerging technique used in many types of networks. It makes the network robust to the packet loss using transmission control protocol. In order to amplify the throughput and robustness in CRNs, NC is a preeminent selection.

Wireless networks are suffering from many problems such as low throughput and dead spots. A variety of techniques were introduced in order to overcome these problems. NC is one of the latest and emerging techniques developed for enhancing the throughput and providing a minimum transmission rate over wireless networks. It is also used to achieve a minimum energy per bit for multicasting in wireless networks. For improvement in energy efficiency, the NC-based scheme has only polynomial time complexity, flouting through the non-deterministically polynomial (NP)-hardness barrier of the conventional routing approaches. Wang et al. [2] briefly discussed about the energy issues and efficiency of NC regarding energy consumption.

NC when applied to CRNs significantly enhances the performance of the network. The use of NC increases the spectrum availability for the SUs in CRNs by improving the estimation of primary users (PUs). A variety of algorithms based on NC has been developed in order to decrease the need of bandwidth in CRNs. NC increases the spectrum utilization for SUs by giving them opportunity to utilize the unused part of the spectrum owned by PUs. By contrast, using traditional techniques, the spectrum usage might be as low as 15%. Wang et al. [3] identified one aspect of CRN as spectrum shaping and the view of NC as a spectrum shaper. Using NC, a number of SUs may use the spectrum at the same time. Furthermore, NC can achieve a potentially lower energy consumption compared to the conventional routing schemes.

Some up-to-date literature on NC is found in [4–8]. Iqbal et al. [4] focussed on NC-aware routing protocols in wireless networks. The physical layer NC (PNC) in wireless networks is discussed in [5,6]. NC for distributed storage is discussed in [7]. Bassoli et al. [8] presented a survey on NC in a theoretical form by focusing on NC theory including information theory and matroid theory. All the aforementioned works focused on wireless networks in general. However, in this chapter, we provide a survey on NC schemes specifically designed for CRNs. To the best of our knowledge, this is the first work that provides such a comprehensive description of NC in the context of CRNs. In this chapter, our major focus will be on the cognitive radio aspect of NC and we recommend the readers to refer [5,8] for more details on NC in wireless networks.

In this chapter, we make the following contributions:

- We provide two simple illustrative examples that help the reader to understand the NC basics.
- We discuss how NC is used in different networks.
- We provide an in-depth discussion of NC in traditional wireless networks.
- We discuss the usage of NC in CRNs by focusing on how it can be beneficial for different types of networks and its classification and advantages.
- We highlight the issues, challenges, and future directions.

The remainder of this chapter is organized as follows: Section 8.2 discusses the NC basics. Section 8.3 discusses NC applied to different networks. Section 8.4 describes NC in traditional wireless networks. Section 8.5 discusses CRNs and NC. Section 8.6 presents open issues, challenges, and future directions. Finally, Section 8.7 concludes the chapter.

8.2 NC Basics

NC is a technique of sending data in an encoded and a decoded form. It is applied to the nodes receiving or sending data packets. This causes less transmission time and thus increases the throughput of the network to a large extent. It enhances the performance of the network by sending more data in limited amount of time. It is also used to handle incoming or outgoing data from a node.

Liew et al. [5] mentioned that PNC is industrialized in 2006 for applications in wireless networks. Its elementary idea is to exploit the mixing of signals that arise naturally. A weighted sum of signals is treated on receiver as an outcome of concurrent transmission at various transmitters. This sum is in the form of NC process by itself. It could also be transmuted in practice of NC. In fact, Liew et al. [5] gave an ephemeral conception of PNC, scrutinized a serious issue in PNC, and anticipated that PNC is not only for wireless networks. Analog NC (ANC) is a version of PNC that is also instigated in this chapter. Liew et al. [5] also attempted to spread the application of PNC to optical networks.

We now discuss some illustrative examples to make the reader understand the basics of NC.

Example 8.1

In Figure 8.1, there are two scenarios of NC. Figure 8.1a shows a multicast with two sources $S1$ and $S2$, two receivers X and Y, and two packets a and b. $S1$ and $S2$ want to transmit these packets holding binary information symbols. R is the intermediate node

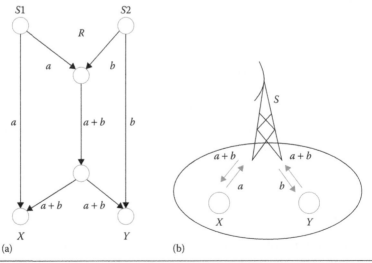

(a)   (b)

Figure 8.1 NC examples: (a) Multicast with two sources and two receivers; (b) wireless point-to-point communication.

that combines a and b and creates a new packet a_b where "$\_$" is the symbol for bitwise exclusive OR. If this is a coded network, R combines a and b; otherwise, R should produce two different packets a and b. Therefore, coding in the network augments the throughput because one transmission will produce less delay. Figure 8.1b shows wireless point-to-point communication in which S is the base station and the circle around it represents the range of base station; X and Y are nodes in the range of base station, but they cannot communicate directly to each other. If X wants to transmit packet a, then it will first send it to base station S, and if Y also sends packet b to S, then S generates new packet a_b and transmits it to both X and Y. In this manner, both examples show the prospective of coding procedures in network nodes.

Example 8.2

Figure 8.2 illustrates an example of ad hoc CRNs in which NC minimizes the number of broadcasts. In this figure, v is the relay node through which S_1 and S_2 relay their data to d_1 and d_2, respectively. Suppose S_1 and S_2 send two packets p_1 and p_2 to d_1 and d_2, respectively, and node d_1 comes to know about packet p_2 and node d_2 comes to know about packet p_1. However, d_1 and d_2 can receive their required packets if relay node v XORs (XOR is a logical operator which gives output 0 when both inputs are same and gives 1 when both inputs are different.) these two packets and transmits the two original packets to the destinations, respectively.

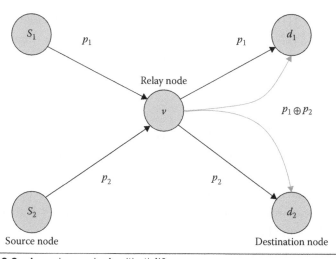

Figure 8.2 A sample example of multipath NC.

8.3 NC in Different Networks

We now describe different networks and how NC is used in these networks.

8.3.1 Orthogonal Frequency-Division Multiple Access Networks

Xu et al. [9] discussed the impacts of the orthogonal frequency-division multiple access (OFDMA) system's parameters on NC gain. They formulated the optimization frameworks and proposed channel-aware coding-aware resource allocation algorithms to exploit the network capacity in slow frequency. They also showed that the node's power and traffic patterns are the dependencies of NC gain.

8.3.2 Underwater Sensor Networks

Guo et al. [10] discussed that in the past, underwater sensor networks faced so many challenges such as the presence of high error probability, long propagation delays, and low acoustic bandwidth. They proposed the use of NC in underwater sensor networks and claimed that NC is a gifted solution for these types of problems in underwater sensor networks.

8.3.3 Vehicular Ad Hoc Networks

Park et al. [11] discussed NC in the context of vehicular ad hoc networks (VANETs). By using NC, they proposed the reliable dissemination of video streaming in case of emergency situations.

8.3.4 Optical Networks

In optical networks, light is used as a medium to transfer information. In this context, optical fiber is used in the network industry as a viable solution to transfer data with high data rates in core networks. How NC can be applied to optical networks is still an open issue and very less work has been done so far. For instance, Kamal and Mohandespour [12] used NC for optical networks.

8.3.5 Delay-Tolerant Networks

Ali et al. [13] proposed the features for the advancement of reliable transport's performance in delay-tolerant network (DTN)'s unicast and multicast flows. Their proposed scheme comprises random linear NC of packets.

8.3.6 Wireless Sensor Networks

NC is widely adopted in wireless sensor networks (WSNs). Table 8.1 shows the parameters used to evaluate NC in WSNs. Stefanovic et al. [14] used a number of encoded packets needed for successful decoding N as a parameter for NC evaluation. Wang et al. [2] evaluated NC using the remaining energy per bit and the remaining energy of cluster head. In [15], the parameters used for evaluation of NC are reliability, number of packets, traffic, energy consumption, and delay.

8.3.7 Wireless Relay Networks

In [16], many kinds of protocols were exposed with the detection of wireless relay network. NC performed its role by changing the wireless relay coding from *store and forward* to *store, process, and forward*.

8.4 NC in Traditional Wireless Networks

As discussed in section 8.1, NC is a latest technique which overcomes the problems of wireless networks by enhancing the throughput and providing a minimum transmission rate over wireless networks. Thus, Table 8.2 presents the parameters used to evaluate NC in traditional

Table 8.1 NC Evaluation Parameters in WSNs

PARAMETER NAME	REFERENCES
Number of encoded packets needed for successful decoding N	[14]
Remaining energy per node and remaining energy of cluster head	[2]
Reliability, number of packs, traffic, energy consumption, and delay	[15]

Table 8.2 NC Evaluation Parameters in Traditional Wireless Networks

PARAMETER NAME	REFERENCES
Protocol handshake duration	[30]
Retransmission rate	[30]
Average decoding delay	[30]
Throughput gain	[9]
End-to-end throughput	[9]
Normalized throughput	[9]
Full decoding percentage	[31]
Total cost	[31]
Node decoding percentage	[31]
Bit error rate	[16,32]
Capacity	[32]
Detection probability	[33]
Average number of transmissions	[34,35]
Transmission time	[36,37]
Throughput	[1,29,36,38–42]
Cumulative fraction of flows	[40]
MAC-independent opportunistic routing (MORE) throughput	[40]
Allocated redundancy	[42]
Cumulative fraction of flows	[43]
Bandwidth overhead	[43]
Signatures per second and latency	[43]
Quality degradation	[44]
Complexity	[8]
Storage per node	[7]
Transmission delay	[37]
Normalized energy consumption	[10]
Successful delivery ratio	[10]
Packet delivery ratio	[39,45]
End-to-end delay	[45]
Resource redundancy degree	[45]
Useful throughput ratio	[45]
Channel capacity	[16]
Average node transmission efficiency improvement (NTEI)	[28]
Coding + MAC gain	[29]
Confliction probability	[1]
Throughput per direction	[46]
Cumulative fraction of flows	[38]
Cumulative distribution	[39]
Packet delivery ratio	[39]
Localization error	[47]

(*Continued*)

Table 8.2 (Continued) NC Evaluation Parameters in Traditional Wireless Networks

PARAMETER NAME	REFERENCES
Average number of linearly dependent packets	[35]
Average broadcast time	[35]
Average number of collisions	[35]
Number of group of pictures (GOPs)	[44]
Percentage of optimal redundancy	[44]
Average peak signal-to-noise ratio (PSNR)	[44]

ESNR, Effective signal-to-noise ratio, ICI, Inter-cell interference, MAC, Media access control, PER, Packet error rate.

wireless networks. In traditional wireless networks, PNC has been studied extensively [17–26].

Mehta and Narmawala [27] concluded that nowadays the use of multimedia applications over wireless networks is at its peak, but there is a huge amount of packets loss and delay in transmission, as the available bandwidth for wireless networks fails to meet the requirements. They suggested that NC may be applied in different layers in order to increase the throughput.

Chi et al. [28] proposed a network coding-based packet forwarding architecture known as 'COPE' for throughput maximization. COPE is based on NC-based packet forwarding architecture. It categorizes packets in small- or large-size simulated queues and then scrutinizes only head packets to bound packet reordering. It familiarizes limited packet reordering when the order of packets arrival is different from the departure. It improves transmission efficiency by around 30%. This improvement is supposed to be further increased by 45% by using a flow-oriented architecture.

Iqbal et al. [4] discussed the basic concepts of applications of the state-of-the-art NC for wireless ad hoc networks in the perspective of routing and recognized demarcation among NC-aware and NC-based routing methods in wireless ad hoc networks. They emphasized on the existing NC-aware routing protocols by providing different assessments and its advantages over traditional routing. In [29], relay-aided NC (RANC) is discussed by developing the physical layer multirate ability in multihop wireless networks.

In [45], RANC is discussed which shows an increase in performance gain of NC by developing the physical layer multirate ability in multihop wireless networks where nodes are acceptable to broadcast at diverse rates according to the channel state. While relayed packets may frequently pass on at a high rate, the liberation of the total packets essential for decoding to each end node can be much earlier than direct transmission. Zhang et al. [45] examined transaction in expanding performance of RANC and systematically offered the solution by dividing the original design problem into subproblems: flow partition and scheduling problems. To enhance the bandwidth utilization in wireless networks by easy operation such as bitwise XOR, NC develops the broadcast nature of wireless medium. The desired operation of NC does not need any advancement of hardware. NC was anticipated by Ahlswede [48] for multicast in wired networks. In COPE proposed by Katti et al. [49], the performance growth in terms of throughput and effectiveness appears from coding and listening and there is also a main role of the number of flows in NC. In COPE, except for end node, the packet sent by a starting node should be effectively overheard by all other nodes, and this node may be the blockage of COPE performance. Developing the physical layer multirate capability allows to fight the crash of poor channel state on the recital of NC. In contrast to COPE, RANC itself broadcasts its native packet above a short range and enhances the performance of NC in which the node has bad channel condition among its far neighbors. In this chapter, Zhang et al. [45] also develops the coding structure for RANC. Substitution in performance gain of RANC is also discussed. RANC protocol is established by separating the original dilemma into two: flow partition and scheduling problems. This supports to diminish the global cost. Replication is used to assess this strategy that RANC can expressively outclass COPE in terms of the throughput of NC.

In WSNs, neighboring sensor nodes have connections of data. The process of sensors data compression in order to provide energy efficiency in WSNs is called distributed source coding (DSC). Using DSC, a network architecture extensively manipulates the compression effectiveness. Dynamic clustering scheme is discussed in [50] in assessment to previous schemes, which signify that this scheme has more efficiency than static clustering schemes.

WSNs are occasional systems based on the combined effort of various sensor nodes to monitor a physical phenomenon and involve spatially dense sensor operation to attain acceptable exposure. In Slepian–Wolf theorem, two connected sources can be encoded, though just encoders have individual admittance to the two supplies, provided that both encoded streams are accessible at the decoder. The partition method in WSNs with DSC is not useful in applications due to manufacturing and financial issues. While winding up, an analytical framework to sculpt the dilemma of partition and dynamic clustering scheme is offered to resolve the problem. It can panel the network vigorously adaptive to the topology and connections of the network with better density performance.

8.5 CRNs and NC

In this section, we first give an overview of CRNs and then discuss NC schemes in CRNs.

8.5.1 Cognitive Radio Networks

CRN is an emerging field and has recently gained a lot of attention from the networking research community [50–54]. This is primarily due to (1) availability of limited spectrum, (2) fixed spectrum assignment policy, and (3) inefficiency in spectrum usage.

CRNs are composed of two types of users: SUs and PUs. The PUs have higher priority over the licensed channels, whereas SUs have lower priority. SUs use the licensed channels opportunistically and are required to vacate the licensed channels as soon as PUs arrive over them.

We now discuss NC schemes proposed for CRNs.

8.5.2 NC in CRNs

Shu et al. [55] proposed NC-aware channel allocation and routing in CRNs by considering the availability of maximum number of channels. Their proposed solution increases the throughput by distributing the channel and link rate at different stages.

Hao et al. [56] anticipated that in CRNs, the need of bandwidth can be amplified by using distributed cooperative spectrum based on the network code. They investigated and entitled the algorithm of the distributed scheme based on the network code, and due to this, the throughput increases outwardly. An approach called gossiping updates for efficient spectrum sensing (GUESS) is anticipated to shrink the protocol overhead. In this work, transmitting data is diminished by NC.

In [1], SUs utilize NC for data transmissions in CRNs. Wang et al. [33] described the use of spectrum sensing that it is used to identify the spectrum holes and detect the presence of PUs. NC is applied to different SUs in order to enhance the spectrum efficiency of cognitive frequency bands.

Figure 8.3 shows an example of butterfly network, in which traditional routing and NC-based routing are presented. S is the source and E and F are the destination nodes. Let the source multicast b_1 and b_2 two unit data, E and F, and the links are SA, AC, CD, DE, and so on. In Figure 8.3a, E obtains only b_1 in one unit time since only one unit data can be broadcast per unit time in excess of link CD, so an utmost broadcast of multicast cannot be obtained. However, if we want to transmit $b_1 \oplus b_2$ to both E and F at the same time, we establish NC

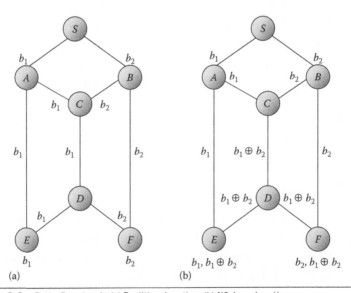

Figure 8.3 Butterfly network: (a) Traditional routing; (b) NC-based routing.

in traditional routing. In this technique, C can XOR the incoming data and broadcast the coded data $b_1 \oplus b_2$ on link CD, and finally, it will decode this data correspondingly. Therefore, E and F can receive b_1 and b_2 in one unit time, through which maximum multicast transmission capacity can be achieved.

8.5.3 Classification of NC Schemes

Figure 8.4 shows the classification of NC [49,57–68]. There are three types of NC schemes available: random NC, vector NC, and linear NC. Linear NC is further classified based on coding with field size equal to one and greater than one. The linear coding with field size equal to one is further classified into two categories, which are then categorized into subcategories. The same is the case with linear coding having field size greater than one. The linear NC in multiple paths is illustrated in Figure 8.5.

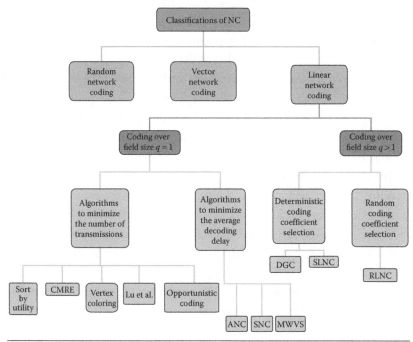

Figure 8.4 Classification of NC. CMRE, cache-based multicast retransmission encoding; DGC, dynamic general-coding; MWVS, maximum weight vertex search; RLNC, random linear network coding; SLNC, sparse linear network coding.

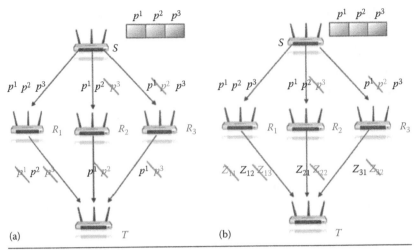

Figure 8.5 (a,b) Random linear NC in multiple paths.

In Figure 8.5, S is the source node and T is the destination, and there are three link-disjoint paths between the source and the destination. The source S sends packets to relay nodes R_1, R_2, and R_3 simultaneously. In Figure 8.5a, S sends packets p^1, p^2, and p^3 to R_1, R_2, and R_3 one by one. Sometimes packet loss occurs; therefore, suppose that R_1 receives all three packets, whereas R_2 and R_3 fail to receive p^2 and p^3, respectively. However, devoid of NC, R_1 forwards p^1, p^2, and p^3; R_2 forwards p^1 and p^2; and R_3 forwards p^1 and p^3 to destination T. Suppose packets p^1 and p^3 sent from R_1, packet p^2 from R_2, and packet p^3 sent from R_3 adopted wrong paths and become lost. Therefore, as a result, only p^1 and p^2 are received at the destination node T. In Figure 8.5b, linear NC for transmitting packets is applied. In this strategy, relay nodes produce the output packet Z_{ij} by linearly merging the received packets. Here, R_1 receives p^1, p^2, and p^3, and produces encoded packets Z_{11}, Z_{12}, and Z_{12}. Subsequently encoded packets Z_{21} and Z_{22} are produced by R_2, and Z_{31} and Z_{32} are produced by R_3. Z_{11}, Z_{13}, Z_{22}, and Z_{32} are lost on their way to T. Therefore, at the destination T receives Z_{12}, Z_{21}, and Z_{31}. T can decode the real packets because these packets are the permutations of linear independent packets by solving following equations:

$$Z_{12} = \alpha_1 p^1 + \alpha_2 p^2 + \alpha_3 p^3 \tag{8.1}$$

$$Z_{21} = \beta_1 p^1 + \beta_2 p^2 \tag{8.2}$$

$$Z_{31} = \gamma_1 p^1 + \gamma_3 p^3 \tag{8.3}$$

where:

$(\alpha_1, \alpha_2, \alpha_3)$, (β_1, β_2), and (γ_1, γ_3) are the encoding vectors generated by R_1, R_2, and R_3, respectively

The parameters used to evaluate NC in CRNs are described in Table 8.3.

Table 8.3 Parameters Used to Evaluate NC in CRNs

PARAMETER NAME	REFERENCES
Symbol error rate	[69]
Average throughput	[50]
Dissemination delay	[70]
Outage probability	[71]
Broadcast cost	[72]
Additional gain on SU's throughput from backoff-based adaptive sensing over random channel sensing	[3]
SU throughput	[3]
PU idle probability	[3]
Probability of success	[73]
Step time overhead	[73]
Normalized packet delay of a message	[74]
PER performance	[75]
Average ICI received by the central cell	[76]
SU goodput	[51]
Spectral efficiency	[77]
Luby transform (LT) decoding error probability	[77]
Spectral efficiency versus Forward Error Correction (FEC)	[78]
Error probability	[79]
Diversity gain	[79]
Cooperative ENSR	[80]
Bit error rate	[81]
Mesh number	[82]
Total bits	[82]
Average capacity	[83]
Uniform capacity	[83]
Batch delay	[84,85]
Decision time	[54]
Energy dissipation	[54]
Throughput	[55,56]
Spectrum efficiency	[86]
Probability of successful control information	[86]

8.5.4 Benefits of NC in CRNs

Wang et al. [2] discussed the efficiency of NC regarding energy issues, which is of great advantageous. The idea of NC was first floated by Ahlswede et al. in 2000. NC allows the node to code the incoming data as well as the transmitting data. It also improves the broadcast efficiency by merging data for data-based WSNs. Wang et al. [2] discussed the benefits of NC applied in WSNs. By performing analytical and simulation evaluation, it shows that NC can significantly lessen the time of information exchange. NC is more appropriate in wireless networks such as wireless ad hoc networks, wireless mesh networks, and CRNs. One of the major assistance in WSNs is energy saving. Other benefits are throughput enhancement and delay minimization. NC also maintains the network load balance and increases the network security.

8.5.5 NC-Based Routing in CRNs

Tang and Liu [34] conferred about the routing scheme with NC for CRNs. We have to propose a routing protocol if we are given some cognitive radio nodes with the geographical locations. This routing protocol should minimize the average number of transmission per packet. The packets in this protocol use temporary nodes hop by hop, until they reach their destinations. Performance can also be enhanced by using the algorithm they proposed. Simulation shows that compared to DSR protocol, this algorithm gains much better performance in terms of lower number of transmission.

8.5.6 Secure Transmission of NC

He and Chen [87] examined the research performed on secure transmission on NC in wireless networks. Security issue is the key problem in wireless networks; to overcome this, NC is used. As there is so much pollution in networks, a scheme was proposed named as secure XOR NC that filters the polluted message. Secure NC scheme uses different coding strategies and different coding methods based on current network safety situation. Adaptive NC scheme is more appropriate for the multifaceted and unstable practical wireless network environment. Kamal and Mohandespour [12] worked on to provide the effective usage of resources and preemptive protection by exploiting NC.

The construction of rectilinear amalgamations of packets, received at diverse inputs links between nodes, was improved by the use of NC in order to improve the capacity. They conveyed by addressing NC-based protection of bidirectional unicast connection and enlightened the use of p-cycles. Diversity coding is also discussed in which mishmashes are molded at special nodes.

8.5.7 NC for Commercial Devices

Paramanathan et al. [88] described the ability of NC that it has the potential to simplify the design with higher throughput and lower energy expenditures. They argued that NC complexity is not an issue for current mobile devices even without hardware acceleration. They provided two design styles of NC, which are intersession and intrasession NC using commercial platforms, supplying real-life measurement results on commercial devices.

8.5.8 NC in Multichannel CRN

It is critical for SUs to have efficient data transmission in order to communicate with each other [85]. Channel fading is the main issue during transmission even if they have full access to that channel. Under practical fading channel conditions, Zheng et al. [85] proposed the random linear coded scheme for an efficient data transmission in multichannel CRNs. They analyzed the performances of two multichannel automatic repeat request (ARQ)-based schemes. Simulations shows that this coded scheme outperforms the ARQ-based scheme in terms of batch delay transmission.

8.6 Issues, Challenges, and Future Directions of NC in CRNs

There are plenty of issues, challenges, and future directions of NC in CRNs. We now highlight the main issues, challenges, and future research directions in Sections 8.6.1 through 8.6.5.

8.6.1 PU Activity

PU activity is the foremost and intrinsic challenge of CRN [89,90]. SUs operate on licensed channels of PUs; therefore, whenever a

PU arrives on licensed channel, then SU has to vacate that licensed channel if it is currently utilizing it because PU has the highest priority to access the channel. The dynamicity of PU activity imposes a big challenge on NC in CRNs. For instance, when the available channel set has diversity, NC scheme will be difficult to implement.

8.6.2 Multiple Channel Availability

Traditional wireless networks and NC for these networks operate on a single channel. But in CRNs, there are multiple channels available and SUs exploit them opportunistically whenever they are available. Since traditional algorithms for NC are developed for a single channel, there is a need to investigate NC algorithms to provide support for multiple channels.

8.6.3 Heterogeneous Channels

Since there are multiple channels available in CRNs, each channel has different characteristics such as bandwidth, data rate, and bit error rate. Therefore, NC has to deal with heterogeneous channels in CRNs, which is also an important challenge that needs investigation.

8.6.4 Contention among SUs

Another challenge may occur when one SU uses some part of spectrum and another SU also accesses the same part of spectrum at the same time. Thus, there might be interference among them. Therefore, it is necessary to develop protocols for channel access among SUs in NC for CRNs.

8.6.5 Spectrum Sensing

Another challenge is to sense the spectrum for SU when PU is not using it. Incorrect spectrum sensing results in either false alarm or miss detection. In false alarm, there is no PU activity, but the spectrum sensing detects that there is PU activity. In this case, there is an opportunity to exploit a spectrum band, but due to false detection, this opportunity is lost. In miss detection, there is PU activity but spectrum sensing detects that there is no PU activity, so SU keeps on using the spectrum band that results in interference among PUs and SUs. This is a very serious issue because in CRNs, we can compromise on losing an opportunity of spectrum access, but we cannot compromise on interference with PUs. Therefore, it also needs further investigation. Sometimes due to incorrect spectrum sensing, there was a spectrum conflict between PUs and SUs, and communication becomes interrupted.

8.7 Conclusion

In this chapter, we presented a comprehensive survey of NC evolving from traditional wireless networks to emerging CRNs. To the best of our knowledge, this is the first novel work that provides a comprehensive survey on taxonomy of NC for CRNs and also provides various examples, benefits, and different mechanisms. In this chapter, first the introduction and basics of NC have been presented with different examples. Moreover, what work has been done in NC and how does it work have been explained in detail. Subsequently, applications of NC have been discussed. After this, NC has been described for different types of networks such as OFDMA networks, underwater sensor networks, vehicular ad hoc networks, optical networks, and WSNs, followed by traditional wireless networks. Then, NC for CRNs has been described followed by its classification, benefits, and different mechanisms. Finally, issues, challenges, and future directions of CRNs in NC have been highlighted.

References

1. W. Mu, L. Ma, X. Z. Tan, L. Li, and L. Liu, Network coding for cognitive radio systems, in *IEEE International Conference on Computer Science and Network Technology*, 1, December 24–26, Harbin, People's Republic of China (2011), pp. 228–231.
2. S. Wang, X. Gao, and L. Zhuo, Survey of network coding and its benefits in energy saving over wireless sensor networks, in *7th International Conference on Information, Communications and Signal Processing*, December 8–10, Macau, People's Republic of China (2009).

3. S. Wang, Y. E. Sagduyu, J. Zhang, and J. H. Li, Spectrum shaping via network coding in cognitive radio networks, in *Proceedings of the IEEE INFOCOM*, April 10–15, Shanghai, People's Republic of China (2011).
4. M. A. Iqbal, B. Dai, B. Huang, A. Hassan, and S. Yu, Survey of network coding aware-routing protocols in wireless networks, in *Journal of Network and Computer Applications*, 34 (2011) 1956–1970.
5. S. C. Liew, S. Zhang, and L. Lu, Physical-layer network coding: Tutorial, survey, and beyond, in *Physical Communication*, 6 (2013) 39.
6. P. Hu and M. Ibnkahla, A survey of physical-layer network coding in wireless networks, in *25th Biennial Symposium on Communications*, May 12–14, Kingston, ON, Canada (2010).
7. A. G. Dimakis, K. Ramchandran, Y. Wu, and C. Suh, A survey on network codes for distributed storage, in *IEEE Proceedings*, 99 (2011) 14.
8. R. Bassoli, H. Marques, J. Rodriguez, K. W. Shum, and R. Tafazolli, Network coding theory: A survey, in *IEEE Communications Surveys & Tutorials*, 15 (2013) 29.
9. Y. Xu, J. C. Lui, and D.-M. Chiu, Analysis and scheduling of practical network coding in OFDMA relay networks, in *Computer Networks*, 53 (2009) 20.
10. Z. Guo, B. Wang, P. Xie, W. Zenga, and J.-H. Cui, Efficient error recovery with network coding in underwater sensor networks, in *Ad Hoc Networks*, 7 (2009) 12.
11. J.-S. Park, U. Lee, and M. Gerla, Vehicular communications: Emergency video streams and network coding, in *Journal of Internet Services and Applications*, 1 (2010) 12.
12. A. E. Kamal and M. Mohandespour, Network coding based protection, in *Optical Switching and Networking*, 11 (2013) 13.
13. A. Ali, M. Panda, T. Chahed, and E. Altman, Improving the transport performance in delay tolerant networks by random linear network coding and global acknowledgments, in *Ad Hoc Networks*, 11 (2013) 21.
14. C. Stefanovic, D. Vukobratovic, V. Stankovic, and R. Fantacci, Packet-centric approach to distributed sparse-graph coding in wireless ad hoc networks, in *Ad Hoc Networks*, 11 (2013) 15.
15. L. Miao, K. Djouani, A. Kurien, and G. Noel, Network coding and competitive approach for gradient based routing in wireless sensor networks, in *Ad Hoc Networks*, 10 (2012) 19.
16. A. H. Mohammed, B. Dai, B. Huang, M. Azhar, G. Xu, P. Qin, and S. Yu, A survey and tutorial of wireless relay network protocols based on network coding, in *Journal of Network and Computer Applications*, 36 (2013) 18.
17. D. Haccoun, M. M'edard, E. Soljanin, and R. W. Yeung, Introduction to the special issue on network coding and its applications to wireless communications, in *Physical Communication*, 6 (2013) 2.
18. K. Xu, Z. Lv, Y. Xu, D. Zhang, X. Zhong, and W. Liang, Joint physical network coding and LDPC decoding for two way wireless relaying, in *Physical Communication*, 6 (2013) 5.

19. J. Manssour, T. ur Rehman Ahsin, S. B. Slimane, and A. Osseiran, Analysis and performance of network decoding strategies for cooperative network coding, in *Physical Communication*, 6 (2013) 14.

20. J. Li, E. Lu, and I.-T. Lu, Joint MMSE designs for analog network coding and different MIMO relaying schemes: A unified approach and performance benchmarks, in *Physical Communication*, 6 (2013) 12.

21. R. Gummadi, L. Massoulie, and R. Sreenivas, The role of coding in the choice between routing and coding for wireless unicast, in *Physical Communication*, 6 (2013) 12.

22. E. Drinea, L. Keller, and C. Fragouli, Real-time delay with network coding and feedback, in *Physical Communication*, 6 (2013) 14.

23. Y. Li, E. Soljanin, and P. Spasojevic, Three schemes for wireless coded broadcast to heterogeneous users, in *Physical Communication*, 6 (2013) 10.

24. K. Cai, K. Letaief, P. Fan, and R. Feng, On the solvability of single rate 2-pair networks—A cut-based characterization, in *Physical Communication*, 6 (2013) 10.

25. N. Karamchandani, L. Keller, C. Fragouli, and M. Franceschetti, Function computation via subspace coding, in *Physical Communication*, 6 (2013) 8.

26. P. Sattari, C. Fragouli, and A. Markopoulou, Active topology inference using network coding, in *Physical Communication*, 6 (2013) 22.

27. T. Mehta and Z. Narmawala, Survey on multimedia transmission using network coding over wireless networks, in *IEEE Nirma University International Conference on Engineering*, December 8–10, Ahmedabad, Gujarat, India (2011), p. 6.

28. K. Chi, X. Jiang, B. Ye, and Y. Li, Flow-oriented network coding architecture for multihop wireless networks, in *Computer Networks*, 55 (2011) 18.

29. J. Qureshi, C. H. Foh, and J. Cai, Online XOR packet coding: Efficient single-hop wireless multicasting with low decoding delay, in *Computer Communications*, 39 (2013) 15.

30. H. Alnuweiri, M. Rebai, and R. Beraldi, Network-coding based event diffusion for wireless networks using semi-broadcasting, in *Computer Networks*, 10 (2012) 15.

31. W. Xuanli, L. Mingxin, J. Laiwei, and S. Lukuan, Distributed cooperative spectrum sensing strategy based on the joint use of network coding and transform-domain processing, in *7th International ICST Conference on Communications and Networking in China*, August 8–10, Kunming, People's Republic of China (2012).

32. D. Bing, Z. Jun, and L. JiangHua, Cooperation via wireless network coding, in *5th International ICST Conference on Communications and Networking in China*, August 25–27, Beijing, People's Republic of China (2010).

33. X. Wang, W. Chen, and Z. Cao, A rateless coding based multi-relay cooperative transmission scheme for cognitive radio networks, in *Global Telecommunications Conference*, November 30–December 4, Honolulu, HI (2009).

34. X. Tang and Q. Liu, Network coding based geographical opportunistic routing for ad hoc cognitive radio networks, in *Global Communications Conference Workshops*, December 3–7, Anaheim, CA (2012).

35. Z. Yang, M. Li, and W. Lou, R-code: Network coding-based reliable broadcast in wireless mesh networks, in *Global Telecommunications Conference*, November 30–December 4, Honolulu, HI (2009).

36. Y. Peng, Q. Song, Y. Yu, and F. Wang, Fault-tolerant routing mechanism based on network coding in wireless mesh networks, in *Journal of Network and Computer Applications*, 37 (2013) 14.

37. K. D. A. Boubacar, Z. Shi-hong, A. Saley, M. Yi-hui, and C. Shi-duan, Cooperative file sharing mechanism with network coding in wireless mesh networks, in *The Journal of China Universities of Posts and Telecommunications*, 16 (2009) 8.

38. J. Dong, R. Curtmola, and C. Nita-Rotaru, Secure network coding for wireless mesh networks: Threats, challenges, and directions, in *Computer Communications*, 32 (2009) 19.

39. S. Chieochan and E. Hossain, Network coding for unicast in a WiFi hotspot: Promises, challenges, and testbed implementation, in *Computer Networks*, 56 (2012) 18.

40. A. Newell, J. Dong, and C. Nita, On the practicality of cryptographic defenses against pollution attacks in wireless network coding, in *ACM Computing Surveys*, 45 (2013) 26.

41. J. Dong, R. Curtmola, C. Nita-Rotaru, and D. K. Y. Yau, Pollution attacks and defenses in wireless inter-flow network coding systems, in *Wireless Network Coding Conference*, June 21, Boston, MA, (2010).

42. H. Wang, S. Xiao, and C.-C. J. Kuo, Random linear network coding with ladder-shaped global coding matrix for robust video transmission, in *Journal of Visual Communication and Image Representation*, 22 (2011) 10.

43. J. Dong, R. Curtmola, and C. N. Rotaru, Practical defenses against pollution attacks in wireless network coding, in *ACM Transactions on Information and System Security*, 14 (2011) 31.

44. H. Wang and C.-C. J. Kuo, Robust video multicast with joint network coding and video interleaving, in *Journal of Visual Communication and Image Representation*, 21 (2010) 12.

45. X. Zhang, H. Zhu, and J. Zhang, Ranc, relay-aided network coding in multihop wireless networks, in *Computer Communications*, 32 (2009) 974–984.

46. L. Lu, T. Wanga, S. C. Liewa, and S. Zhang, Implementation of physical layer network coding, in *Elsevier, Physical Communication* (2011) 14.

47. Z. Li and W. Wang, Node localization through physical layer network coding: Bootstrap, security, and accuracy, in *Elsevier, Ad Hoc Networks*, 10 (2012) 11.

48. R. Ahlswede, N. Cai, S. Y. R. Li, and R. W. Yeung, Network information flow. in *IEEE Transactions on Information Theory* (2000), 46(4), pp. 1204–1216. http://en.wikipedia.org/wiki/Rudolf_Ahlswede.

49. S. Katti, H. Rahul, W. Hu, D. Katabi, M. Medard, and J. Crowcroft, XORs in the air: Practical wireless network coding, in *IEEE/ACM Transactions on Networking*, 16 (2008) 14.

50. J. Jin, H. Xu, and B. Li, Multicast scheduling with cooperation and network coding in cognitive radio networks, in *IEEE INFOCOM* Piscataway, NJ (2009) p. 9.

51. A. Asterjadhi, N. Baldo, and M. Zorzi, A distributed network coded control channel for multihop cognitive radio networks, in *IEEE Networks*, 23 (2009) 7.

52. I. F. Akyildiz, W.-Y. Lee, and K. R. Chowdhury, CRAHNs: Cognitive radio ad hoc networks, in *Ad Hoc Networks*, 7 (2009) 27.

53. Z. Wang, Y. E. Sagduyu, J. H. Li, and J. Zhang, Capacity and delay scaling laws for cognitive radio networks with routing and network coding, in *IEEE Networks, Military Communications Conference*, October 31-November 3, San Jose, CA (2010), p. 6.

54. V. Abrol and P. Sharma, Sector coded transmission protocol for cognitive radio networks, in *International Conference on Communication, Information & Computing Technology*, October 19–20, Mumbai, Maharashtra, India (2012).

55. Z. Shu, J. Zhou, Y. L. Yang, H. Sharif, and Y. Qian, Network coding-aware channel allocation and routing in cognitive radio networks, in *IEEE Global Communications Conference*, December 3–7, Anaheim, CA (2012), p. 6.

56. Z.-H. Hao, Y.-X. Tang, and J.-X. Xia, A distributed cooperative spectrum sensing based on network code in cognitive radios, in *Apperceiving Computing and Intelligence Analysis* (2009) 5.

57. Y. E. Sagduyu and A. Ephremides, On network coding for stable multicast communication, in *IEEE MILCOM*, October 29–31, Orlando, FL (2007).

58. J. Heide, M. V. Pedersen, F. H. P. Fitzek, and T. Larsen, Network coding for mobile devices—Systematic binary random rateless codes, in *IEEE ICC Workshops*, June 14–18, Dresden, Germany (2009).

59. C. W. Sung, K. W. Shum, and H. Y. Kwan, On the sparsity of a linear network code for broadcast systems with feedback, in *IEEE NetCod*, July 25–27, Beijing, People's Republic of China (2011).

60. H. Y. Kwan, K. W. Shum, and C. W. Sung, Generation of innovative and sparse encoding vectors for broadcast systems with feedback, in *IEEE ISIT*, May 24, St. Petersburg, Russia (2011).

61. K. Chi, X. Jiang, and S. Horiguchi, Network coding-based reliable multicast in wireless networks, in *Computer Networks*, 54 (2010) 4.

62. J. Barros, R. A. Costa, D. Munaretto, and J. Widmer, Effective delay control in online network coding, in *IEEE INFOCOM*, April 19–25, Rio de Janeiro, Brazil (2009).

63. P. Sadeghi, D. Traskov, and R. Koetter, Adaptive network coding for broadcast channels, in *IEEE Workshop on Network Coding, Theory and Applications*, June 15–16, Lausanne, Switzerland (2009).

64. S. Sorour and S. Valaee, Minimum broadcast decoding delay for generalized instantly decodable network coding, in *IEEE Global Communications Conference*, December 6–10, Miami, FL (2010).

65. C. Zhan, Y. Xu, J. Wang, and V. Lee, Reliable multicast in wireless networks using network coding, in *IEEE MASS*, October 12–15, Macau, People's Republic of China (2009).

66. E. Rozner, A. Padmanabha, L. Qiu, Y. Mehta, and E. M. Jafry, Efficient retransmission scheme for wireless LANs, in *ACM CoNEXT*, New York (2007).

67. L. Lu, M. Xiao, M. Skoglund, L. Rasmussen, G. Wu, and S. Li, Efficient network coding for wireless broadcasting, in *IEEE Wireless Communications and Networking Conference*, April 18–21, Sydney, Australia (2010).

68. W. Fang, F. Liu, and Z. Liu, Reliable broadcast transmission in wireless networks based on network coding, in *IEEE INFOCOM Computer Communications Workshop*, April 10–15, Beijing, People's Republic of China (2011).

69. M. H. Islam, Y.-C. Liang, and R. Zhang, Robust precoding for orthogonal space-time block coded MIMO cognitive radio networks, in *IEEE 10th Workshop on Signal Processing Advances in Wireless Communications*, June 21–24, Perugia, Italy (2009).

70. A. Asterjadhi and M. Zorzi, Jenna: A jamming evasive network-coding neighbor-discovery algorithm for cognitive radio networks, in *IEEE Wireless Communications*, 17 (2010) 6.

71. V. A. Bohara, S. H. Ting, Y. Han, and A. Pandharipande, Interference-free overlay cognitive radio network based on cooperative space time coding, in *Proceedings of the 5th International Conference on Cognitive Radio Oriented Wireless Networks & Communications*, June 9–11, Cannes, France (2010).

72. Y. Liu, Z. Feng, and P. Zhang, A novel ARQ scheme based on network coding theory in cognitive radio networks, in *IEEE International Conference, Wireless Information Technology and Systems*, August 28-September 3, Honolulu, HI (2010).

73. S. H. Alnabelsi, A. E. Kama, and T. H. Jawadwala, Uplink channel assignment in cognitive radio WMNs using physical layer network coding, in *IEEE International Conference on Communications* (2011).

74. B. Shahrasbi and N. Rahnavard, Rateless-coding-based cooperative cognitive radio networks: Design and analysis, in *8th Annual IEEE Communications Society Conference on Sensor, Mesh and Ad Hoc Communications and Networks*, June 27–30, Salt Lake City, UT (2011).

75. T. Khomyat, P. Uthansakul, and M. Uthansakul, Hybrid-MIMO receiver with both space-time coding and spatial multiplexing detections for cognitive radio networks, in *International Symposium on Intelligent Signal Processing and Communications Systems*, December 7–9, Chiang Mai, Thailand (2011).

76. K. Yang, W. Xu, S. Li, and J. Lin, A distributed multiple description coding multicast resource allocation scheme in OFDM-based cognitive radio networks, in *Wireless Communications and Networking Conference*, April 7–10, Shanghai, People's Republic of China (2013).

77. H. Kushwaha, Y. Xing, R. Chandramouli, and H. Heffes, Reliable multimedia transmission over cognitive radio networks using fountain codes, in *IEEE Proceedings*, 96 (2008) 11.

78. A. Chaoub and E. Ibn-Elhaj, Multiple description coding for cognitive radio networks under secondary collision errors, in *16th IEEE*

Mediterranean on Electrotechnical Conference, March 25–28, Yasmine Hammamet, Tunisia (2012).

79. C. Wang, M. Xiao, and L. Rasmussen, Performance analysis of coded secondary relaying in overlay cognitive radio networks, in *Wireless Communications and Networking Conference*, April 1–4, Shanghai, People's Republic of China (2012).

80. I. Stupia, L. Vandendorpe, R. Andreotti, and V. Lottici, A game theoretical approach for coded cooperation in cognitive radio networks, in *5th International Symposium on Communications Control and Signal Processing*, May 2–4, Rome, Italy (2012).

81. R. E. Bardan, E. Masazade, O. Ozdemir, and P. K. Varshney, Performance of permutation trellis codes in cognitive radio networks, in *Sarnoff Symposium*, May 21–22, Newark, NJ (2012).

82. Q. Zhang, Z. Feng, and P. Zhang, Efficient coding scheme for broadcast cognitive pilot channel in cognitive radio networks, in *Vehicular Technology Conference*, May 6–9, Yokohama, Japan (2012).

83. Y. Yang, and S. Aissa, Cross-layer combining of information-guided transmission with network coding relaying for multiuser cognitive radio systems, in *IEEE Wireless Communications Letters*, 2 (2013) 4.

84. C. Zheng, E. Dutkiewicz, R. P. Liu, R. Vesilo, and Z. Zhou, Efficient network coding transmission in 2-hop multi-channel cognitive radio networks, in *International Symposium on Communications and Information Technologies*, October 2–5, Gold Coast, QLD, Australia (2012).

85. C. Zheng, E. Dutkiewicz, R. P. Liu, R. Vesilo, and Z. Zhou, Efficient data transmission with random linear coding in multi-channel cognitive radio networks, in *Wireless Communications and Networking Conference*, April 7–10, Shanghai, People's Republic of China (2013).

86. L. Yang, F. Zhi-yong, and Z. Ping, Optimized in-band control channel with channel selection scheduling and network coding in distributed cognitive radio networks, in *The Journal of China Universities of Posts and Telecommunications*, 19 (2012) 9.

87. M. He and L. Chen, Survey on secure transmission of network coding in wireless networks, in *International Conference on Computer Science & Service System*, August 11–13, Nanjing, People's Republic of China (2012).

88. A. Paramanathan, M. V. Pedersen, D. E. Lucani, and F. H. P. Fitzek, Lean and mean: Network coding for commercial devices, in *IEEE Wireless Communications*, 20 (2013) 8.

89. M. H. Rehmani, A. C. Viana, H. Khalife, and S. Fdida, SURF: A distributed channel selection strategy for data dissemination in multi-hop cognitive radio networks, in *Elsevier Computer Communications Journal*, 36(10–11) (2013) 1172–1185.

90. M. H. Rehmani, A. C. Viana, H. Khalife, and S. Fdida, Activity pattern impact of primary radio nodes on channel selection strategies, in *Proceedings of the 4th International Workshop on Cognitive Radio and Advanced Spectrum Management in Conjunction with ISABEL*, Barcelona, Catalonia, Spain, October 26–29, 2011.

Biographical Sketches

Muhammad Zubair Farooqi is currently doing BS in computer science at COMSATS Institute of Information Technology, Wah Cantt, Pakistan. His research interests include cognitive radio networks, network security and protection, and wireless sensor networks.

Salma Malik Tabassum is currently doing BS in computer science at COMSATS Institute of Information Technology, Wah Cantt, Pakistan. Her research interests include cognitive radio networks, network security and protection, and wireless sensor networks.

Mubashir Husain Rehmani obtained his BE from Mehran UET, Jamshoro, Pakistan, in 2004. He received his MS and PhD from the University of Paris XI and the University of Paris VI, Paris, France, in 2008 and 2011, respectively. He was a postdoctoral fellow at the University of Paris Est, Paris, France, in 2012. He is an assistant professor at COMSATS Institute of Information Technology, Wah Cantt, Pakistan. His current research interests include cognitive radio networks and wireless sensor networks. He served in the technical program committee of Association for Computing Machinery (ACM) Conference on emerging Networking Experiments and Technologies (CoNEXT'13 SW), IEEE International Conference on Communications (ICC'14), and IEEE International Wireless Communications & Mobile Computing (IWCMC'13). He currently serves as a reviewer of *IEEE Journal of Selected Areas in Communications* (*IEEE JSAC*), *IEEE Transactions on Wireless Communications* (*IEEE TWireless*), *IEEE Transactions on Vehicular Technology* (*IEEE TVT*), *Computer Communications* (*ComCom*) (Elsevier), and *Computers and Electrical Engineering* (*CAEE*) (Elsevier) journals.

Yasir Saleem received his four-year BS in information technology from National University of Sciences and Technology, Islamabad, Pakistan, in 2012. He is currently doing MS in computer science (by research) under a joint program of Sunway University, Selangor, Malaysia, and Lancaster University, Lancaster, UK. He is also a reviewer of journals and conferences such as *Computers & Electrical Engineering*

(Elsevier), *International Journal of Communication Networks and Information Security* (*IJCNIS*), and IEEE International Conference on Communications (ICC) 2013. His research interests include cognitive radio networks, cognitive radio sensor networks, cloud computing, and wireless sensor networks.

9

Toward an Autonomic Cognitive Radio Platform

YOUSSEF NASSER, MARIETTE AWAD, ALI YASSIN, AND YOUSSEF A. JAFFAL

Contents

9.1 Introduction

Data traffic is currently growing exponentially. By 2015, the mobile traffic is expected to reach 7000 PB per month, with an increase of 92% from 2010, whereas the global IP traffic will reach 110.3 EB by 2016 [1] with devices 3 times the number of the global population. Around 65% of the mobile data are expected to be mobile video, whereas the rest are mainly web content and related data. The video stream delivered to tablets and smart handheld devices will require a minimum of 1 Mbit/s per video stream, while the mobile network operators have not been planning for this. While this problem mainly affects the wireless access networks, the core wired network can be built to support high bit rates by using fiber networks. For instance, content delivery network providers such as Akamai, Google, or own solutions (e.g., Netflix) can support technology that provides content to proxies located near the final customer; however, they cannot support the final wireless link.

In dense populated areas, the available spectrum for mobile operators is usually not enough. Thus, mobile operators are looking for a spectrum allocated to more traditional services, such as TV broadcasting, for their networks. TV operators, however, are not eager to let go of the spectrum. TV broadcasting technology has been developed, thanks to efforts by the digital video broadcasting (DVB) technological groups, to a performance level that it will be hard to develop further (8 bit/s/Hz). Indeed, when the digital video broadcasting operators deploy using the latest DVB technology, the spectrum usage is on a level that mobile operators can never reach. A number of high definition (HD)-quality video streams can reach millions of viewers using a single 8 MHz channel. However, the video viewing habits today, with interactive video services such as YouTube, do not support the broadcasting model. Therefore, the broadcasting model is not a solution to the scarce spectrum.

Since 1906, the spectrum allocation has normally been assigned to licensees, via auctions, for a certain time by the International Telecommunication Union (ITU) Radio Regulations. This process, well performed in practice, can result not only in an effective use of spectrum but also result in bad ones, as it might be very hard to predict the real value of a spectrum allocation far in future. Nowadays advocates of spectrum policy call for free spectrum access since interference-avoidance technology has matured enough to render interference a benign issue. As such, once the main cause for spectrum regulation is resolved, there would be no licenses, which would remedy the *bottleneck* problem of spectrum and thus allow any technology to be used any time. However, the interference level can still create concerns: Each new spectrum terminal affects others through its interference, and it is very probable to encounter overcrowded spectrum utilization. This might limit the technological innovations: If an entrant knows that someone else can use the same band at the same time and immediately offer the same service, this may limit the innovation. Supporters of the common approach believe that operators have interest in supporting interference-limiting solutions despite the fact that, for the time being, they are still limited.

The problem of spectrum shortage is not a new dilemma. In 1952, a report of the United States Joint Technical Advisory Committee of 1952 (Federal Communications Commission [FCC] 1952) states, "Wireless communication is plagued by a shortage of space for new services. As new regions of the radio spectrum have been opened to practical operation, commerce and industry have found more than enough uses to crowd them." In today's communications infrastructure, the spectrum issue is even more obvious due to the rapidly increasing number of services and the quality of service (QoS) required. This increasing rate will surely lead a high spectrum use requiring dynamism in innovation and technology as well as an effective spectrum management.

It is evident that a rigid solution for spectrum allocation is not optimal in terms of innovation, revenues, and services. Hence, the quite logical next step, even if at the moment it is only discussed in research projects, is to go for a real spectrum market with quasi-real-time allocation. The idea of flexible coexistence of broadcast and cellular networks is often mentioned as a solution. It is based on the utilization of the available bandwidth, for example, currently unused TV spectrum, widely called as white space (WS), for services other than broadcast services. Cognitive radio refers to the technologies that enable flexible use of spectrum. The spectrum could be available on a frequency exchange market, where users could buy, sell, and trade frequency bands. These allocations could be traded for certain frequency ranges or for certain geographical areas for a certain time. Mobile operators, broadcasters, Internet service providers, and event organizers could acquire frequencies for their services from this spectrum market with real-time or almost real time. These available bands could be allocated and divided as the services requirements varying from short-range communications with small and large bandwidths to long-range communications with medium and large bandwidths. Hence, the partitioning of the spectrum into small or large pieces or equivalently into units of bands will follow the services. The cost of a unit or a piece, either small or large, will also be following the supply and demand, and, finally, the consumers of radio telecommunications systems would pay the price of services following the market. It remains to define a unit of bandwidth or a piece of sharing so that all stakeholders, including end users, are satisfied with their incomes seen in terms of the technology adopted, the QoS requirements, and the financial cost and revenue.

Surely, such a cognitive radio (CR) market as described earlier is not yet available for real-life implementation, but this market would and should be available in future as its advantages are widely acknowledged. From software and hardware points of view, some CR principles are currently implemented on different platforms such as CORAL, ORBIT, CORNET, and CREW [2,3,4]; however, these platforms are still lacking a lot of properties and tools. Dynamic and opportunistic spectrum access will enable the technology for CR-oriented wireless networks to allow low-priority secondary users (SUs) to communicate over licensed bands when there is no full usage of the spectrum by legacy users. To foster the development of this concept, it is necessary to improve the existing technologies and methodologies for a harmonized spectrum access in both license-exempt bands and licensed spectra, and to design a new interface, software tools and platforms for a more market share the use of radio spectrum resources.

Six main bricks can assess the concept of real spectrum access and help move it from a rigid configuration to a flexible and open one as portrayed in Figure 9.1. Their maturity levels are shown in Figure 9.2.

- *Adaptive software platform, radio, and metrics design*: Software-defined radio (SDR) technologies provide the flexibility in the sensing technology as well as in the learning and adaptive components. The necessary metrics of the SDR interfaces are

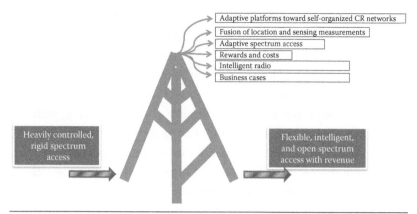

Figure 9.1 Toward a flexible spectrum market.

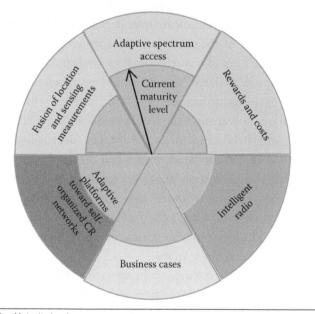

Figure 9.2 Maturity level.

diverse such as the signal-to-noise ratio (SNR), the interference level, and spectrum efficiency. Currently, even though a lot of research efforts have been provided on the theoretical aspects of adaptive CR engines, there are still huge gaps in the proof of concept through platforms, test beds, and metrics for test beds. Although some standardization activities such as IEEE 802.22 and research initiatives are running toward a unified definition and classification of these metrics, the gap to cover is still huge.

- *Fusion of location and sensing measurements*: When the CR concept was first proposed in literature, radio-frequency (RF) sensing measurements based on signal processing techniques were prioritized. Later on, new technologies based on the location and radio environment mapping (REM) databases have been proposed as a complement of the RF sensing techniques. Nowadays, the fusion of both approaches is seen as a promising concept for the near future since it combines the pros and cons of each approach. This opens up new research doors for the estimation of spectrum availability.

- *Adaptive and opportunistic access*: Dynamic and opportunistic spectrum access is presented as an enabling technology for CR-oriented wireless networks. This technology permits low-priority SUs to communicate over licensed bands when there is no full usage of the spectrum by legacy users.

- *Rewards and costs*: Because policies of traditional static spectrum assignment have been to award each wireless service with an exclusive utilization of specific frequency bands, they leave many spectrum bands unlicensed to be used by scientific, industrial, and medical subtasks. Because spectrum scarcity among those bands is attained due to the fast spread of low-cost wireless applications operating in unlicensed spectrum bands, sharing of underused licensed spectrum between unlicensed devices within the framework of an incentive of rewards and costs provides an encouraging solution to the problem of spectrum scarcity.

- *Advanced learning techniques*: In spectrum allocation where sharing is allowed in either licensed or licensed-free bands, radio access technology recognition and learning become two important approaches for reducing interference and facilitating cooperation among CRs. For instance, in the licensed bands, the cognitive nodes need to be able to differentiate between transmissions of licensed users and other unlicensed users, and they should only free a band when the licensed primary user (PU) starts to transmit. Therefore, transmission technology classifications will have a vital role in such shared spectrum bands for coexistence/cooperation purposes.

- *Business cases*: The introduction of broadband services is increasingly putting pressure on traditional spectrum allocation procedures, leading regulation authorities and governments to revise their assignment approaches. Hence, exploring the value of a band and setting up an economic valuation of the available or shared spectrum could not be feasible without the initiation of successful business cases. The reciprocal approach also holds. The successful driven business cases would also lead to new ways for spectrum assignment different than the traditional ones. In literature, different business cases have been analyzed in different European projects (QOSMOS, ACROPOLIS, QUASAR), but the future is still open for more initiatives. Some examples of possible future applications include but not limited to small cell extension, smart grids, machine-to-machine (M2M) communications, and so on.

9.2 General Concepts of CR and Learning

CR was introduced by Mitola and Maguire in 1999 [5] as a fusion between model-based reasoning and software radio. It is expected to play a major role toward meeting the exploding traffic demand over cellular systems because it provides the radio system with some intelligence to maintain a highly reliable communication with efficient utilization of the radio spectrum. Considered by Haykin in 2005 [6] as a brain-empowered wireless communication, CR is aware of its RF environment and adapts, after learning from the interactive experiences with its surroundings, to the statistical variations in the input signals. Basically, CR senses the environment, analyzes the outdoor parameters, and decides on a dynamic time–frequency–power resource allocation as well as on an optimal management of the spectrum utilization.

Figure 9.3 shows a wireless communications system with some PUs or networks that own and have higher priority to use the spectrum and some SUs that may use the spectrum only upon availability and without causing any harmful interference. For best resource allocation and network performance, CR network (CRN) should be able to sense the environment and analyze the outdoor parameters in order to propose a dynamic resource allocation and management scheme that

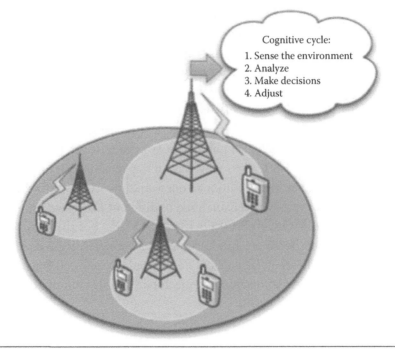

Cognitive cycle:
1. Sense the environment
2. Analyze
3. Make decisions
4. Adjust

Figure 9.3 Generic CRNs with PUs and SUs.

could improve the utilization of the radio electromagnetic spectrum between the PUs and the SUs. Thus, it is important for the CRN to know few parameters [7,8] such as channel characteristics between the base station and the users of the spectrum, spectrum holes, user and application requirements, power availability, and local policies or other limitations, if any. CR-sensed parameters feed as inputs into the resource management and resource allocation objectives that include but are not limited to minimizing the bit error rate (BER), power consumption and interference while at the same time maximizing the throughput, spectrum efficiency, QoS, and quality of user experience (QoE).

Although CR aims to satisfy the aforementioned objectives, combination of some objectives such as simultaneously minimizing power consumption and BER, often creates conflicting solutions, which calls for some trade-off solution instead of an optimal one. Constrained optimal power and bandwidth solutions can be formulated in a closed-form optimization as a mixed-integer nonlinear programming problem that is nonconvex and NP-hard as shown in

[8] and can be solved using centralized and distributed algorithms using the Lagrangian dual method. However, such an optimization formulation would result in significant computational complexity that could render sometimes the solution inappropriate for real-time decision making. As alternatives to the closed-form formulation, machine learning (ML) and artificial intelligence (AI) approaches are emerging as promising potentials to reduce the solution complexity and meet the real-time resource allocation requirements of CR, mainly in fast-varying environments. A branch of AI and ML uses statistics as well as heuristics to build mathematical models whose core aim is to infer and generalize based on a sample population. Aside than being predictive, an ML model can also be descriptive.

The main ML steps in CR—shown in Figure 9.4—are not different from a generic ML workflow that typically consists of offline training and testing phases. The ML workflow starts with data acquisition and data preprocessing before being fed into the ML learning engine. Data preprocessing consists of identifying outliers or noisy

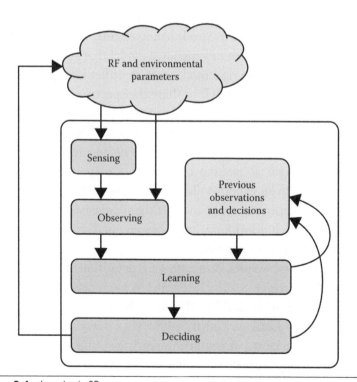

Figure 9.4 Learning in CR.

measurements as well as sometimes performing feature extraction or reduction to only retain the most relevant parameters. The model is then learned and iterated few times for acceptable accuracy before its characteristics are locked. This iterative learning often consists of performing grid search on some meta-parameters and cross-validation so that the ML model can generalize well on yet-to-be-seen observations not included in the training phase. Learning can be supervised, transductive, unsupervised, or reinforced. Once new data are acquired, it is processed—as need be—before being applied to the learned model and a decision is made. Therefore, for the CRN, it first needs to sense the RF parameters such as channel quality, observe the environment, and analyze its feedback such as ACK responses. A cognitive device would acquire channel-state information as well as its physical and spectral environment utilization. It would learn the current situation while continuously sensing the status of the channel and storing it for archive purposes. The history of the spectrum usage information will be used for predicting the future profile of the spectrum. This would help a lot in the analysis and decision of secondary devices to operate on the vacant hole. Thus, based on many collected observations as inputs to the core-learning engine, CR can then build an intelligent discriminant or generative model that should be able to learn after iterative training a model that identifies idle spectrum and opportunistic scenarios for SUs to share the spectrum.

Learning in CR has been proposed based on fuzzy logic, genetic algorithms (GAs), artificial neural networks (ANNs), game theory, and reinforcement learning (RL) [9–19], to name a few. These methods aim in general at making cognitive nodes perform tasks in a manner similar to an expert. While each approach has its advantages and limitations, unfortunately, most of them are based on theoretical background with a priori knowledge of some performance metrics, a scenario far away from practice.

In what follows, we provide a high-level definition of some AI tools used in CR. However, readers interested in more details on ML concepts are invited to check AI textbooks such as [9].

Introduced by Lotfi A. Zadeh in 1965 [15], fuzzy set theory solves and models uncertainty, ambiguity, imprecisions, and vagueness using mathematical, approximate reasoning, and empirical models, in a similar manner to a human expert. Different from binary modeling, fuzzy logic variables are not limited to only two values (True or False); they are defined in crisp sets [16], and each element has a degree of membership or compatibility with its own and negation sets. In general, the inputs to the fuzzy inference system (FIS) need to be fuzzified using linguistic hinges and, after being applied to some application-specific if–then rules, will determine the output of the system.

GA originated from the work of Friedberg (1958), who attempted learning by mutating small FORTRAN programs; therefore, by making an appropriate series of small mutations to a machine code program, one can generate a program with good performance for any particular simple task [20]. GA simply searches the solution space, with the goal of finding the one that maximizes the fitness function by evolving a population of initial solutions, chromosomes basically, toward a better one. The chromosomes are represented as a string of binary digits. This string grows as more parameters are used by the system. The search is parallel instead of processing a single solution because each element can be seen as a separate search. GA formulations can provide awareness in processing, decision making, and learning in CR.

Developed as an attempt to replicate human intelligence by modeling the brain's basic learning process, ANNs have had a history riddled with highs and lows since their inception. On a nodal level, ANNs started with highly simplified neural models such as the McCulloch–Pitts neurons (1943) and then evolved into Rosenblatt's perceptrons (1957) and a variety of more complex and sophisticated computational units. Many topologies and arrangements of these units have been suggested. From single-layered to multilayered networks, such as self-recurrent Hopfield networks (1986) and self-organizing maps (SOMs or Kohonen networks) (1986), ANNs have seen much iteration in structure. Similar to the biological neural network, ANNs are formed of nodes, also called neurons or processing elements, which are connected together to form a network. ANNs get information from all neighboring neurons and propagate the outputs depending on some weight and activation functions. To accomplish the learning process, these weights are adjusted till some cost function, often the error between actual output and desired one, is minimal. ANNs are used to make the CR learn from the environment and take decisions in order to improve the QoS of the communication system [21].

Game theory provides solutions for decentralized multiagent systems (MASs), similar to the players in a game, under full or partial observability assumptions. As it is a branch of discrete mathematics and linear algebra, it is used as a decision-making technique where several players take actions and consequently affect the interests of the other players. Each player can decide on his or her actions based on the history of actions selected by other players in previous rounds of the game. Game theory reduces the complexity of adaptation algorithms in large cognitive networks where the CRNs are modeled as the players in the game. The set of actions players can choose from are the RF parameters such as transmit power and channel selection. These actions taken by CRNs are based on observations that are represented by environmental parameters such as channel availability, channel quality, and interference. Therefore, each CR network will learn from its past actions by observing the actions of the other CR networks, and modify its actions accordingly [13,22] to reach the game equilibrium.

Finally, RL plays a key role in the multiagent domain, as it allows the agents to discover the situation and take actions using trial and error to maximize the cumulative reward. The basic RL model consists of environment states, actions, rules for transition between states, immediate reward of transition rules, and agent observation rules. In RL, an agent needs to consider the immediate rewards and the consequences of its actions to maximize the long-term (LT) system performance [12,23]. In the multiagent system, the cognitive nodes are considered as agents. An agent needs to share information (e.g., decisions of neighbors) to make a good decision. The self-adaptation of the MAS is gradually reached through a direct interaction with the environment after correctly factoring in previous system's experiences. Therefore, the decision of MAS agent will be based on a distributed or centralized learning strategy from the environment so that one or multiple objective functions can be reached. While the multiagent bandit (MAB) is based on decision making by each bandit that maximizes the cumulative gains by pulling the arms of different slot machines, the MAS is mainly based on modeling each cognitive node as an agent and on exchanging information in a distributed manner between these agents.

9.3 Literature Review

Recently, the QUASAR project reported in [24,25] on the overall system aspects of CR technologies with a particular attention to the economic viability of different use cases. The project findings reclaimed that secondary sharing could be utilized in larger scenarios than a simple detection of spectrum holes. It also claimed that from a business perspective, secondary spectrum usage through TV white spaces does not provide real opportunities for either cellular incumbents or new market entrants in the cellular domain due to the high cost of necessary deployment at the ultra high frequency (UHF) bands and the requirement of large band needs. Moreover, the commercial success of secondary access turned out not viable due to the fact that the secondary network is not scalable, and most importantly, the secondary spectrum usage is not an attractive method for most of the commercially interesting scenarios except for some very specific scenarios such as short-range indoor communications.

Although the outcomes of the QOSMOS project highlighted some interesting economic use cases such as femtocells in cellular networks, cellular coverage expansion, mobile to mobile in cellular systems, all of which constitute an important step for the evolution and success of the cognitive sharing technologies, the status of the state of the art in CR research is a topic worth further exploration. In what follows, the most relevant research work on CR according to the six pillars identified earlier is detailed.

9.3.1 Opportunistic Access

Recent research works have proposed some novel schemes for open spectrum access, aiming at maximizing the overall throughput, minimizing delays, insuring fairness among participants, and improving the power efficiency of the network [23]. In general, the different spectrum allocation mechanisms among participants can be categorized as centralized and distributed spectrum access. In the centralized approach, the decision is taken at a central node that controls the resource allocation among different participants in the area, where each participant sends its information to the central node, such as channel conditions, data demand, and possible interferers. This approach could be seen as a competition between operators to get a specific spectrum, whereas the central broker controls the dynamic allocation of the spectrum among the different operators. The centralized approach can achieve the optimal performance since it has full information and control of the network, but it is impractical to implement due to the backhaul requirements and the complex computation at the central node [26,27]. In the distributed approach, each participant takes its own decision based on its observation and interactions with other participants. The main techniques found in literature focus on game theoretic approaches, where each participant tries to maximize its throughput [28]. Spectrum access can also be divided into cooperative and noncooperative. In the cooperative case, the cognitive nodes share their information or a portion of it in order to improve the overall performance and to avoid collisions. The different participants negotiate until they reach a deal to divide the available resources, or they use common etiquettes that ensure a stable network behavior. Bater et al. [29] proposed a cooperative scheme based on Nash bargaining solution. In [30], a reassignment algorithm assumed that all the participants follow a common etiquette, and when a new participant tries to get some resources, the existing participants creates a gap in the spectrum so that the new entrant can use it. If the cognitive nodes decide not to share, then their trivial solution is to keep sensing the spectrum to have full information about the spectrum utilization. Some research works proposed some RL methods in order to reduce the sensing requirements [31,32,33]. To guarantee that the network will reach equilibrium without conflicts between different participants, the involved participants should follow some common protocols or etiquettes. Bae et al. [34] proposed a game model where each participant pays some fees to other inactive participants when it sets up an access point before converging to a Nash equilibrium. It is also worth mentioning that the spectrum could be shared for the utilization of a single radio access network (RAN) or multiple RANs. In the case of single RAN, the spectrum might be rented or released from other RANs or from a spectrum pool. In the case of multiple RANs, centralized units would be needed in collaboration with local decision engines for multiple spectrum access.

Finally, some research works have focused on the proposition, design, and analysis of physical (PHY) layer waveforms with flexible reconfiguration and reduced out-of-band emissions [35] using filter bank multicarrier techniques [36], noncontiguous orthogonal frequency division multiplexing (OFDM) [37], and multicarrier code division multiple access, to name a few.

9.3.2 Rewards and Costs

Pandit and Singh [22] approached the problem of the fixed spectrum allocation policy from an economic point of view where a simulation model that improves the allocation of bandwidth between the PUs and the SUs of a CRN was proposed. They developed an algorithm that minimizes the cost of this bandwidth while maximizing the effectiveness of the SUs in the CRN. They utilized a game theory utility to model the payoffs between the SUs and the PUs, which they consider to be the players. The main aim of the PUs was to maximize their revenue, whereas the SUs have an objective of improving QoS satisfaction at an acceptable cost. Alrabaee et al. [38] addressed the spectrum management in CR. They studied the spectrum trading spectrum management without game (SMWG) theory and the spectrum competition spectrum management with game (SMG) theory. Considering the SMWG theory first, they introduced a novel factor called competition factor that models the spectrum competition between the different PUs and also introduced a new QoS-level function that relates to the spectrum availability and is variable according to SU requirements. In both cases, they assumed that the trade-off is between the PUs' desire to maximize the revenue and the SUs' aim to obtain the required QoS level. RL is also another tool proposed in literature for intelligent CRNs as well as for intelligent rewarding system. For instance, Busoniu et al. [19] considered the immediate rewards of an agent's needs and the consequences of its actions to maximize the LT system performance. Barve and Kulkarni [39] incorporated a system of errors and rewards based on each decision, and hence every agent tried to maximize its own rewards.

It is worth mentioning that this concept of rewards and costs could (or should) be combined with ML algorithms and the auction-based CR concept for a better optimization of the spectrum utilization and resource allocation satisfying the end users.

9.3.3 Spectrum Databases

Basically, spectrum sensing and WS databases are the two typical methodologies proposed in literature to detect the available channels in the vicinity of the users. In spectrum sensing, the SU detects the availability of the free band using, for instance, signal processing techniques. In the case of database-driven technologies, the SU should query a central database to attain a spectrum availability information at its location [40].

Even though a lot of research works was focused on the spectrum sensing, this concept was excluded from the latest regulations of FCC. As an alternative, the WS database-driven technologies were adopted. In general, geolocation databases are proposed for CR queries. They hold information about the PU signal (parameters, transmit power, position) and the transmission pattern as well as information about the geographic environment (geography) and the positioning of PU and SU transmitters. Therefore, it is fundamentally based on field strength estimates of the primary service obtained using radio propagation models. However, those models have limited estimation capabilities to areas of hundreds of square meters. To overcome the limitations of propagation models and suboptimal field sensing algorithms, a geostatistical procedure is proposed using spatial simulated annealing to minimize the mean universal kriging variance that improves the precision of local field strength estimates [41].

REM was also introduced as a database-driven technology in CRNs. Defined as an abstraction of a real-world environment storing multidomain information (e.g., PUs, policies, and terrain data), it can also be considered more generally as an intelligent network entity that can further process the gathered information, inspect the spatiotemporal characteristics, and derive a map of the RF environment [42]. REMs act as cognition engines by building LT knowledge via processing spectrum measurements collected from sensors to estimate the state of locations, which do not have any measurement data. Wei et al. [43] proposed a generic top-down approach for CRs to obtain situation awareness by exploiting REM through a general framework for CR learning algorithms that incorporates both high- and low-level learning loops [42]. Global situation awareness and coordination help CRs to make desired adaptations beyond single node's capability.

They showed that global REM information can significantly improve the performance of both PUs and SUs, reduce the CRN adaptation time, and mitigate the hidden node problem. They [43] discussed the REM construction technique and analyzed the trade-off between the number of measurements (sensors) and REM accuracy. REM design relies on data gathering/representation, data processing/fusion, and data retrieval/query. In [43], REM is built by considering the coverage of all networks based on the cognitive pilot channel technology. Sensors are randomly deployed in a region without taking into account mesh boundaries. The majority of the sensor measurements in a mesh determine the radio environment and these values are used to construct REM for the region. Huseyin Birkan and Tugcu [44] presented REM as a *cognitive engine* and compared their active transmitter location estimation-based (LIvE) REM construction technique with kriging and inverse distance weighted interpolation in shadow and multipath fading channels. Simulation results showed that the LIvE REM construction outperformed the compared methods because it utilized additional channel parameters' information. When deploying REM, some factors should be taken into consideration as stated in [45]:

- *Signaling overhead*: It is related to the amount of relevant data transmitted between the CR and the REM nodes as well as some limitations of capacity in the connection between these two nodes.
- *Time of validity of data*: It acts as a lifetime for the information that REM stores. This information should be regularly updated to stay relevant and accurate.
- *Mobility and location*: It should be updated as the user location is changing to benefit from the spectrum opportunities in the respective area.
- *Wireless devices*: They use the same WS channels that the REM is trying to use, but they are not always knowledgeable to the REM. This will cause a problem when allocating some spectrum spacing to other users.
- *Bootstrapping*: It happens when a new device wants to connect to a base station. Usually, the REM has not yet given the device a spot so the device might interfere with other PUs when trying to connect to the base station and transmit its location.

It is worth mentioning that the literature is not limited to location-based database technologies. Höyhtyä et al. [46] presented the utilization of LT and short-term (ST) database utilization. While LT database helps in the functioning of the CR system and decreases its time of sensing via channel prioritization, ST database permits prediction and classification of the interested bands. This methodology enhances the performance of the system by improving the secondary system throughput and decreasing the interference triggered toward PUs. Thus, the use of LT and ST databases introduced a smart method of selecting channel for data and control transmission. It was also shown in [46] that combining the utilization of LT and ST databases leads to more advantages than using either alone. This unique combination in CR is independent of the utilized frequencies and CR types. LT database collaborates with CRs during the normal functioning, whereas CRs can continue functioning independently if the connection to LT database is vanished.

Another pioneering work in this area is presented in [47]. Saeed et al. [48,49] introduced the distributed resource map (DRM) as a framework for collaboration and knowledge accumulation in CR ad hoc networks. DRM is a database-driven knowledge base intended to exploit distributed sensing accomplished by heterogeneous CR nodes and internodes cooperation to sustain a network-wide support architecture. Similarly, some authors proposed the dual-mode scheme where the secondary device should have sensing and geolocation capabilities to enhance its detection capabilities [49].

From an application point of view, different corporations have proposed database services. For instance, Google Inc. and Comsearch were designated by the FCC as TV bands device database administrators. Recently, Koos Technical Services Inc. [50] and Telcordia Technologies Inc. [51] presented their designed database systems.

Even though database-driven CRNs present an encouraging methodology by following the traditional technique of location-based services, they suffer from privacy intimidations specifically on the feature of location privacy. For instance, an attacker can geolocate the SU by tracing its database queries. Hence, this will generate a severe privacy leak if the sensitive data of the SU are strictly correlated to its location. Moreover, the user in database-driven CRN has to register his/her utilized spectrum in the database. The work presented in [40] introduces a privacy preserving spectrum retrieval and utilization architecture for database-driven CRs, claimed as PriSpectrum. Their novel private channel utilization protocol diminishes the leaking of the location privacy in spectrum usage phase by choosing the most stable channel with the minimum number of channel switch events.

Finally, it is worth mentioning that there is no doubt that leveraging database and sensing could overcome the database backhaul problems by adding an over-the-air sensing, especially, for the short-range coexistence or when the backhaul is absent [52]. However, even though sensing is not sufficient for avoiding the incumbent licensed users, it is compulsory for all TVWS secondary systems. The idea of having the geolocation database monitor the use of the available spectrum and assist in coordinating that usage is actually gaining ground as can be seen from [53,54], but the focus so far has largely been on general frameworks for database-assisted coordination of secondary TVWS spectrum use and not on the underlying decision-making mechanisms. Moreover, there are very little researches on combining spectrum sensing techniques and location databases to enhance the decision mechanisms of the channel utilization and allocation.

9.3.4 On Learning Algorithms

As mentioned earlier, an intelligent machine should be able to deduce, reason, solve problem, learn, and represent knowledge adequately [9]. Within the context of CR, few researchers have investigated ML.

Kaur et al. [55], Aryal et al. [56], Matinmikko et al. [57] and Qin et al. [58] used fuzzy logic theory in CR to optimize bandwidth allocation, interference and power management, spectrum availability assessment methods, and resource allocation, respectively. Aryal et al. [56] proposed a centralized FIS that can allocate the available bandwidth among cognitive users considering traffic intensity, type, and QoS priority. They [56] researched power management while reducing interference and maintaining QoS. Taking into consideration the number of users, mobility, spectrum efficiency, and synchronization constraint, their results checked the power consumption and interference price while adjusting the power values. For detecting the available bandwidth, Qin et al. [58] fuzzified the probability of detection, the operational SNR, the available time for performing the detection, and the a priori information before comparing their results to the fixed spectrum schemes. Chen et al. [11] proposed the fuzzy inference rules for resource management in distributed heterogeneous wireless environments. The fuzzy convergence was designed in a hierarchical manner: First, the local convergence calculation was based on local parameters such as interference power, bandwidth of a frequency band, and path loss index, and then the local convergence calculations were collected from all nodes and aggregated to generate a global control for each node.

GA was proposed in [11] to enhance the CR system performance by minimizing the BER and maximizing the throughput. Researchers proposed encoding operating variables for different numbers of subcarriers into a chromosome, including the center frequency, transmission power, and modulation type. In [59], spectrum optimization in CRs was addressed using elitism and four parameters for representing the chromosome structure, which are frequency, power, BER, and modulation scheme. It is to be noted that elitism is often used for selecting the best chromosomes and transferring them to the next generation, before performing crossover and mutation, which prevent the loss of the most likely solutions in the available pool of solutions. Hauris et al. [60] used GA for RF parameter optimization in CR. Modulation and coding schemes, antenna parameters, transmit and receive antenna gains, receiver noise figure, transmit power, data rate, coding gain, bandwidth, and frequency were used as genes, while the fitness function related link margin, carrier to interference (C/I), data rate, and spectral efficiency.

While Tan et al. [12] proposed the back propagation ANN to solve the inefficiency in current communication networks by using ANN. Abdulghfoor et al. [13] tried to improve the performance of spectrum sensing in CRNs by learning an ANN at every SU in order to predict the sensing probabilities of these units. Their various simulations, with considerably high differences in SNR between the different SUs, reported quite impressive results with a global false-alarm probability of .01 for the CRN. Yang et al. [61] combined GA and radial basis function network in order to effectively adapt the parameters of the CR system as the environment changes. Using the data collected from the network simulator 2 (NS2) platform, their results showed superior reliability and accuracy performance compared to belief propagation (BP) ANN.

Abdulghfoor et al. [13] suggested game theory (GT) to model resource allocation in ad hoc CRNs and showed through simulations that GT can be used to design efficient distributed algorithms, whereas its application at the media access control (MAC) layer proved to be most challenging. Li et al. [62] proposed a single framework for cooperative spectrum sensing of CRNs as well as for self-organization of femtocells by relying on the data generated by the CR users to better understand the environment. Researchers suggested the creation of a large spectrum database that relies on both macrocell and femtocell networks with the flexibility of linking local decisions to the centralized fusion center. Alrabaee et al. [38] compared spectrum trading and management with a Bertrand–Stackelberg game and without game theory. In both cases, they assumed that the PUs aim to maximize revenues, while SUs desire to reach the required QoS level. Yau [10] incorporated RL to correctly complete the cognition cycle in centralized and static mobile networks. The RL approach was applied at the level of the SUs where they rank channels dynamically according to PU utilization levels and packet error rate during data transmission to select the best one to communicate with other SU receivers.

Barve and Kulkarni [39] considered routing in CRNs and proposed a new RL system that jointly works on channel selection and routing for a multihop CRN. The RL model incorporated the errors and rewards on each decision, and hence every agent tried to maximize its own rewards after few trials and errors. They also used the feedback obtained from the environment and modeled the problem using the Markov decision process. Zhou et al. [63] addressed the problem of the high-power consumption in CRNs that is generated due to overhead communication between different CR users. They proposed a new design for a power control algorithm that is simple to implement by relying on RL, which eliminates the need for information sharing about interference channels and power strategy between the different CRN elements. Various simulations showed that the proposed model was much more realistic than the current models and pointed out that the case of Bush–Mosteller reinforcement with Lagrange multipliers provided the best answers. Introduced by Robbins in 1952 as a simple ML approach, MAB [64] aims at maximizing the resource allocation gains in CR by modeling each available band as a bandit machine, based on an analogy faced by a gambler trying to get the biggest rewards from a *K* number of machine slots. Similarly, MAS is another approach partially explored in literature and deserves further investigation, mainly in cooperative CRNs with uncertainties.

9.3.5 *Adaptive Radio, Software, and Metrics*

Because experimentation facilities play an important role in the transition of technology from concept to prototype, requirements for an open platform allowing experimentation with CRs have been outlined in [2]. Several large CR test beds such as CORAL, ORBIT, CORNET, and CREW [3] have been deployed in the United States and Europe. These are often built using off-the-shelf equipment such as Tmote Sky1 and Universal Software Radio Peripheral but can also be based on custom-made hardware platforms such as CORAL [4], the imec sensing engine3, and the versatile platform for sensor network applications (VESNA). The majority of the experiments performed on these platforms to date focus on spectrum sensing.

Some practical initiatives for CR test beds have taken place [65,66] including improvements to the fourth-generation (4G) LT evolution (LTE) technology augmented with cognitive capabilities.

While theoretical frameworks and computational simulations abound, it is clear that the experimental investigation of the behavior and performance of resource allocation algorithms, intelligent engine, and RF adaptation is still scarce and far away from the needs.

9.4 Toward Autonomic and Self-Organizing CRNs

It is clear from the state of the art presented in Section 9.3 that CR is still far from realistic network deployment due to different constraints, challenges, and open issues toward autonomic and self-organized CRNs. Indeed, the demand to control the interference between various operators motivated the utilization of licensed spectrum for wireless communication. Nevertheless, there is no significant efficiency in spectrum usage with this mode of regulation. Also, the evolution of wireless networks is delayed by the scarcity of free frequency bands and the huge investments for acquisitioning a license. Here, a different growth path for the evolution of wireless networks and the efficient utilization of the spectrum should be addressed. The path should be based on the utilization of unlicensed spectrum, while the users are allowed to have an open access to all public (unlicensed) networks. To do so, alternative interference mitigation strategies and network planning mechanisms based on the information offered by end users and their information exchange are very promising solutions within this context of self-adaptive and self-organizing networks. Many challenges such as spectrum sharing dimensions, uncertainties, opportunistic versus open spectrum access, centralized versus distributed spectrum access, cooperative versus noncooperative spectrum sharing, AI-based modeling of spectrum sharing, and RF front-end designs to name a few are still to be closely investigated.

In what follows, we discuss some of the main challenges foreseen in this context:

9.4.1 Challenges in Spectrum Sensing

The challenges in spectrum sensing discussed in this chapter are as follows:

- *Optimal threshold setting*: Spectrum sensing is a binary detection challenge of setting the right threshold value. The key challenge for the case of energy detector (ED)-based sensing is observed in the trade-off between the probabilities of missed detection and false alarm while setting the threshold spectrum sensing. Basically, the probability of false alarm designates the probability that the detection technique falsely decides that PU is present in the scanned frequency band when it actually is absent; however, the probability of missed detection designates the probability of not detecting the presence of a PU signal when there is actually a PU transmission.
- *Noise uncertainty problem*: There is a strong dependence on the value of the threshold set and the actual noise variance that might alter with location and time.
- *Primary receiver uncertainty problem*: The PU transmitter detection-based sensing technique cannot detect the PU activity when CR lies outside the range of the primary transmitter. Thus, CR generates an inevitable interference to the primary receivers especially when CR lies close to them.

- *Detecting hidden primary transmitter*: A problem may arise when CR lies inside the primary transmitter range but observes heavy shadowing or deep fading. Also, destructive interference can be imposed by secondary transmission on primary receivers.
- *Cooperative sensing among CRs*: The challenge is to solve the problem of multipath fading and shadowing based on the underlying principle that SUs are distributed spatially with different channel conditions.
- *Spectrum sensing in a multiuser environment*: In such an environment, interference between SUs may occur causing the detection of PUs to be a difficult task.
- *Sensing/communication trade-off*: Even though this topic has been investigated in some works, however, it is still not clear how to optimize the resource sharing between sensing and communications, especially in scenarios with uncertainties.

9.4.2 Challenges in SDR Platforms

In this section, the challenges are quite high due to the different requirements of a suitable yet easy implementation and utilization methodology for an experimental evaluation of a simple yet efficient resource allocation on a real-world experimental test bed. The latter should be updated toward a real CR test network where cognitive nodes are implemented on a software radio platform operating on different frequencies such as TVWS. The test network should consist of devices, which can be configured to act as cognitive nodes or base stations within the test bed. The constraints imposed by the software and testbed platforms, to overcome, are as follows:

- *Interface and metrics*: The challenge here is to propose general and transparent interfaces and metrics to develop a common testbed platform within a self-organizing framework. The interface should be scalable and available to the open-source platform.
- *Transmission power levels*: The testbed power scenarios should be developed to match the optimization techniques proposed in the framework of self-organized networks to address some

green aspects and proof of concept in such a way that some algorithms could be easily tested.

- *Co-use of geolocation and sensing*: The main question here is how to address the access to the databases from the developed SDR devices, knowing that this requires some specific licenses that may be costly.
- *Interoperability*: The main issue to solve in this context is how to propose a suitable platform solution with interoperability capabilities in terms of hardware, systems, specifications, and so on.

9.4.3 Challenges in Hybrid Data Fusion of Location Information and Spectrum Sensing Measurements

Given that Zhang and Banerjee [67] have recently (November 2013) shown through measurements collected at over 1 million locations across a 100 km^2 area that the commercial databases tend to overpredict the coverage of certain TV broadcasts by almost 42% measured locations leading to a blocking of the WS usage; developing tools and techniques for a hybrid data fusion between sensing and location measurements is necessary for an improved spectrum access. The output of this fusion should reflect the fundamental performance limit of location-aided CRs in future cognitive access networks. Specifically, the following issues should be addressed.

- Location-enabled cooperative spectrum sensing techniques that improve the reliability of spectrum sensing by allowing distributed CRs to share their spectral estimates (data fusion). For instance, using the location information of cognitive nodes, we can exclude those unreliable estimates in the data fusion process to improve sensing performance with reduced communication overhead and computational cost.
- Location-based cooperation among heterogeneous networks in the license-free bands, which improve the sensing capabilities of cognitive devices.
- Enhanced algorithms for the spectrum sensing and utilization in future wireless systems. For instance, it is of interest to continuously perform a statistical analysis of radio observables

coming from the cognitive nodes. These learnt parameters will be then stored for utilization, for instance, in small cells and/or coordinated for the usage of M2M.

- Hybrid data fusion of spectrum sensing measurements and location information in a single database and at the cognitive nodes for an improved utilization and sharing of the licensed bands (TVWSs) and unlicensed bands (e.g., the 2.4 GHz).
- Adaptive path-loss models and geostatistical approaches for an improved location-based spectral utilization.

9.4.4 Challenges in Uncertainties and Context Awareness

One step beyond the state-of-the-art context-aware CR engine in heterogeneous and/or cooperative networks would consist in its extension to the case of dynamic network topology. Indeed, an M2M scenario, for instance, the cognitive nodes could be moving along a certain track, thus creating strong time correlations between the different positions along that track. By applying tracking filters that include mobility models on these time correlations, new context-aware location-based and spectrum sensing (i.e., time-based intelligent algorithms) could be adapted to reduce the complexity of the cognitive engine.

A thorough study in this research item consists of analyzing the impact of the error measured between the estimated and actual distributions of the cognitive nodes as well as the spatial and temporal correlations of fading channels and the activity pattern of the nodes.

It is well known that the noise level could be time variant, and hence prone to noise uncertainty [68]. The effect of other systems may also lead to some changes in the received detected signal yielding interference uncertainty. Other uncertainties could also come from the channel variations in a dynamic environment, hidden primary nodes, and nonreliable secondary nodes in the distributed cooperative scenarios. An interesting investigation would be to evaluate the impact of the uncertainty and/or measurement procedures on the hybrid, heterogeneous, and cooperative sensing algorithms in terms of sensing performance, overhead, and latency (including algorithm convergence and delivery times). This investigation could propose the models and scenarios of realistic interference with consistent operating conditions such as imperfect synchronization, dynamic channel, and RF impairments and assess their performance. For instance, depending on the level of collaboration between PUs and SUs, the deployment of CR solutions over unused frequencies should avoid creating disturbances over the existing technologies. To afford this, the analysis and modeling of the potential co-channel interferences are critical issues that should be covered. With this knowledge, it is possible to take a more thoughtful decision on the transmission parameters (e.g., power, bandwidth, etc.) to be used by the new opportunistic service.

9.4.5 Challenges in Optimization through Rewards and Costs toward Self-Organizing CRNs

The resource allocation problem in a dynamic environment using capacity, power, or any other metric, as the optimization function, is in general a nonconvex optimization problem. In literature, researchers proposed to deal with the convexity of these metrics using some suboptimal solutions through the dual decomposition theory [69]; however, the complexity of these algorithms is still too complex to be implemented in a real scenario. In a CR scenario, the resource allocation becomes even more complex as it requires information on interference levels, neighborhood, locations, and so on. In general, the solutions of the dynamic resource allocation could be based on two main categories: utility-based functions [70] and graph-based theory approaches [71].

Some key questions that are interesting to investigate are as follows:

- What are the main suitable utility functions for cooperation and coordination between cognitive nodes with improved QoS and reduced overhead?
- What are the main parameters and how to develop reliable methods for relaying and exploiting geolocation information and spectrum sensing measurements in MAS?
- How to offer reliable and efficient PHY/MAC cross-layer design techniques exploiting the availability of positioning information?
- How to relate the rewards and costs between the PHY and MAC layer indicators?

9.4.6 Challenges in Advanced Learning Algorithms

Learning techniques may have many advantages; however, their implementation faces many challenges. We will focus on support vector machine (SVM), ANN, and GA techniques mostly. SVM offers a principled approach to ML problems due to its mathematical foundations in statistical learning theory. It constructs its solution as a weighted sum of support vectors (SVs), which are only a subset of the training input. SVs are considered most influential in deciding the learning model. Beyond minimizing some error cost functions based on the training data sets similarly to what other discriminant ML techniques do, SVM imposes an additional constraint to the optimization problem that forces the optimization step to find a model that would eventually generalize better since it is situated at an equal and maximum distance between the classes. SVs and their corresponding weights are found after an exhaustive optimization step that uses Lagrange relaxation and solves Karush–Kuhn–Tucker. For linearly nonseparable problems, SVM uses the kernel trick, which consists of a kernel function satisfying Mercer's theorem to map the data to the feature space where the data would become at worst pseudo-linearly separable. Despite the robustness and optimality of the original SVM formulation, SVMs do not scale well computationally. Suffering from slow training convergence on large data sets, SVM online testing time can be suboptimal because it writes its model as a weighted sum of SVs whose number cannot be estimated ahead of time and can total as much as half the original training data set. Thus, it is with larger data sets, though, that SVM fails to efficiently deliver, especially in the nonlinear classification case. Large data sets impose great computational time and storage requirements during training, rendering SVM, in some cases, slower than ANN, already known for their slower convergence. Another challenge for SVM is kernel selection and grid search for its meta-parameters.

As a biologically inspired ML technique by excellence, GA has been widely adopted. However, similar to other ML approaches, GA faces a few technique-specific challenges such as the design of an efficient and data-aware fitness function that rewards better solutions during evolution to insure faster convergence. In the presence of class skewness and linearly nonseparable feature space, the *vanilla flavor* GAs tend to produce classifiers biased toward the majority class at the expense of the minority class accuracy due to the overwhelming majority of one class in the training examples. Another challenge specific to GA is solution suboptimality as GA is reputed for being nonreliable in consistently and systematically reaching the desired final solution given all its heuristically selected parameters. Compared to GA and ANN, SVM training returns a uniquely defined model for a given training set, whereas ANN and GA classifier models are different each time training is initialized. It is because GA explores, in an evolutionary fashion, the search space of many possible solutions that maximizes a selected fitness function. The search space is traversed using a combination of selection, crossover, and mutation mechanisms applied to the chromosome population. The fittest chromosomes are allowed to propagate through generations few times, and when the termination conditions are achieved the best chromosome is chosen. Given its heuristic nature, GA solution is thus not unique, instead it is often suboptimal. Its robustness and repeatability are easily affected by the choice of the initial population, fitness function, and evolution parameters.

Similar to GA, the perception aim is to only minimize the error during training, which translates into many solutions meeting this requirement. If many models can be learned during the training phase, only the optimal one is best to retain because training is practically performed on some samples of the population, while the test data might not exhibit the same distribution as the training set. ANN suffers from the classical multilocal minima, the curse of dimensionality, and overfitting because it somewhat heuristically moved from applications to theory while SVM evolved from a robust theory to implementation. Relying instead on the empirical risk model, ANN topology needs to be designed for optimal model complexity, whereas SVM automatically determines the model complexity by selecting the number of SVs and by minimizing the structural risk.

GA, ANN, and SVM share the same challenge of labeled data availability in supervised learning scenarios. Table 9.1 shows some of the advantages and limitations of the most promising AI approaches.

Table 9.1   Strengths and Limitations of AI Techniques Currently Proposed in CR

LEARNING TECHNIQUE	SPECTRUM SENSING	DECISION MAKING	STRENGTHS	LIMITATIONS
Neural networks	x	x	Adaptation ability to minor changes	Require training data labels
			Construction using few examples, thus reducing the complexity	Poor generalization Overfitting
SVM	x	x	Generalization ability	Requires training data labels and previous knowledge of the system
			Robustness against noise and outliers	
				Complex with large problems
GAs		x	Multiobjective optimization	Require prior knowledge of the system
			Dynamically configure the CR based on environment changes	Suitable fitness function High complexity with large problems
Game theory	Related to spectrum sensing	x	Reduction of the complexity of adaptation Solutions for MASs	Requires prior knowledge of the system and labeled training data
Reinforcement learning	x	x	Learning autonomously using feedback	Needs learning phase of the policies
			Self-adaptation progressively in real time	
Fuzzy logic	Related to spectrum sensing	x	Simplicity; decisions are directly inferred from rules	Needs rule derivation Accuracy is based on these rules
Entropy approach	x	x	Statistical model	Requires prior knowledge of the system

9.4.7 *Challenges in RF Front-End Design*

Besides the challenges related to intelligence distribution and implementation, decision making, sensing algorithms and learning process, delay/protocol overhead, cross-layer design, security, geolocation, and flexible hardware design, major difficulties lie in the implementation of such systems, mainly related to antennas, amplifiers, and oscillators, for which hardware implementation and system development are progressing at a slower pace, because of the complexities involved in designing and developing CR systems [72]. Several works in literature addressed some of these

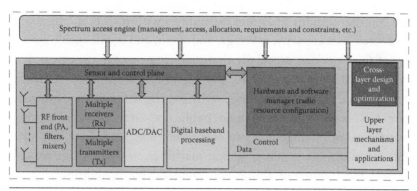

Figure 9.5 Full CR platform design.

challenges facing CR [72,73,74,75,76]. Figure 9.5 presents a schematic block diagram of a CR system for ST and mid-term implementation. It is to be noted that parallel processing is employed for the analog-to-digital interfaces in order to operate on a very wide band or multiband simultaneously. Multiantennas are designed for multiple input multiple output (MIMO) operation and/or multiband operation, while passive modules used for switching or duplexing, RF filtering, and impedance matching between antennas and power amplifiers (PAs) are located after the antennas. Then multireceiver (Rx) and multitransmitter (Tx) are utilized before a multi-analog to digital converter/digital to analog converter (ADC/DAC) module. Since high performance is expected for CR, digital filtering, dc offset cancellation, digital automatic gain control, calibration and correction of analog errors, and nonlinearities are required besides the conventional processing for modulation and demodulation, coding and encoding, and so forth. This calls for a feedback from the baseband to RF front end and transceiver combined with control plane and sensor in order to boost the performance of the analog part. The main challenges of RF front end and transceiver in the ST/midterm are to reduce the off-chip and passive components, increase their frequency agility, minimize the power dissipation, and reduce the chip area.

9.5 Conclusion

This chapter reported on the state-of-the-art research in CR and ML toward a flexible spectrum usage. We expect that intelligent techniques derived from ML and AI will move CR toward more autonomic and self-organizing CRNs such that M2M will improve both the communications links and capabilities with respect to efficiency, data rates, user capacity, coverage, and so on, and it will enable a variety of new services, especially that requiring context awareness. For instance, M2M networks can play essential roles in data exchange between machines without necessarily passing through an infrastructure. CRNs will also enable more public and safety services, making *smarter*, more intelligent, more energy efficient, and more comfortable our transport systems, homes, machines, as well as all personal devices and Internet of Things with connectivity capabilities.

Acknowledgments

We thank Miss Nadine Abbas for her initial input and contribution to an earlier draft of this work. This work was supported by the Lebanese National Council for Scientific Research, Beirut, Lebanon and the University Research Board of the American University of Beirut, Beirut, Lebanon.

References

1. Cisco, "Cisco Visual Networking Index: Forecast and Methodology 2011–2016," *White Paper*, May 2012.
2. L. A. DaSilva, A. B. MacKenzie, C. da Silva, and R. W. Thomas, "Requirements of an Open Platform for Cognitive Networks Experiments," pp. 1–8, October 2008.
3. J. Sydor, A. Ghasemi, S. Palaninathan, and W. Wong, "Cognitive, Radioaware, Low-Cost (Coral) Research Platform," pp. 1–2, 2010.
4. "LOG-a-TEC, Testbeds with Sensor Platforms," 2013. Available: http://www.log-a-tec.eu/.
5. J. Mitola and G. Q. Maguire, "Cognitive Radio: Making Software Radios More Personal," *IEEE Personal Communications*, vol. 6, no. 4, pp. 13–18, 1999.
6. S. Haykin, "Cognitive Radio: Brain-Empowered Wireless Communications," *IEEE Journal on Selected Areas in Communications*, vol. 23, no. 2, pp. 201–220, 2005.
7. L. Wang, W. Xu, Z. He, and J. Lin, "Algorithms for Optimal Resource Allocation in Heterogeneous Cognitive Radio Networks," *Proceedings of the 2nd International Conference on Power Electronics and Intelligent Transportation System*, December 2009.
8. F. Khan and K. Nakagawa, "Comparative Study of Spectrum Sensing Techniques in Cognitive Radio Networks," *Proceedings of the World Congress on Computer and Information Technology*, June 2013.

9. S. Russels and P. Norvig, *Artificial Intelligence: A Modern Approach.* Prentice Hall, Upper Saddle River, NJ, December 2009.
10. K.-L. A. Yau, "Reinforcement Learning Approach for Centralized Cognitive Radio Systems," *Proceedings of the International Conference on Wireless Communications and Applications*, pp. 1–5, October 2012.
11. S. Chen, T. R. Newman, J. B. Evans, and A. M. Wyglinski, "Genetic Algorithm-Based Optimization for CR Networks," *Proceedings of the IEEE Sarnoff Symposium*, pp. 1–6, April 2010.
12. X. Tan, H. Huang, and L. Ma, "Frequency Allocation with Artificial Neural Networks in CR System," *Proceedings of the IEEE TENCON Spring Conference*, April 2013.
13. O. B. Abdulghfoor, M. Ismail, and R. Nordin, "Application of Game Theory to Underlay Ad-Hoc CR Networks: An Overview," *Proceedings of the IEEE International Conference on Space Science and Communication*, pp. 296–301, July 2013.
14. B. Wang, W. Wu, and K. J. R. Liu, "Game Theory for CR Networks: An Overview," *Computer Networks*, vol. 54, pp. 2537–2561, 2010.
15. L. A. Zadeh, "Fuzzy Sets," *Information and Control*, vol. 8, pp. 338–353, 1965.
16. L. Gavrilovska, V. Atanasovski, I. Macaluso, and L. DaSilva, "Learning and Reasoning in Cognitive Radio Networks," *IEEE Communications Surveys and Tutorials*, no. 99, pp. 1–17, 2013.
17. R. Rojas, *Neural Networks A Systematic Introduction.* Springer, New York, 1996.
18. M. Bkassiny, Y. Li, and S. K. Jayaweera, "A Survey on Machine-Learning Techniques in Cognitive Radios," *IEEE Communications Surveys and Tutorials*, vol. 15, no. 3, pp. 1136–1159, 2013.
19. L. Busoniu, R. Babuska, and B. De Schutter, "A Comprehensive Survey of Multiagent Reinforcement Learning," *IEEE Transactions on Systems, Man, and Cybernetics, Part C: Applications and Reviews*, vol. 38, no. 2, pp. 156–172, 2008.
20. S. J. Russell and P. Norvig, *Artificial Intelligence: A Modern Approach.* Prentice Hall, Englewood Cliffs, NJ, 1995.
21. T. Zhang, M. Wu, and C. Liu, "Cooperative Spectrum Sensing Based on Artificial Neural Network for Cognitive Radio Systems," *Proceedings of the 8th International Conference on Wireless Communications, Networking and Mobile Computing*, September 2012.
22. S. Pandit and G. Singh, "Spectrum Sharing in Cognitive Radio Using Game Theory," *Proceedings of the 3rd International Advance Computing Conference*, February 2013.
23. A. C. V. Gummalla and J. O. Limb, "Wireless Medium Access Control Protocols," *IEEE Communications Surveys and Tutorials*, vol. 3, pp. 2–15, 2000.
24. http://www.quasarspectrum.eu/.
25. J. Zander, L. K. Rasmussen, K. Sung, P. Mähönen, M. Petrova, R. Jäntti, and J. Kronander, "On the Scalability of CR: Assessing the Commercial Viability of Secondary Spectrum Access," *IEEE Wireless Communications*, vol. 20, no. 2, pp. 28–36, 2013.

26. R. D. Yates, C. Raman, and N. B. Mandayam, "Fair and Efficient Scheduling of Variable Rate Links via a Spectrum Server," *Proceedings of the IEEE International Conference on Communications*, vol. 11, pp. 5246–5251, 2006.

27. M. M. Buddhikot, P. Kolodzy, S. Miller, K. Ryan, and J. Evans, "DIMSUMNet: New Directions in Wireless Networking Using Coordinated Dynamic Spectrum Access," *Proceedings of the IEEE International Symposium on a World of Wireless Mobile and Multimedia Networks*, pp. 78–85, 2005.

28. K. Akkarajitsakul, E. Hossain, D. Niyato, and D. I. Kim, "Game Theoretic Approaches for Multiple Access in Wireless Networks: A Survey," *IEEE Communications Surveys & Tutorials*, vol. 13, no. 3, 2011.

29. J. Bater, H. Tan, K. N. Brown, and L. Doyle, "Maximising Access to a Spectrum Commons Using Interference Temperature Constraints," *Proceedings of the 2nd International Conference on CR Oriented Wireless Networks and Communications*, pp. 441–447, August 2007.

30. L. Wang, X. Xu, W. Xu, Z. He, and J. Lin, "A Nash Bargaining Solution Based Cooperation Pattern for Open Spectrum CR Networks," *Proceedings of the IEEE International Conference on Wireless Information Technology and Systems*, pp. 1–4, September 2010.

31. X. Chen and J. Huang, "Evolutionarily Stable Open Spectrum Access in a Many-Users Regime," *Proceedings of the IEEE Global Telecommunications Conference*, pp. 1–5, 2011.

32. H. Li, D. Grace, and P. D. Mitchell, "Multiple Access with Multi-Dimensional Learning for CR in Open Spectrum," *Proceedings of the 11th International Symposium on Communications and Information Technologies*, pp. 298–302, October 2011.

33. H. Li, D. Grace, and P. D. Mitchell, "CR Multiple Access Control for Unlicensed and Open Spectrum with Reduced Spectrum Sensing Requirements," *Proceedings of the 7th International Symposium on Wireless Communication Systems*, pp. 1046–1050, September 2010.

34. J. Bae, E. Beigman, R. Berry, M. L. Honig, and R. Vohra, "Incentives and Resource Sharing in Spectrum Commons," *3rd IEEE Symposium on New Frontiers in Dynamic Spectrum Access Networks*, 2008.

35. www.ict-phydyas.org/.

36. B. Farhang-Boroujeny, "OFDM versus Filter Bank Multicarrier," *IEEE Signal Processing Magazine*, vol. 28, no. 3, pp. 92–112, 2011.

37. P. Gao, J. Wang, and S. Li, "Non-Contiguous CI/OFDM: A New Data Transmission Scheme for CR Context," *Proceedings of the IEEE International Conference Wireless Information Technology and Systems*, pp. 1–4, August 2010.

38. S. Alrabaee, A. Agarwal, N. Goel, M. Zaman, and M. Khasawneh, "Comparison of Spectrum Management without Game Theory and with Game Theory for Network Performance in CR Network," *Proceedings of the 7th International Conference on Broadband, Wireless Computing, Communication and Applications*, pp. 384–355, November 2012.

39. S. S. Barve and P. Kulkarni, "Dynamic Channel Selection and Routing through Reinforcement Learning in CR Networks," *Proceedings of the International Conference on Computational Intelligence and Computing Research*, pp. 1–7, December 2012.

40. Z. Gao, H. Zhu, Y. Liu, M. Li, and Z. Cao, "Location Privacy in Database-Driven CR Networks: Attacks and Countermeasures," *Proceedings of the INFOCOM*, pp. 2751–2759, April 2013.

41. J. Ojaniemi, J. Kalliovaara, A. Alam, J. Poikonen, and R. Wichman, "Optimal Field Measurement Design for Radio Environment Mapping," *Proceedings of the Annual Conference on Information Sciences and Systems*, pp. 1, 6, March 2013.

42. Y. Zhao, J. Gaeddert, K. K. Bae, and J. H. Reed. "Radio Environment Map Enabled Situation-Aware CR Learning Algorithms," *Proceedings of the SDR Forum Technical Conference*, 2006.

43. Z. Wei, Q. Zhang, Z. Feng, W. Li, and T. A. Gulliver, "On the Construction of Radio Environment Maps for CR Networks," *Proceedings of the Wireless Communications and Networking Conference*, pp. 4504–4509, April 2013.

44. Y. Huseyin Birkan and T. Tugcu, "Location Estimation Based Radio Environment Map Construction in Fading Channels," *Proceedings of the Wireless Communications and Mobile Computing*, pp. 1–5, 2013.

45. J. van de Beek, T. Cai, S. Grimoud, B. Sayrac, P. Mahonen, J. Nasreddine, and J. Riihijarvi, "How a Layered REM Architecture Brings Cognition to Today's Mobile Networks," *IEEE Wireless Communications Magazine*, vol. 19, no. 4, pp. 17–24, 2012.

46. M. Höyhtyä, J. Vartiainen, H. Sarvanko, and A. Mammela, "Combination of Short Term and Long Term Database for CR Resource Management," *Proceedings of the International Symposium on Applied Sciences in Biomedical and Communication Technologies*, pp. 1, 5, 7–10, November 2010.

47. S. N. Khan, M. A. Kalil, and A. Mitschele-Thiel, "Distributed Resource Map: A Database-Driven Network Support Architecture for Cognitive Radio Ad Hoc Networks," *4th International Congress on Ultra Modern Telecommunications and Control Systems and Workshops*, pp. 188, 194, October 3–5, 2012.

48. R. A. Saeed and S. J. Shellhammer, *TV WS Spectrum Technologies: Regulations, Standards, and Applications*, CRC Press, December 2011.

49. R. A. Saeed, R. A. Mokhtar, J. Chebil, and A. H. Abdallah, "TVBDs Coexistence by Leverage Sensing and Geo-Location Database," *Proceedings of the International Conference on Computer and Communication Engineering*, pp. 33–39, July 2012.

50. FCC Encyclopedia, "Chairman Genachowski Announces Approval of First Television WSs Database and Device," December 22, 2012. Available: http://www.fcc.gov/encyclopedia/white- space-database-administration.

51. FCC, "Office of Engineering and Technology Announces the Approval of Telcordia Technologies, Inc.'s TV Bands Database System for Operation," March 26, 2012. Available: http://www.fcc.gov/encyclopedia/white-space-database-administration.

52. V. Goncalves and S. Pollin, "The Value of Sensing for TV WSs," *Proceedings of the IEEE Symposium on Frontiers in Dynamic Spectrum Access Networks*, pp. 231–241, May 2011.

53. S. Probasco and B. Patil, "Protocol to Access WS Database: PS, Use Cases and RQMTS," http://tools.ietf.org/html/draft-ietf-paws-problem-tmtusecases-rqmts-08, August 2012.

54. G. P. Villardi, Y. D. Alemseged, C. Sun, C.-S. Sum, T. H. Nguyen, T. Baykas, and H. Harada, "Enabling Coexistence of Multiple Cognitive Networks in TV WS," *IEEE Wireless Communications*, vol. 18, no. 4, pp. 32–40, 2011.

55. P. Kaur, M. Uddin, and A. Khosla, "Fuzzy Based Adaptive Bandwidth Allocation Scheme in Cognitive Radio Networks," *Proceedings of the 8th International Conference on ICT and Knowledge Engineering*, November 2010.

56. S. R. Aryal, H. Dhungana, and K. Paudyal, "Novel Approach for Interference Management in Cognitive Radio," *Proceedings of the 3rd Asian Himalayas International Conference on Internet*, November 2012.

57. M. Matinmikko, J. Del Ser, T. Rauma, and M. Mustonen, "Fuzzy-Logic Based Framework for Spectrum Availability Assessment in Cognitive Radio Systems," *IEEE Journal on Selected Areas in Communications*, vol. 31, no. 11, pp. 1136–1159, 2013.

58. H. Qin, L. Zhu, and D. Li, "Artificial Mapping for Dynamic Resource Management of Cognitive Radio Networks," *Proceedings of the 8th International Conference on Wireless Communications, Networking and Mobile Computing*, September 2012.

59. M. Riaz Moghal, M. A. Khan, and H. A. Bhatti, "Spectrum Optimization in Cognitive Radios Using Elitism in Genetic Algorithms," *Proceedings of the 6th International Conference on Emerging Technologies*, October 2010.

60. J. F. Hauris, D. He, G. Michel, and C. Ozbay, "Cognitive Radio and RF Communications Design Optimization Using Genetic Algorithms," *Proceedings of the IEEE Military Communications Conference*, October 2007.

61. Y. Yang, H. Jiang, C. Liu, and Z. Lan, "Research on Cognitive Radio Engine Based on Genetic Algorithm and Radial Basis Function Neural Network," *Proceedings of the Spring Congress on Engineering and Technology*, May 2012.

62. Y. Li, H. Zhang, and T. Asami, "On the Cooperation between Cognitive Radio Users and Femtocell Networks for Cooperative Spectrum Sensing and Self-Organization," *Proceedings of the Wireless Communications and Networking Conference*, April 2013.

63. P. Zhou, Y. Chang, and J. A. Copeland, "Reinforcement Learning for Repeated Power Control Game in Cognitive Radio Networks," *IEEE Journal on Selected Areas in Communications*, vol. 30, no. 1, pp. 54–69, 2012.

64. W. Jouini, C. Moy, and J. Palicot, "On Decision Making for Dynamic Configuration Adaptation Problem in CR Equipments: A Multi-Armed Bandit Based Approach," *Proceedings of the of 6th Karlsruhe Workshop on Software Radios*, pp. 21–30, March 2010.

65. A. Sanchez, I. Moerman, S. Bouckaert, D. Willkomm, J. Hauer, N. Michailow, G. Fettweis, L. Dasilva, J. Tallon, and S. Pollin, "Testbed Federation: An Approach for Experimentation-Driven Research in CRs and Cognitive Networking," *Proceedings of the Future Network Mobile Summit*, pp. 1–9, 2011.

66. J. Xiao, R. Hu, Y. Qian, L. Gong, and B. Wang, "Expanding LTE Network Spectrum with CRs: From Concept to Implementation," *IEEE Wireless Communications*, vol. 20, no. 2, pp. 12–19, 2013.

67. T. Zhang and S. Banerjee, "Inaccurate Spectrum Databases? Public Transit to Its Rescue!" *ACM HotNets Conference*, November 2013.

68. H. Kobeissi, O. Bazzi, and Y. Nasser, "Wavelet Denoising in Cooperative and Non-Cooperative Spectrum Sensing," *Proceedings of the 20th International Conference on Telecommunications*, pp. 1–5, May 2013.

69. W. Yu and R. Lui, "Dual Methods for Nonconvex Spectrum Optimization of Multicarrier Systems," *IEEE Transactions on Communications*, vol. 54, no. 7, pp. 1310–1322, 2006.

70. W.-H. Kuo and W. Liao, "Utility-Based Resource Allocation in Wireless Networks," *IEEE Transactions on Wireless Communication*, vol. 6, no. 10, pp. 3600–3606, 2007.

71. L. Narayanan, *Channel Assignment and Graph Multicoloring*, Wiley, pp. 71–94, 2002.

72. P. Pawelczak, K. Nolan, L. Doyle, S. Oh, and D. Cabric, "Cognitive Radio: Ten Years of Experimentation and Development," *IEEE Communications Magazine*, vol. 49, no. 3, pp. 90–100, 2011.

73. V. T. Nguyen, F. Villain, and Y. L. Guillou, "Cognitive Radio RF: Overview and Challenges," Hindawi Publishing Corporation, 2012.

74. B. Razavi, "Cognitive Radio Design Challenges and Techniques," *IEEE Journal of Solid-State Circuits*, vol. 45, no. 8, pp. 1542–1553, 2010.

75. E. P. Tsakalak, O. N. Alrabadi, A. Tatomirescu, E. de Carvalho, and G. F. Pedersen, "Antenna Cancellation for Simultaneous Cognitive Radio Communication and Sensing," *International Workshop on Antenna Technology*, pp. 215–218, 2013.

76. J. I. Choi, M. Jain, K. Srinivasan, P. Levis, and S. Katti, "Achieving Single Channel, Full Duplex Wireless Communication," *16th Annual International Conference Mobile Computing and Networking*, pp. 1–12, 2010.

Biographical Sketches

Youssef Nasser obtained his master's and PhD degrees in signal processing and communications from the National Polytechnic Institute of Grenoble, Grenoble, France, in 2003 and 2006, respectively. From 2003 to 2006, he worked with the Laboratory of Electronics and Information Technologies in Grenoble [Laboratoire d'Electronique et de Technologies de l'Information (LETI)] as an

R&D engineer. From 2006 to 2010, he worked as a senior research engineer/assistant professor with the Institute of Electronics and Telecommunications of Rennes (IETR), Rennes, France. During the same period, he was a key research collaborator with different European R&D centers such as BBC R&D, Panasonic (Germany), DLR (Germany), and Turku University (Finland). He is currently with the department of electrical and computer engineering at the American University of Beirut, Beirut, Lebanon.

Since 2007, Nasser has been involved in the standardization of European broadcasting systems including DVB-T2, DVB-T2 Lite, and DVB-NGH. He is the technical program committee chair and co-organizer of many conferences. He is a member of the Radio Communications Committee (RCC) and the Ad Hoc and Sensor Networks (AHSN) Committee of the Communications Society (ComSoc). He is a voting member of the Institute of Electrical and Electronics Engineers (IEEE) Standards Association and member of the DVB Project. He is a senior member of the IEEE.

Nasser was a guest editor-in-chief of many international journals. Nasser has published more than 80 papers in international peer-reviewed journals and conferences. He is an active technical program committee member of different IEEE conferences such as ICC, GlobeCom, SPAWC, MC-SS, ISWPC, and ICSPCS. He has coordinated tasks and/or packages of many European projects (4MORE, B21C, NewComm++, ENGINES, etc.) on cooperative networks, broadcasting systems, and cross-layer design for cellular communications systems. These projects received the Golden and Silver awards from the European Union for their achievements. Nasser is currently coordinating two projects, "Localization in Heterogeneous Networks" and "Convergence between DVB and LTE Systems."

Mariette Awad (mariette.awad@aub.edu.lb) received her PhD in electrical engineering from the University of Vermont, Burlington, Vermont, in 2007. Prior to her academic position, she was with IBM System and Technology group in Vermont as a wireless product engineer. Over the years, her technical leadership and innovative spirit have earned her management recognition, several business awards, and multiple patents at IBM. She was invited by CVC labs, Google, and Qualcomm to present her work on machine learning and image processing. She has been a visiting professor at Virginia Commonwealth University, Richmond, Virginia; Intel Mobile Group, and Massachusetts Institute of Technology (MIT), Cambridge, Massachusetts. She is an assistant professor in the electrical and computer engineering department of the American University of Beirut, Beirut, Lebanon. She is an active member of the Institute of Electrical and Electronics Engineers (IEEE) and a reviewer for IEEE transactions. She has published in numerous conferences and journals and is managing few grants. Her current research interests include machine learning, game theory, energy-aware computing, control, and cybernetics.

Ali Yassin was born in A'adchite, Lebanon, in 1989. He received his BE and ME degrees in 2012 and 2014, respectively, from the American University of Beirut, Beirut, Lebanon. Since then, he has been a research assistant at the American University of Beirut, and he is now willing to start the PhD program. His research interests are in cooperative positioning and localization techniques in HetNet and cognitive network, in addition to different tracking filters.

Youssef A. Jaffal received his BEng degree in electrical and computer engineering (with distinction) from the American University of Beirut, Beirut, Lebanon, in 2012. He worked as a research assistant at the German Aerospace Center for 4 months in 2013. Currently, he is working on his master thesis at the American University of Beirut. His current research interests include wireless communications, MIMO systems, resource allocation, mobile radio positioning, and digital implementation of communication systems.

PART IV
ROUTING

10

INTELLIGENT ROUTING IN GRADED COGNITIVE NETWORKS

T. R. GOPALAKRISHNAN NAIR
AND KAVITHA SOODA

Contents

10.1 Introduction

In recent years, it has become increasingly difficult to operate large networks for performing diagnostics and preventing cascading failures. Hence, there is a requirement for networks to think and learn in a nondeterministic way. This is where cognitive approach plays a vital role to overcome the shortcomings. Cognition is the ability to effectively *self-regulate*, *learn*, and *evolve*. Cognitive network (CN) aspects have become crucial when a system is subject to a complex and varied set of stimuli, which is certainly the case of fast-evolving Internet of Things. CNs require each node to cooperate with the data distribution process and make use of information about the network scenario. To get CN working, there is a need to rethink on the architecture and protocols of the components in the global communication infrastructure. A prominent research direction looks into how to mimic nature-like mechanisms to realize smarter communication networks, which in turn can make sense of the hidden communication patterns and do the self-regulation of the topology.

The routing decisions in the existing networks are generally based on the table-driven systems provided at the node level. The current system usually has less awareness of the environment around it. This is where intelligent networks can play a significant role. Many researchers have been working on the area of autonomic network (AN) since 2003. The companies such as Motorola and IBM were some of the initial players in this domain. The aspects of game theory, probability, linear programming, evolutionary algorithm, genetic algorithm, artificial immune system, artificial intelligence, and many more stochastic approaches have been applied to achieve the awareness and learning capabilities in the network. Once there is awareness about the network, routing can be performed effectively. In order to meet the demand for improved network, the nodes need to be intelligent and capable of making decisions on their own. The current nodes of the communication systems usually do not have the information regarding the topology achieved that might have been formed during the routing. In this situation, it is always desirable for an advanced system to learn the environment of routing and remember it for future operations. Hence, the nodes need to have a mechanism to collect the data and become aware of the vital parameters of the network and, if required, communicate this intelligently to other participating nodes. In addition to this, advance networks need to achieve the ability to learn, remember, and reason out in a way as presented in Figure 10.1, which was initially proposed by Mitola [1].

CN implements an approach whereby each node cooperates with others by distributing quality-of-service (QoS) metric parameters and other information-rich data set in the network. In order to transform the existing system to CNs, there is a need to reorient the architecture and protocols of the subsystems in the CN [2,3]. The main orientations for achieving a CN are to be directed toward learning and reasoning. The available learning methods include reinforcement learning, Q-learning, foraging algorithms, evolutionary algorithms, and neural

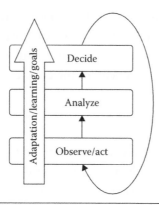

Figure 10.1 Cognitive cycle.

networks. The net result of the learning process happening over the network will later be made available as an information database of all required nodes. The knowledge thus formed facilitates a useful set of information on network status for further reference. This concept of further reference is depicted in Figure 10.2. The reasoning techniques include proactive, reactive, inductive, deductive, one-shot, sequential, centralized, and distributed methods. They help the extraction of quality nodes while exploring the optimal path. Routing is a fundamental function to create dynamic interconnections between the end nodes that are not directly linked. Current networks and nodes have only a rudimentary mechanism to build paths and respond to congestion. These networks are not able to adjust to different types of stimuli and contextual conditions with their deterministic way of functioning.

The remainder of the chapter is organized as follows: Section 10.2 deals with the work done in the field of CNs that are discussed along with the developed products. Section 10.3 presents learning

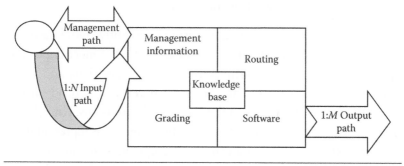

Figure 10.2 Knowledge base design.

and reasoning for CNs. Section 10.4 presents the bioinspired intelligent networks. Sections 10.5 through 10.7 discuss the graded cognitive network, simulation results, and future research direction, respectively. Section 10.8 concludes the chapter.

10.2 Background

The section presents a detailed description on the existing technology depicting the concept of CN in the cognitive domain. One of the initial incorporation of intelligence in the communication domain took place in the development of cognitive radio in order to detect the unused bandwidth. More or less in the same way several improvements were brought into networking domain, which also uses the principles of CN such as autonomic networking, Motorola foundation–observation–comparison–action–learn–reason (FOCALE), and software-defined networking (SDN), to name a few, which have tried to infuse better logic and decision-making process into the network.

10.2.1 Cognitive Radio

Cognitive radio is an intelligent wireless communication system that is capable of sensing its surrounding signals and organizing its required bandwidth resources. It adopts internal states to the incoming RF stimuli by making corresponding changes in certain operating parameters (e.g., transmit power, carrier frequency, and modulation strategy) in real time. The two main purposes for making changes in operating parameters are as follows:

- Reliable communications whenever and wherever needed
- Efficient utilization of the radio spectrum

We can see that 50% of the nomenclature in cognitive radio development has led to further development in CNs [1]. The most important aspects of this could be learning from the past decisions and using them for future behavior. Both are goal driven and depend on the observations and knowledge of the node to take decisions. Knowledge in cognitive radio is represented by radio knowledge representation language (RKRL). The two attributes of cognitive radio are goal orientation and context-awareness. For this to exist in CNs, a network level equivalence needs to be realized. The optimization space requires tunable parameters in cognitive radio provided by software-defined radio (SDR), which is comparable to software-adaptable network (SAN). Therefore, it could be assumed that both technologies utilize a software tunable platform that is controlled by the cognitive process.

10.2.2 Autonomic Network

An AN is realizable in a situation where states are fairly predictable, and the search space or paths can be handled fairly well using an algorithm. When the environment is larger, the cognitive approach will play its role more globally with learning and reasoning [4]. IBM was one of the players to propose the AN concept. It described the agents in three dimensions. The first dimension is the agency where it determines the degree of autonomy and authority assigned to the agent. The second dimension is about the intelligence describing the reasoning and the concurrent learned behavior. The third dimension is the mobility of the cognition enabling agents that travel through the network. The current routing approaches are based on packet delivery. This type of operation is asynchronous and loosely coupled to a complex network that satisfies the lowest degree of autonomy. The intelligent axis is not considered in the current routing along with user preference and is not given enough importance. In the case of CNs, it is supposed to utilize intelligent agents for knowledge plane management. On the agency axis, the agent can act on behalf of user requirements especially based on data, application, or services. On the intelligent axis, the agent can hold a model and perform reasoning, planning, and learning on it. These agents can be run statically or as a dynamic mobile script, which is almost similar to the cognition loop, and thus, it provides a chance for an intelligent approach as presented in Table 10.1.

Table 10.1 IBM's View of ANs

INTELLIGENCE X-AXIS	AGENCY Y-AXIS	MOBILITY Z-AXIS
Preferences	Service interactivity	Static
Reasoning	Application interactivity	Mobile scripts
Planning	Data interactivity	Mobile objects
Learning	Representation of user	
	Asynchrony	

10.2.3 *Motorola FOCALE*

The autonomic management platform developed by Motorola is referred to as FOCALE for core networks. This development was based on the human nervous system that performs unconscious actions. The system actually tries to figure out the functions not known to the human. This system contains all the elements of cognitive approaches. It has two control loops: one for maintaining the current state and the other for reconfigurations. The idea behind this development was to consider the business-level requirement in simple language. Transformation to the autonomic configured network

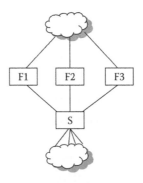

Configuration I			
	P	T	Action
	In-port match	Packet type	
S	1,2		Fwd F1
	3,4		Fwd F2
	5,6		Fwd F3
F1		A,B	Monitor
			Allow
F2			Allow
F3			Allow

Figure 10.3   SDN access control flow. Here S, F1, F2 and F3 are node names of which P is the port number and T is the type of data. A,B are the names for packet type and Fwd refers to forward which is the action taken.

takes place where it gets converted to a form that would sense and act according to the requirement of the end user.

10.2.4 Software-Defined Networking

The technique used in SDN is based on a centralized controller that directly dictates where the packet should move. Here, there is some support for multiswitch updates, which makes it strong without asking the user to program with low-level commands. In this case, we do not have to worry about the step-by-step updates. But programmers who use this must be able to distinguish between the multiple abstract updates that are useful to one another as presented in Figure 10.3. This will require a consistency class, mechanism, and language to code for proper setup of SDN [5].

10.3 Learning and Reasoning for CNs

Learning and reasoning plays an important role in the CN. Reasoning helps in the immediate decision process using the historical knowledge as well as the current state of the network. The primary goal of reasoning is to choose a set of actions. Learning is a long-term process that is based on the accumulation of knowledge on the perceived result of past actions. CN nodes learn by enriching the knowledge base so that the efficacy of reasoning improves in future.

In general, any CN structure will require some assistant to accomplish a certain amount of accounting for the optimal solution. This is where agents will assist in performing the job of optimization. A cognition agent in general will work on a particular problem space, where it tries to achieve its goal and remember it for future purpose. To take proper action, it looks into what was learned along with the current state and takes a decision based on belief and goal.

Most of the CN systems need to work with the support of a generalized agent mechanism. The agent extracts the required parameters from the information base and operates on the network topology. The network topology supports the reasoning to obtain the optimal path and to choose an action to be executed in the environment. The reasoning process is driven by the data structures in the information base. Agents are tuned based on a set of training data set, and probabilistic reasoning is the best approach as errors are inevitable.

Another approach for reasoning includes learning from how the human brain works. The brain helps in the movement, sensing, and regulation of involuntary body processes such as breathing. The study of brain has helped in understanding human behavior, thoughts, and feelings. All these actions take place based on the reasoning involved in the brain, which map to the properties required for implementing a CN.

10.4 Bio-Inspired Intelligent Network

The evolution of the Internet has seen a paradigm shift from static, hierarchical network structures to highly distributed, reconfigurable systems without any form of significant intelligent control over its operation. For networking nodes, the ability to self-adapt and self-organize in a changing environment has become a key issue. In conventional Internet usage, there is a hierarchical order with centralized and static control. But with the progressive growth of Internet, there has been a drift in static control. The working mode has shifted to dynamic control in an ad hoc mode of connectivity among the participating nodes. For these types of new dynamic networks, the following network controls are considered mandatory:

- *Expandability (or scalability)*: This feature of network defines the increasing number of users on the network such as Internet and the devices attached to the Internet.
- *Mobility*: The mobility of the device moving from one to another network space must be handled in a probabilistic way.
- *Diversity*: The network must be able to handle new software and hardware as it creates new challenges in the environment.

The above characteristics can be met if the nodes or end hosts are equipped to adapt to the current network status dynamically. This is where bio-inspired approaches can facilitate a promising direction as they are capable of self-adaptation. The methods we apply can be slow to adapt to change in the environment. The research results in bio-inspired algorithms, although they are not so new, were applied mainly for optimization of network control successfully. There is further scope to explore and get inspired from the nature and focus on scalability, adaptability, self-organization, and robustness properties of networks.

Though networks can be tuned to the properties derived from the biological systems, limitations do exist in the application of properties as adaptable to the Internet. A trade-off exists while we apply the concept on distributed and centralized control systems. Though scalability has been realized well on the Internet, the approaches applied for this distributed system come with a performance cost. There can be a lack of global view of the entire distributed system. Methods applied with a pure local view would only yield suboptimal solutions, which probably could not be accepted. Therefore, a trade-off must be incorporated into the distributed system as to what is required by the end user in terms of network performance. A distributed system needs to manage the scalability of the nodes as well as preserve the advantages of the centralized control during critical situations.

Self-organization can be achieved by molecular processing, adaptive response, and artificial immune system. The bio-inspired architecture includes swarm intelligence, ant colony optimization, AntNet, and AntHocNet, to name a few. The adaptive systems can be developed based on the principles of how species evolve. Genetic algorithms [6,12], particle swarm optimization [7], artificial bee colony (ABC) [8,14], and memetic [9,13] have been the major application algorithms in this direction of research.

10.5 Graded Cognitive Network

A grade is like an index, which represents the satisfied status of node QoS parameters. The status helps in estimating the selection of the node quickly once the network is realized. The realization of such parameters is done once in half an hour, and thus, faster convergence to the optimal path takes place when the learning for grade is done [10].

With higher bandwidth available in forward channels, the nature of the parameter is *good*, and also the grade assigned will be *best*. Resource availability is mainly related to the availability of the buffer. If the buffer is available for taking the excess demand from all channels, the grade is the maximum. If the buffer is completely not available, then the grade is low. Delay is the maximum time for a packet to reach destination. If packet loss is high, contribution to the grade is very low. If packet loss is low, then the grade is better.

The nodes participating in routing are studied under various parameters and evaluated based on Equation 10.1. The grade value obtained by this equation for a given node is thus stored in the memory of the node for further reference.

Grade value = $(a1 \times$ contribution from $B) + (a2 \times$ contribution from $RA)$

$+ (a3 \times$ contribution from $De) + (a4 \times$ contribution from $PL)$

$$\text{Grade value} = (a1 \times B) + (a2 \times RA) + (a3 \times De) + (a4 \times PL) \quad (10.1)$$

where:
   B is the bandwidth of the link
   RA is the resource availability
   PL is the packet loss
   De is the delay
   $a1, a2,\ldots, a4$ are the grade coefficients and full grade $= 1$

A higher grade value indicates better satisfaction of the parameters considered. Figure 10.4 depicts the grade value graphically with p5 indicating all parameters being satisfied and p1 indicating none satisfied.

A graded model is a centralized abstract entity that is usually an intelligent algorithm and does not require a central controller as in SDN.

$$\text{Grade value} = 0.6 \times B + 0.2 \times RA$$
$$+ 0.1 \times De + 0.1 \times (1 - PL) \quad (10.2)$$

Here, the channel depends on the bandwidth. We have not considered the congestion scenario for the node's grade calculation.

This can be shown mathematically with lumped polynomial and coefficient estimations heuristically. Staggered levelling is practiced in most of the network companies. Here bandwidth is considered as the major contributor to determine the capacity of the link for forwarding data. The grade value swings between 0 and 1 with a right parameter of the model equation stated above.

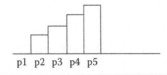

p1 p2 p3 p4 p5

Figure 10.4 Parametric grading.

In this work, we perform two types of grading operation: fixed time grading and variable time grading. If an optimal path is not obtained, it means to say that grading conditions are not satisfied. If quality deteriorates in the topology considered, then we go for fixed time grading, that is, we obtain new topology and assign new random values to the nodes where at least 70% of them are of good quality. Once the new setup is performed, we keep on trying to obtain the paths that undergo variable time grading. In variable time grading, continuous paths are obtained for which the quality gets updated automatically based on the situation. If most of the nodes fail to forward the data at a given point of time, then we go for fixed time grading.

Here we consider the coordinate position of the source and destination nodes to determine the reference line. Once the reference line is obtained, we fix the cone angle to be 15°, which is depicted as B in Figure 10.5. This angle is drawn on either side of the reference line drawn from the coordinate $S(x1,y1)$ and $D(x2,y2)$. Given the coordinates of two points and the angle known, we can get the slope of the lines that can be determined on either side of the reference line to obtain the points $(x3,y2)$ and $(x4,y2)$. This cone now is reduced to a triangular search space as shown in Figure 10.5.

$$m = \frac{y2 - y1}{x2 - x1} \tag{10.3}$$

$$m = \tan A \tag{10.4}$$

$$A = \tan^{-1} m \tag{10.5}$$

$$\tan(A + B) = m1 \tag{10.6}$$

$$\tan(A - B) = m2 \tag{10.7}$$

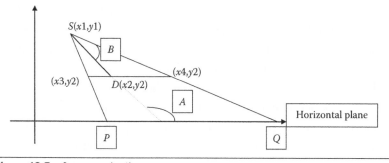

Figure 10.5 Cone approximation.

Therefore, the equation of *SQ* from (*x*1,*y*1) having slope *m*1 is

$$(y2 - y1) = m1(x3 - x1) \qquad (10.8)$$

Therefore, the equation of *SP* from (*x*1,*y*1) having slope *m*2 is

$$(y2 - y1) = m2(x4 - x1) \qquad (10.9)$$

The reason for this structure is to reduce the search space. This cone is based on the direction of the destination node where the source acts like the starting vertex. If the path to the desired node is not determined by the cone, then the initial angle is expanded and cone expansion takes place. The procedure now again repeats for path determination. Within this cone structure, the nodes that belong to this are graded. This grade index for the nodes helps in determining the quality of the node. The quality is prioritized based on the number of criteria each individual node meets. The grade signifies the quality of the router that is nothing but the knowledge of the environment.

In the design of network model, we have all the nodes connected either directly or indirectly with one another. It is referred to as a connected graph. Suppose that the network has a set of nodes *N*. The neighbors of each node $i \in N$ are in the corresponding set Ne_i. All links in the network belong to the set *L*. In this model, we send data to a single destination *D* at a time. For the given source and destination nodes, we draw a conical structure for determination of optimal path. The cone is drawn with 15° angle from the reference line drawn between the source and the destination on either side.

In this model, in every region, the node information is collected by the management information base. The agent then does grading that is referred to as parametric level assignment. The bio-inspired algorithm is executed by the routing software module and the output is brought into action by the routing hardware that would select one of the many output links to forward the data.

In this model, depending on the algorithm chosen, we select the forward hop at every hop or choose a complete path from the source to the destination. Figure 10.6 shows the hop-level selection after grading, which is actually done by the chosen bio-inspired algorithm.

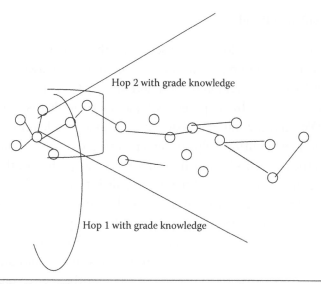

Figure 10.6   Hop-level jump.

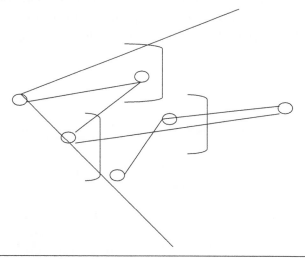

Figure 10.7   Zigzag formation of path.

Here is a likely chance that the path can go in a zigzag fashion as it depends on the parametric grade level and also the bandwidth available on the link as shown in Figure 10.7.

Search will take place at every hop and will form a zigzag pattern depending on the topology setup. Sometimes the data path will be almost full, and hence, its grade value will be low. The nodes will be queried for their quality and through the management path. Depending on the grade value obtained, the data will be sent to the data path successfully.

10.6 Simulation Results

In the simulation developed, the user enters the number of nodes for a random run. Depending on the number entered by the user, the nodes are spread across the region of view and links get generated as shown in Figure 10.8 [11]. The required QoS parameter is also assigned randomly. We enter the source and destination nodes for determining the path. The Learn button helps in remembering the path already executed on the run. This button also carries out the grading of the node, which signifies the quality of the node. This information is useful for future, that is, if such a pair of source and destination nodes is again sought, there is no need for the path to be computed again by the algorithm. Instead, the path can be determined from the information base. We choose the desired algorithm for viewing the output, which is selected with the help of logical reasoning for the graded node.

Here, the data are assigned based on random function. Table 10.2 gives a detailed description of the experimental runs carried out for node numbers varying from 32 to 8192, which are all a power of 2.

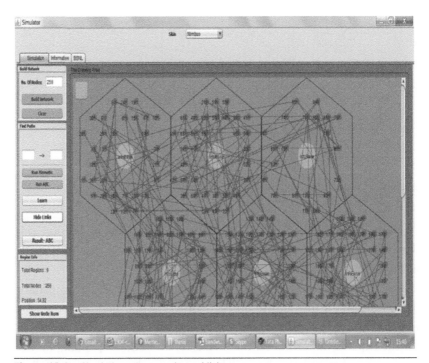

Figure 10.8 Simulation setup—nodes and links.

Table 10.2 Simulation Setup Time

NUMBER OF NODES	MEMETIC		ABC	
	FITNESS VALUE	BANDWIDTH	FITNESS VALUE	BANDWIDTH
32	0.8284	130	0.974	125
64	0.87782	469	0.821	362
128	0.8105	277	0.8107	271
256	0.8205	548	0.827	627
512	0.8595	573	0.861	549
1024	0.8153	470	0.8143	505
2048	0.864	531	0.811	419
4096	0.816	415	0.826	422
8192	0.857	730	0.8553	741

Table 10.3 Simulation With and Without Cone

NUMBER OF NODES	REDUCED SPACE	AGENT	SIMULATION SETUP TIME	
			USING CONE	WITHOUT CONE
32	18	3	133	83
64	44	4	182	35
128	58	5	259	18
256	114	8	407	59
512	159	18	676	160
1024	252	34	1542	392
2048	326	69	2178	572
4096	432	141	1506	443
8192	270	275	11700	1477

Using the cone approach, we reduce the search space. Each region has one agent that collects information about the nodes belonging to the region and also assigns the grade value to each node. We observe that the simulation setup time using cone is higher than that without using the cone. This is because learning aspect takes place within the cone. The learned knowledge is made available in the information base which is nothing but the agent. The number of nodes that participate without using the cone is the complete search space, while there is a considerable reduction in the search space when cone is used for determination of optimal path as shown in "Reduced space" column of Table 10.3.

Table 10.2 shows the way in which two algorithms perform on the graded network while using the with and without cone concept. ABC algorithm is roughly 20% better than memetic algorithm. The total

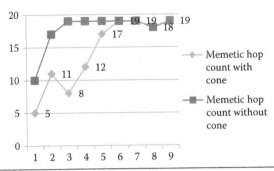

Figure 10.9 Memetic algorithm—hop count comparison.

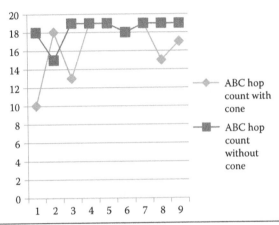

Figure 10.10 ABC algorithm—hop count comparison.

throughput provided by memetic is more in five out of nine cases compared to ABC.

The graphs in Figures 10.9 and 10.10 show the hop count obtained when memetic and ABC algorithms were carried out with cone and without cone, respectively. While considering the hop count as the parameter for success, the cone actually helped in obtaining the shorter path with required quality providing a complete path.

The simulation studies were carried out for a data set obtained from an actual network operating center, an Internet Service Provider (ISP) in Karnataka, India, and it was recorded for 1 month. The data set considered here are spread across a few days with 51 nodes active in its topology.

The network delay depends on the location from which the client is trying to access the Internet. There exists an upper limit for it from a given location to the server that we are trying to reach. For example, from India we can get connected to Yahoo server at Singapore with a maximum of 120 ms, the United Kingdom with 200 ms, and the United States with 350 ms approximately. On the Internet, if 80% of the link is occupied, then congestion is definite. Depending on the Internet usage and service-level agreement made with the client, the network operator automatically deviates the packets with a little extra delay. The output that we see in the simulation is one of the paths from Bangalore to Yahoo server at Singapore with a connectivity of 66 ms and 11 hops. The other nodes have been designed at random and a value is assigned to them using the data set obtained. On average, it is known that a minimum of 7 hops and a maximum of 19 hops are required to connect to the server. Therefore, the simulation has considered the upper limit for hops as 19. It was observed that the real network output that has 11 hops to get connected to Singapore was also the same when we applied a cone and network grading approach.

10.7 Future Research Direction

It calls for substantial improvements in our understanding of how the intelligence is built into the systems in nature. There is a need for deep insight into information processing systems to achieve successful CNs of significant size. The designs for new computing elements and data networks need to be more mimicking natural brains than the rigid work patterns employed in the current systems of narrow outlook. Future research should utilize stronger techniques, and perhaps use intelligent databases to store past performances and outcomes. It must engage powerful learning schemes and stochastic meta-heuristics such as genetic algorithms or improved neural networks for reasoning. While game theory can provide a fair model for selfish behaviors, it does not represent the cooperative schemes well, in support of expanded networks. In order to realize, large cooperative networks with intelligence parallel reasoning schemes need to be investigated.

10.8 Conclusion

The application of intelligent approaches in network environments has been growing steadily over the past few decades. The development of better property optimization techniques in computational intelligence along with improved hardware employable in CNs will enable future network systems to work more intelligently. Several bio-inspired algorithms may find their place in future network construction based on a variety of applications that need to be managed in them.

In this chapter, we have presented the successful implementation of a network approach for optimal path determination by accounting the quality levels of the nodes. Here, the grading process done over the network promotes self-awareness among the nodes in the network. The optimal path from source to destination in a typical communication is obtained using intelligent algorithms along with the quality of the nodes. In order to achieve geographical binding, we engage an abstract a cone structure with vertex at source and the base capturing the destination node.

Out of all the intelligent algorithms, one significant aspect observed was related to ABC algorithm in the network. It outperformed the memetic technique in terms of throughput and fitness for optimal path determination. One significant achievement was the reduction of hop count that originated as a result of introducing the geographically constraint cone.

In this chapter, various details about the development and implementation of CNs have been presented along with sufficient emphasis given on intelligent routing techniques.

References

1. Joseph Mitola, Cognitive radio: An integrated agent architecture for software defined radio, PhD thesis, Royal Institute of Technology (KTH), Stockholm, Sweden (2001).
2. Ryan W. Thomas, Luiz A. DaSilva, and Allen B. Mackenzie, Cognitive networks, *Proceedings of the IEEE,* Baltimore, MD (November 2005) 352–360.
3. Carolina Fortuna and Michael Mohorcic, Trends in the development of communication networks: Cognitive networks, *Journal of Computer Networks,* 53 (2009) 1354–1376.
4. David Raymer, Sven van der Meer, and John Strassner, From autonomic computing to autonomic networking: An architectural perspective, *5th IEEE Workshop on Engineering of Autonomic and Autonomous Systems,* Washington DC (2008) 174–183.
5. Mark Reitblatt, Nate Foster, Jennifer Rexford, and David Walker, Consistent updates for software-defined networks: Change you can believe in! *HotNets,* 7, ACM, Cambridge, MA (2011).
6. T. R. Gopalakrishnan Nair and Kavitha Sooda, Application of genetic algorithm on quality graded networks for intelligent routing, *World Congress on Information and Communication Technologies,* Mumbai, India (December 11–14, 2011), pp. 558–563, doi:10.1109/WICT.2011.6141306.

7. T. R. Gopalakrishnan Nair and Kavitha Sooda, Particle swarm optimization for realizing intelligent routing in networks with quality grading, *7th IEEE International Conference on Wireless Communications, Networking and Computing*, Wuhan, People's Republic of China (September 23–25, 2011) 1–4.

8. Kavitha Sooda and T. R. Gopalakrishnan Nair, Optimal path selection in graded network using artificial bee colony algorithm with agent enabled information, *12th IEEE International Conference on Hybrid Intelligent Systems*, Pune, India (December 4–7, 2012), doi: 10.1109/HIS.2012.6421356.

9. Mohammed El-Abd, Performance assessment of foraging algorithms vs. evolutionary algorithms, *Information Sciences*, 182 (2012) 243–263.

10. T. R. Gopalakrishnan Nair and Kavitha Sooda, An intelligent routing approach using genetic algorithms for quality graded network, *International Journal of Intelligent Systems Technologies and Applications*, 1(3/4) (2012) 196–211, doi:10.1504/IJISTA.2012.052495.

11. Kavitha Sooda and T. R. Gopalakrishnan Nair, Competitive performance analysis of two evolutionary algorithms for routing optimization in graded network, *3rd IEEE International Advance Computing Conference IACC*, Ghaziabad, India (February 22–23, 2013), pp. 666–671, doi:10.1109/IAdCC.2013.6514306.

12. Aluizio F. R. Araujo and Maury M. Gouvea, Jr., Multicast routing using genetic algorithm seen as a permutation problem, *AINA Proceedings of the IEEE Computer Society*, 1 (2006) 6.

13. Xianshun Chen, Yew-Soon Ong, Meng-Hiot Lim, and Kay Chen Tan, A multi-facet survey on memetic computation, *IEEE Transactions on Evolutionary Computation*, 15(5) (2011) 591–607.

14. Dervis Karaboga and Celal Ozturk, A novel clustering approach: Artificial bee colony algorithm, *Applied Soft Computing*, 11 (2011) 652–657, doi:10.1016/j.asoc.2009.12.025.

Biographical Sketches

T. R. Gopalakrishnan Nair holds an MTech in electronics from the Indian Institute of Science (IISc), Bangalore, Karnataka, and PhD in computer science from the Indian Institute of Science (IISc), Bangalore, Karnataka. He has three decades of experience in computer science and engineering through research, industry, and education. He has published several papers and holds patents in multidomains. He has won the PARAM Award for technology innovation. Currently, he is the Saudi ARAMCO Endowed Chair of Technology, Prince Mohammad Bin Fahd University, Al Khobar, Saudi Arabia, and the vice president (sabbatical) DSI, Bangalore, India.

Kavitha Sooda holds an MTech in computer science and engineering from BMSCE, Bangalore. She is pursuing her PhD from Jawaharlal Nehru Technological University, Hyderabad, India. She has 12 years of teaching experience. She has published 13 papers. Her interest includes routing techniques, quality-of-service application, cognitive networks, and genetic algorithms. Currently, she is a research scholar at Advanced Networking Research Group, Research and Industry Incubation Center (RIIC), Dayananda Sagar Institutions, and an associate professor of the department of computer science and engineering, Nitte Meenakshi Institute of Technology, Bangalore, India.

11

ENERGY-EFFICIENT ROUTING PROTOCOL FOR COGNITIVE RADIO AD HOC NETWORKS

A Q-Learning Approach

LING HOU, K. H. YEUNG, AND ANGUS K. Y. WONG

Contents

11.1 Introduction

In recent years, a rapid increase of wireless devices and services has led to a dramatic increase in the demand for radio spectrum. But radio spectrum suitable for wireless communications is very limited. Thus, spectrum scarcity is becoming more and more serious. In parallel with that, many spectrum bands are extremely underutilized under traditional fixed spectrum assignment policy, as reported by the Federal Communications Commission (FCC) in [1]. Cognitive radio (CR) is then proposed as a promising technique to fully utilize the spectrum resource. CR technique enables unlicensed users or secondary users (SUs) to exploit the unutilized spectrum bands assigned to licensed users or primary users (PUs) without harmfully interfering PUs' services [2].

CR ad hoc networks (CRAHNs), networking of SUs, can provide an ultimate spectrum-aware and infrastructure-free communication paradigm [2]. They have a wide spread of applications, such as military surveillance deployment network and disaster response network. A large amount of research has been conducted on CRAHNs in lower layers, for example, spectrum sensing in physical layer [3–6] and spectrum management and nondisruptive spectrum handoff in media access control (MAC) layer [7–10]. In CRAHNs, due to the limited radio range of the CR link, it may be necessary for one SU to enlist the aid of other intermediate SUs in forwarding information to its destination. Thus, routing problem in the network layer should be well investigated to provide end-to-end communications for CRAHNs.

CRAHNs have some specific characteristics that differentiate them from traditional ad hoc networks. First, the channel availability in CRAHNs is significantly different from that in traditional ad hoc networks. The set of available channels at a SU is of time varying and changes in correlation with the PU activities [11,12]. Second, the mobility of SU and the changes in PU activity result in dynamically changing network topologies of CRAHNs. The changes in PU activity may affect several links at a same time [12]. Consequently, the routing protocols for traditional ad hoc networks cannot be applied directly to CRAHNs. Novel routing algorithms are required for CRAHNs to cope with the time-varying spectrum channel availabilities in lower layers.

Energy efficiency has also been a major design concern for CRAHNs. The energy consumption of many CR functions—observe, orient, decide, and act—is inherently high. The devices in CRAHNs are typically battery driven. Current battery techniques cannot support the CR devices work for a long period of time. Meanwhile, many application environments of CRAHNs, such as military surveillance, are harsh for battery replacement or energy harvesting. Therefore, energy efficiency becomes much more critical and challenging for CRAHNs. For a CRAHN to perform its function for as long period of time as possible, network lifetime awareness should be considered in the early design stage of CRAHNs [13]. In this chapter, network lifetime is the time span from the deployment of a network to the instant when the first CR node dies. Energy-efficient routing will be able to reduce the energy consumption of the relay SUs, balance the residual energy distribution in the network, and thus prolong the network lifetime of CRAHNs.

In this chapter, we address the problem of energy-efficient routing in CRAHNs. A trade-off should be made between finding minimum energy cost path and balancing energy usage. To address the issues mentioned earlier, we propose an energy-efficient and lifetime-aware routing protocol based on Q-learning technique, which is a reinforcement learning (RL) technique that solves decision problems in an adaptive and distributed manner. Each SU in the network is seen as a learning agent. By observing the network environment and evaluating a reinforcement function, which gives the direct reward of taking a forwarding action in a given state, the distributed learning agents are able to find the optimal or near-optimal routing path.

The major contribution of this work is that we apply the Q-learning technique in a routing protocol for CRAHNs to reduce the energy cost for packet transmission, balance residual energy distribution among SUs for longer network lifetime, and learn the environment effectively for better adaptability to dynamic underlying environment. The Q-learning-based routing protocol is easy to implement and computationally efficient.

The remainder of this chapter is organized as follows: Section 11.2 reviews the related work on routing protocols for CR networks and some RL-based routing protocols. Section 11.3 gives a detailed introduction on Q-learning technique. Section 11.4 presents the assumptions and system model adopted. Section 11.5 discusses the energy-efficient routing problem that is formulated as a Q-learning model. Section 11.6 provides the simulation results. Section 11.7 concludes the chapter.

11.2 Related Work

Some previous works have studied the routing algorithms and protocols for CR wireless networks. Recently, some works have started investigating the possibilities of applying RL techniques into routing protocols for wireless networks. In this section, we briefly review the related work on routing protocols for CR networks and RL-based routing protocols.

The most commonly used approach for CR networks is to define proper routing metrics to assess the route quality and incorporate them into some types of a reactive or an on-demand routing protocol [12,14]. These routing protocols could be categorized into different sets according to the specific metric used: interference minimizing protocols, delay minimizing protocols, energy consumption minimizing protocols, throughput maximizing protocols, and route stability maximizing protocols [14,15]. A brief summary of these routing protocols is given in [14], and some detailed description is given in [15]. As an example, a minimum power routing for cognitive wireless ad hoc networks is proposed in [16]. The routing metric is defined to be the transmission power that SUs use along the routing path. The weight of a CR link is calculated as a function of the transmission power. The proposed routing protocol tries to discover minimum weight paths between a source and a destination. As analyzed in [15], the model does not take into account the impact of PU activities. Another example in [13] also addresses the routing issue for multihop CR networks. It reviews an on-demand routing protocol applicable to CRAHNs, which is based on dynamic source routing (DSR). Inspired by this, a routing protocol called Energy based DSR (E-DSR) was proposed for CR networks with end-to-end energy consumption as a routing metric. The proposed algorithm aims at finding the routing path with minimum energy consumption. However, minimum energy consumption for CR nodes does not necessarily lead to energy efficiency or longer lifetime. In Section 11.6, we will perform comparison experimentally between our proposed routing protocol and the E-DSR protocol.

Recently, there have emerged some routing protocols based on RL techniques. As described in [17], Q-routing is one of such early works based on Q-learning technique. The routing metric of Q-routing incorporates not only the number of hops to the destination but also the effects of traffic jams. The simulation results show that the algorithm performs well under changing network topologies and high network workload scenario. Q-routing could be easily deployed into wireless networks because it uses only local information and works in a distributed manner. An improved algorithm proposed in [18] converges faster than Q-routing. Another extended algorithm is named bio-inspired quality-of-service (QoS) routing [19]. The proposed algorithm is based on a Dijkstra multipath routing approach combined with the Q-routing. Bio-inspired QoS routing enables to optimize multiple QoS criteria in routing decisions. Collaborative RL (CRL) extends RL with different feedback models, including a negative feedback model that decays an agent's local view of its neighborhood and a collaborative feedback model that allows agents to exchange the effectiveness of actions they have learned with one another [20]. The CRL technique is implemented in the routing protocol for mobile ad hoc networks. It enables the distributed nodes to find throughput maximizing routes online based on local information.

Relatively, few studies [21–23] have been done to integrate RL techniques into routing for CR networks. A combined framework of channel selection and route selection is proposed in [21] for multihop CR networks. The proposed routing protocol is designed based on the model-based RL technique. Channel and route selection is modeled as Markov decision process (MDP). The increased spectrum utilization is designed as a reward or reinforcement signal for each action. The routing strategy aims to explore different state-action pairs to come up with various routing solutions, which are ranked according to their reinforcement signal. The simulation shows that the protocol could significantly improve spectrum utilization. A CR Q-routing (CRQ-routing) scheme [22] is designed to find the routes minimizing interference to PUs. Basically, CRQ-routing is an RL-based approach with which SUs can learn to make dynamic and efficient routing decisions on the fly while addressing the important characteristics of CR networks, namely, dynamicity and unpredictability of channel availability, and SUs' interference to PUs. According to [23], routing in a dynamic CR networks is essentially a problem of walking in a random maze. RL-based solutions proposed for the task of random maze achieved good performance. Motivated by this, Zhang proposed an RL-based routing for CR networks. All these machine learning-based routing protocols for CR networks show their superiority in easy implementation and ability to make dynamic and efficient routing decisions online.

11.3 Q-Learning Technique

Q-Learning is a representative model-free RL technique. To better understand the definition of Q-learning, we will first give a brief introduction to RL technique in this section. Then we will give a detailed description of Q-learning technique.

11.3.1 Reinforcement Learning

An RL technique provides learning agents a framework to determine a sequence of actions or a policy, which maps the state of an unknown stochastic environment to an optimal action plan [14]. RL problems are usually be modeled as a three-tuple $\{S, A, R\}$, where S is the set of system states, A is the set of available actions, and R is the reinforcement function. The observation of the states and rewards are obtained from the operating environment. An agent in RL chooses its actions $a_t (a_t \in A)$ according to its observed state of the system environment $s_t (s_t \in S)$ at time t and the outcome of the action a_t is observed as a reinforcement $r_t (r_t \in R)$ (called reward for positive reinforcement and punishment for negative reinforcement). The observed reinforcement in turn causes an update to the agent's action in the following decision process. The goal of RL is to maximize the total reinforcements received over a time horizon.

RL algorithms can be either model based or model free. Model-based learning strategies build a model of the environment (typically in the form of an MDP) and calculate the optimal policy based on this model, whereas model-free methods do not use an explicit model and learn directly by mapping environmental states to actions. In a particular MDP-based RL, a four-tuple $\{S, A, P, R\}$ is used to describe its characteristics, where P is the state transition distribution function. $P: S \times A \to \Pi(S)$, where $\Pi(S)$ is the set of probability distributions over the set S. With an MDP model in hand, the resultant next state and next reward can be predicted based on the current state and action. Hence, a future course of action can be contemplated by considering possible future situations before they actually happen [20]. Typically, dynamic programming-related algorithms of value iteration are used to find the optimal policy function for model-based RL. Thus, the model-based RL techniques usually require a lot of computational resources in finding the optimal policy. Besides, these methods require greater storage memory but show slower execution times, especially when the state size grows [20]. Due to constrained storage resources and energy resources of CR nodes, we choose model-free Q-learning techniques to solve the routing problem for CRAHNs.

11.3.2 Q-Learning

Q-learning as a popular model-free RL was first introduced by Watkins [14] in 1989. This technique enables learning agents to act optimally with limited computational requirements. In this technique, agents do not have explicit model for the underlying environment and the reward transition probability. Instead, Q-learning operates by iteratively approximating the value of an action (Q-value) with experienced outcomes using an idea known as a temporal difference approach. Q-learning is proceeded based on incrementally improving approximation of the Q-values, which reflect the quality of particular actions in particular states. We use $Q(s_t, a_t)$ to denote the expected reward that can be received by taking action a_t in state s_t and thereafter following the same optimal policy. According to [17], $Q(s_t, a_t)$ can be calculated as follows:

$$Q(s_t, a_t) = r_t + \gamma \sum_{s_{t+1} \in S} P^{a_t}_{s_t s_{t+1}} \max_a Q(s_{t+1}, a) \tag{11.1}$$

where:

r_t denotes the direct reward received after taking an action a_t and $P^{a_t}_{s_t s_{t+1}}$

$P^{a_t}_{s_t s_{t+1}}$ is the probability of going to the state s_{t+1} from state s_t under the action a_t

γ denotes the discount factor ($0 \leq \gamma < 1$) used to discount the future rewards

The value of γ determines the importance of trade-off between current rewards and future rewards. The Q-value can also be approximated by iterations [14]:

$$Q(s_t,a_t) \leftarrow (1-\alpha)Q(s_t,a_t) + \alpha[r_t + \gamma \max_a Q(s_{t+1},a)] \qquad (11.2)$$

where:

α ($0 < \alpha \le 1$) is the learning rate that determines the rate of updating Q-values

The Q-value approximation is proceeded in every learning step and a new policy is generated based on which drives the next action to execute. The Q-learning algorithm could find the optimal action if every state is sufficiently visited.

One critical issue for Q-learning is the trade-off between exploration and exploitation. Exploitation would yield immediate reward, whereas exploration would require tolerating momentary cost of not using the currently known optimal policy for the opportunity of potential information about better policies [14]. One popular method to balance the ratio of exploration and exploitation is to choose probabilistically among the set of possible actions using the Boltzmann distribution. The Boltzmann distribution function is defined as

$$P(s_t,a_t) = \frac{e^{-(Q(s_t,a_t)/T)}}{\sum_{a \in A} e^{-(Q(s_t,a)/T)}} \qquad (11.3)$$

where:

T is the temperature that determines the ratio of exploitation to exploration

The higher the temperature, the more likely exploration action will be chosen. In general, the networks with more dynamic topologies and few stable links should have a higher temperature than those with stable topologies.

11.4 System Model

We consider a time-slotted CRAHN, within which PUs and SUs are randomly deployed. Figure 11.1 illustrates an example network topology. PUs hold license for specific spectrum channels, whereas SUs can only sense and temporally utilize idle channels. Let N_{ch} denote the number

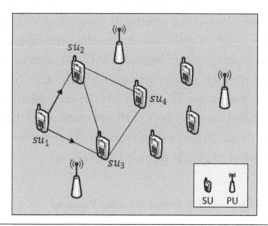

Figure 11.1 An example of CRAHN topology.

of licensed spectrum channels for PUs and ch_i $(i = 1, 2, \ldots, N_{ch})$ denote the spectrum channel indexed with i in the group. Let N_{su} denote the number of SUs in the network and su_i $(i = 1, 2, \ldots, N_{su})$ denote the SU indexed with i in the group. Though routing focuses on the network layer, routing in CRAHNs is also dependent on spectrum availability in the lower layers. We give several assumptions on the lower layers in the following text.

At the physical layer, we assume that SUs execute spectrum sensing at the beginning of each timeslot to detect idle spectrums. If a channel ch_j is detected to be idle at su_i, then su_i could use this channel ch_j in this timeslot. Let Ch_i denote the group of idle channels detected by su_i. There may be some errors in the sensing results. If a channel is busy but detected to be idle by a SU, then utilizing this channel would cause collision between PU and SU.

At the MAC layer, we assume that the spectrum management function is readily available [21]. The function promises to select the best available common channel for data transmission among a SU node pair and avoid channel collisions among SUs' transmissions. For example, the detected idle channel set for su_i and su_j are Ch_i and Ch_j, respectively. Then the channel for data transmission between Ch_i and Ch_j is selected from the common set $Ch_i \cap Ch_j$. Since collision is avoided among SUs, a collision occurs to a SU only when a PU emerges to the channel using by this SU. During a collision event, SU's packets are lost. To retransmit this lost packet, SU must detect and switch to another idle channel in the next timeslot. The process of spectrum switch is called spectrum handoff. A SU transmission may encounter multiple collisions due to PU interruptions, and thus experience multiple spectrum handoffs. We also assume that each SU communicates through a local interaction with other SUs within its radio transmission range. The SUs with the transmission range of a given SU are called the neighbors of this particular SU. Each su_i keeps a record of its dynamic set of neighbors Γ_i. As shown in Figure 11.1, the neighbor set of su_1, Γ_1, is recorded as $\{su_2, su_3\}$.

11.5 Proposed Q-Learning-Based Routing Protocol

The energy-efficient routing problem aims at selecting proper intermediate nodes from source to destination resulting in an energy-efficient forwarding path. In this section, each SU is modeled as a learning agent. Q-learning model is deployed into the learning agents to build an energy-efficient routing protocol for CRAHNs.

11.5.1 Problem Description

In CRAHN routing, we consider the whole network as a system. The energy-efficient routing problem for a source node is essentially to find a sequence of proper intermediate nodes from source to destination such that the energy efficiency is improved. CRAHN routing is a challenging problem due to the specific characteristics of CRAHNs, such as time-varying channel availability, lack of global network information, and frequently changing network topology due to SU mobility and PU activities. To promise the system properties of the routing protocol, routing agents should make their routing decisions based on the frequently changing local environment. In our proposed scheme, the Q-learning model is embedded in each SU, and it enables each SU to make its local routing decision (i.e., next-hop node selection).

We consider routing from the point of view of an individual packet. Thus, the system states are defined to be related to individual packets. For example, in the network shown in Figure 11.1, if a data packet is generated at node su_1 and its destination is node su_4, then the start state related to this packet is su_1 and the termination state is su_4. At state su_1, the forwarding action needs to be determined by choosing one of its neighbors. The set of available actions at state su_1 is related to the neighbor set of node su_1. As shown in Figure 11.1, the neighbor set of node su_1 is $\Gamma_1 = \{su_2, su_3\}$, and thus, the set of actions available at state su_1 is $A_1 = \{a_2, a_3\}$, where a_2 indicates selecting node su_2 as the next hop and a_3 indicates choosing node su_3 as the next hop. Assume that action a_2 is taken at state su_1. Then the system state related to this packet transits from su_1 to su_2 only if the packet is successfully forwarded from su_1 to su_2; otherwise, the system state will stay in state su_1. Each forwarding action may not always succeed due to channel quality variation, interference from PU activities, or SU mobility. The possible outcomes of a forwarding action are described as reinforcement function in Section 11.5.2.

11.5.2 Energy Efficiency-Related Reinforcement Function

The reinforcement function is critical to Q-learning, as it determines the desirability of an agent's action at a specific state within the environment. The goal of the routing algorithm employing Q-learning is to get the packet delivered to the destination with maximum reinforcement [13]. In the proposed routing protocol, we attempt to minimize the energy consumption per packet and balance the residual energy distribution, and thus prolong the network lifetime. The reinforcement should be closely related to the energy consumption and residual energy.

In order to build the reinforcement function model, we first analyze some system events that will induce energy cost. In the proposed protocol, we sample the number of these system events within some timeslots into the past. The events are listed as follows:

1. *Attempted forwarding*: When a SU node forwards a packet to the next hop, some energy will be consumed at this SU. We use E_{tr} to denote the energy cost induced by each attempted forwarding. The number of sample forwarding attempts is denoted by N_a.
2. *Successful transmissions*: If a packet is successfully delivered to the next hop, ACK will be sent from the next-hop node. Thus, the number of successful transmissions can be obtained as N_s at the sender node. We denote the energy cost consumed by receiving a packet as E_{rc}.

3. *Collision due to PU Interference*: Since error may exist in spectrum sensing results, SU may transmit a packet on some channel, which is being used by some PUs. Then a collision occurs between PU and SU transmissions. Both the PU and SU transmissions fail. The first important principle of SU is to utilize the channel without interfering PU transmission. Therefore, we impose a fixed high penalty E_p to reduce the number of PU interference. The spectrum handoff should be made for the interrupted packet in avoidance of interference to PU service. Thus, the total energy cost is the sum of the energy cost induced by the spectrum handoff (denoted by E_h) and the energy cost induced by retransmission (denoted by E_{tr}). The number of collisions can be detected and is denoted by N_c.

4. *Transmission failure due to channel unreliability*: The inherent unreliability of a wireless channel results in received signal variations, which may cause error in decoding some received bits of a data packet. The packet is successfully received only if all bits of the packet are successfully decoded. We assume that another attempted transmission will be induced if a transmission fails due to an unreliable channel. Then the energy cost induced is also E_{tr}. In this case, the packet arrives but cannot be decoded successfully. The energy cost E_{rc} of receiving the packet is also induced in this event. ACK will also be sent back to the sender node. Thus, the number of these events can be obtained as N_{fu}.

5. *Transmission failure due to SU Mobility*: SU mobility may cause the transmission link disconnected. Then, the data packet could not be relayed to the selected next hop. The learning agent needs to make a new decision. We impose a punishment, E_a, to this case due to the extra energy consumption of selecting new forwarding action. Then the total energy cost is calculated as the sum of E_a and E_{tr}. In this case, the packet or feedback is lost during transmission. The number of these events cannot be obtained through direct sample or detection. The number of these events is calculated as $N_{fm} = N_a - N_s - N_c - N_{fu}$.

With the rate of these system events, we could estimate the probability of every possible outcome of an attempted transmission. Then, the expected energy cost $\bar{E}$ that will be induced by an action execution could be calculated as follows:

$$
\begin{aligned}
\bar{E} = {} & \frac{N_s}{N_a} E_{rc} + \frac{N_c}{N_a} (E_p + E_b + E_{tr}) \\
& + \frac{N_{fu}}{N_a} (E_{tr} + E_{rc}) + \frac{N_{fm}}{N_a} (E_a + E_{tr})
\end{aligned}
\tag{11.4}
$$

To guarantee energy efficiency and long lifetime, residual energy distribution should also be considered in the reward function model. Consider an example that an action a_j is chosen at state su_i. If the forwarding attempt a_j fails, then the state will stay in su_i. The reinforcement function $r^{a_j}_{su_i,su_i}$ for the state-action pair (su_i, a_j) is defined as

$$
r^{a_j}_{su_i,su_i} = -E_{tr} - \bar{E} + E_{res}(su_i)
\tag{11.5}
$$

where:

E_{tr} is the energy cost of one attempt rewarding
$E_{res}(su_i)$ is the residual energy of node su_i

$\bar{E}$ is the extra punishment imposed to the reward because the packet is still at node su_i after one attempt forwarding, and it may experience the possible event again at node su_i. With the extra punishment $\bar{E}$, the learning agent is compelled to choose the next hop with which the connection link is more stable, and hence the number of retransmissions will be reduced. The transmission delay and energy cost will be reduced accordingly. $E_{res}(su_i)$ is incorporated into the reinforcement function. A larger $E_{res}(su_i)$ will encourage the learning agent to choose su_i thereafter, while a lower $E_{res}(su_i)$ will encourage learning agent not to choose su_i thereafter.

If the forwarding attempt a_j is successful, then the state will transit to su_j. The reinforcement function $r^{a_j}_{su_i,su_j}$ for the state-action pair (su_i, a_j) is defined as

$$
r^{a_j}_{su_i,su_j} = -E_{tr} - E_{rc} + E_{res}(su_j)
\tag{11.6}
$$

where:

E_{tr} is the energy cost of one attempt rewarding
E_{rc} is the energy cost of receiving one packet
$E_{res}(su_j)$ is the residual energy of node su_j

With E_{tr} and E_{rc} as penalties, the leaning agents will choose the relatively shorter path to maximize the reward. With the residual energy $E_{res}(su_j)$ as a direct reward, the learning agents have to choose the node with higher residual energy and avoid the lower one in order to maximize the reward. In this way, the energy usage will be relatively fair among the networks and the residual energy will be more balanced. Thus, the network lifetime could be prolonged.

Based on the reinforcement function model given above, Q-values of the source and intermediate nodes could be updated using value iteration method in Equation 11.2.

11.5.3 The Proposed Protocol

In Section 11.5.2, the routing problem is modeled as a Q-learning model, with which the system states, available actions, and reward functions are defined. The detailed description of our proposed protocol is given in this section. We use the example in Figure 11.1, where node su_1 has a data packet to transmit to node su_4, to illustrate how our protocol works. Our protocol is operated in terms of five stages: (1) initialization, (2) selection and transmission, (3) reception and ACK, (4) detection, and (5) Q-value update.

(1) Initialization stage

Node su_1 has a data packet for node su_4. The system state related to this packet is su_1. The set of possible system states are the set of the network nodes. The set of available actions is $A_1 = \{a_2, a_3\}$. The Q-values of all the state-action pairs are initialized to be 0, and $N_a = N_s = N_c = N_{fu} = N_{fm} = 0$.

(2) Selection and transmission stage

The action is selected based on the rule given by Equation 11.3. Once the action is determined, the packet is transmitted to the selected node (next hop). In our example, since the initial Q-values are identical, node su_1 will randomly select an action, say a_2, from its action set. Then the packet will be transmitted to node su_2. Then the sender su_1 goes to detection stage to observe the outcome of executing action a_2, whereas the next-hop node su_2 goes to reception and ACK stage.

(3) Reception and ACK stage

If the next hop receives the packet, then it will decode the packet. If the packet could be successfully decoded, then it should put its residual energy in an ACK packet and send the ACK packet back to the previous hop. If the packet could not be decoded successfully, then an ACK packet indicates that the transmission failure should be sent back to the previous hop.

(4) Detection stage

In this stage, the sender node detects whether the transmitted packet is interrupted by PU arrival and whether ACK is received. If the received ACK indicates that the packet is successfully decoded, then N_s will be incremented by 1. The system state transits to the next hop and the reinforcement is calculated through Equation 11.6 with the residual energy information contained in the ACK packet. Otherwise, the state does not change. If the packet transmission is interrupted, then the node will make spectrum handoff and switch to a new idle channel for transmission, and N_c will be incremented by 1. If the received ACK indicates that the packet is not successfully decoded, then N_{fu} will be incremented by 1. If no collision or ACK is received within a predefined time frame, then the packet is assumed to be lost during transmission, and N_{fu} will be incremented by 1. If the transmission is not successful, the reinforcement is determined through Equation 11.5.

(5) Q-value update stage

Each learning agent maintains a routing table, which stores the Q-value for each state-action policy. With the reward obtained in the detection stage, Q-value could be updated in this stage through Equation 11.2. The updated Q-value replaces the old value in routing table. It in turn affects the following action selection in the selection and transmission stage.

The evaluation of the Q-values will be incrementally improved, and thus, our routing algorithm proceeds by an ongoing operation in these stages.

11.6 Simulation Results

In this section, we evaluate our routing protocol using Java programming. We perform comparisons between our routing protocol and E-DSR, which was introduced in Section 11.2. The E-DSR protocol uses the DSR approach with end-to-end energy consumption as a route selection metric. The network performance is evaluated in terms of packet delivery ratio, energy consumption per packet sent, and network lifetime.

11.6.1 Simulation Setup

The parameters used in our simulation are summarized in Table 11.1. The simulated environment is a two-dimensional (2D) area of 1000 m × 1000 m. PUs and SUs are randomly deployed in the simulated area. The underlying channels are Rayleigh fading, which vary randomly according to a Rayleigh distribution with parameter $\delta^2 = 10$. PUs can only access to the spectrum channel assigned to them, whereas SUs could detect idle spectrums and use them in an opportunistic manner. The PU arrival rate of the assigned channel is set to be 20 packets per second. We use a constant packet size of 512 bytes for data packets. SUs are randomly moving in the 2D space according to the random walk-based mobility model, based on which a node moves with a direction uniformly and a speed uniformly from 0 to $v_{\max}$ with exponentially distributed epochs [24]. The maximum

Table 11.1 Simulation Parameters

PARAMETERS	VALUE
Number of channels, N_{ch}	30
Number of SUs, N_{su}	20
Maximum moving speed, $v_{\max}$	10 m/s
Initial energy of SU	100 J
Energy cost per transmission, E_{tr}	0.0033 J
Energy cost per reception, E_{rc}	0.0017 J
Energy cost per spectrum handoff, E_h	0.0005 J
Energy cost by selecting new action, E_a	0.0001 J
Penalty imposed by collision, E_p	0.005
SU learning rate, α	0.7
Discount factor, γ	0.5

transmission range of each node is set to be 300 m. According to [25], we set the energy cost per transmission E_{tr} as 0.0033 J and the energy cost per reception E_{rc} as 0.0017 J.

11.6.2 Packet Delivery Ratio

We compare the performance in terms of packet delivery ratio between our Q-learning-based protocol and E-DSR. The packet delivery ratio used here is defined to be the ratio of the number of packets successfully delivered to the destination node to the total number of packets generated in the network. The results of the packet delivery ratio are shown against the varying arrival rate of SU packets in Figure 11.2.

From Figure 11.2, we can see that our proposed scheme maintains a higher delivery ratio even at high packet arrival rate compared to that of E-DSR. As the data packet arrival rate increases, the competition for spectrum channels or transmission links is getting more and more

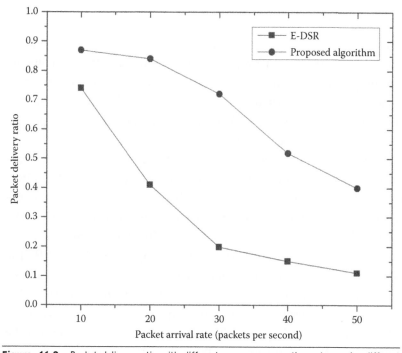

Figure 11.2 Packet delivery ratio with different message generation rates under different routing protocols.

intense. The waiting time for each packet is becoming longer relatively. Thus, the observed packet delivery ratio falls down. As shown in the figure, the rate of decline is much lower for our proposed protocol than that of E-DSR. The packet delivery ratio of our proposed protocol could measure up to 0.4 even when the data packet rate is 50 packets per second. In our proposed approach, adaptive Q-learning in next hop selection enables SUs to choose more stable next hop in the dynamic environment. Our proposed approach will choose the route with rich residual energy, and thus indirectly lead to a traffic load distribution balance. Thus, our proposed scheme maintains a higher delivery ratio even at high packet arrival rate. However, E-DSR uses the DSR approach with end-to-end energy consumption as a route selection metric. The next hop chosen is only based on the estimated energy consumption. It cannot adapt to varying network conditions in terms of intense spectrum resource competition and transmission link competition. Hence, the delivery ratio of E-DSR goes down to 0.1 as the data packet rate increases to 50 packets per second.

11.6.3 Energy Consumption per Packet

We compare the performance in terms of average energy consumption per packet between our proposed protocol and E-DSR. The energy consumption per packet is defined to be the total energy consumption divided by the number of packets successfully delivered during the simulated period. Figure 11.3 shows the average energy cost per SU packet against varying arrival rates of packets.

From Figure 11.3, we can see that the average energy cost per packet of our proposed scheme is much lower than that of E-DSR, especially when the packet arrival rate becomes high. As shown in the figure, the energy consumption per packet is about 0.011 when the data packet rate is 10 packets per second. But for E-DSR it is about 0.017. The energy cost is reduced by 35% compared with E-DSR. The energy cost increases as the packet arrival rate increases. As shown in Figure 11.2, the increase rate of E-DSR is much higher than that of our proposed protocol. Since our proposed adaptive approach can select a stable next hop, the PU interval and retransmission can be much lower than that using E-DSR. Thus, the energy consumption is much lower. Even at high arrival rate with severe competition for

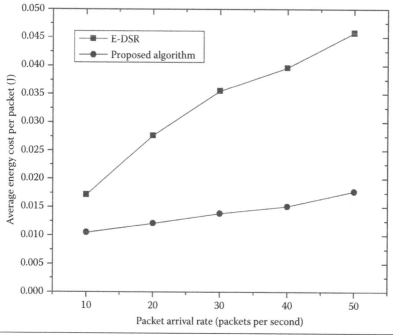

Figure 11.3 Average energy cost per packet with different message generation rates under different routing protocols.

spectrum channels, SUs can adapt to select less busy channels to reduce retransmissions and thus reduce energy consumption.

11.6.4 Network Lifetime

In this section, we evaluate the network lifetime of our protocol and that of E-DSR. Network lifetime is defined as the time span from the deployment of a network to the instant when the first CR node dies. Figure 11.4 plots the network lifetime with different message generation rates under different routing protocols.

The figure reveals that the network lifetime decreases as the message generation rate increases. From Figure 11.4, we can also observe that our protocol performs much better than E-DSR when the message generation rate is high. As shown in Figure 11.4, the network lifetime is about 300 for E-DSR when the data packet rate is 10 packets per second. But it falls down to 100 when the data packet rate increases to 50 packets per second. The rate of decline is about 67% for E-DSR. As for our proposed approach, the network lifetime achieves

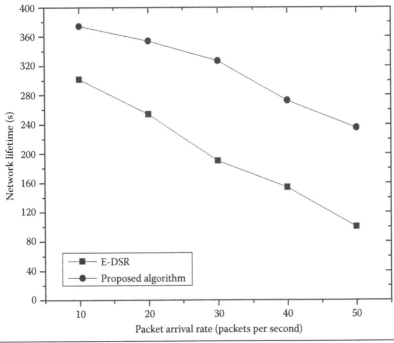

Figure 11.4   Energy efficiency measured in network lifetime with different message generation rates under different routing protocols.

about 375 when the data packet rate is 10 packets per second. It yields 25% longer lifetime compared with E-DSR. The network lifetime falls down to 240 when the data packet rate increases to 50 packets per second. The rate of decline is only 36% for our proposed protocol, which is much lower than that of E-DSR (67%). E-DSR focuses on the reduction of total energy consumption in transmitting data packets, but does not consider the balance of energy usage among the networks. By contrast, our proposed approach considers both the energy consumption per hop and the balance of residual energy distribution.

11.7 Conclusions

In this chapter, we have proposed a novel distributed routing protocol for CRAHNs based on Q-learning. In the proposed protocol, the reinforcement function is defined to estimate the energy consumption and residual energy. With the reinforcement signal received, each agent stores and updates a table of Q- values on the fly. Thus, a routing path with which energy consumption is relatively low and the energy is fairly used could be built. Compared to the most routing protocol of CRAHNs, our proposed protocol is simpler to be implemented. Performance evaluation shows that our routing protocol outperforms E-DSR significantly in the reduction of energy consumption per packet and the growth of the network lifetime under the situation that the data packet arrival rate is larger than 10 packets per second. Meanwhile, it has been shown that the delivery ratio is also improved significantly.

Acknowledgment

This work was supported by City University Strategic Research Grant No. 7002734.

References

1. FCC, "Spectrum policy task force report," FCC 02-155, Washington, DC, 2002. http://wireless.fcc.gov/auctions/31/releases/fc020155.pdf.
2. Akyildiz, I. F., Lee, W. Y., and Chowdhury, K., "CRAHNs: Cognitive radio ad hoc networks," *Ad Hoc Networks*, Vol. 7, No. 5, pp. 810–836, 2009.
3. Liang, Y. C., Zeng, Y., Peh, E. C. Y., and Hoang, A. T., "Sensing-throughput tradeoff for cognitive radio networks," *IEEE Transaction on Wireless Communications*, Vol. 7, No. 4, pp. 1326–1337, 2008.
4. Jiang, H., Lai, L., Fan, R., and Poor, V., "Optimal selection of channel sensing order in cognitive radio," *IEEE Transaction on Wireless Communications*, Vol. 8, No. 1, pp. 297–307, 2009.
5. Li, Z., Yu, F. R., and Huang, M., "A distributed consensus-based cooperative spectrum sensing in cognitive radios," *IEEE Transaction on Vehicular Technology*, Vol. 9, No. 4, pp. 1370–1379, 2010.
6. Cacciapuoti, A. S., Akyildiz, I. F., and Paura, L., "Optimal primary-user mobility aware spectrum sensing design for cognitive radio networks," *IEEE Journal on Selected Areas in Communications*, Vol. 31, No. 11, pp. 2161–2172, 2013.
7. Su, H. and Zhang, X., "Cross-layer based opportunistic MAC protocols for QoS provisionings over cognitive radio mobile wireless networks," *IEEE Journal on Selected Areas in Communications*, Vol. 26, No. 1, pp. 118–129, 2008.
8. Cormio, C. and Chowdhury, K. R, "A survey on MAC protocols for cognitive radio networks," *Ad Hoc Networks*, Vol. 7, No. 7, pp. 1315–1329, 2009.

9. Si, P., Yu, F., Ji, H., and Leung, V., "Optimal cooperative internetwork spectrum sharing for cognitive radio systems with spectrum pooling," *IEEE Transaction on Vehicular Technology*, Vol. 59, No. 4, pp. 1760–1768, 2010.

10. Zhang, Y. and Lazos, L., "Vulnerabilities of cognitive radio MAC protocols and countermeasures," *IEEE Network*, Vol. 27, No. 3, pp. 40–45, 2013.

11. Yuan, Z., Song, J. B., and Han, Z., "Interference aware routing using network formation game in cognitive radio mesh networks," *IEEE Journal on Selected Areas in Communications*, Vol. 31, No. 11, pp. 2494–2503, 2013.

12. Filippini, I., Ekici, E., and Cesana, M., "A new outlook on routing in cognitive radio networks: Minimum-maintenance-cost routing," *IEEE/ACM Transactions on Networking*, Vol. 21, No. 5, pp. 1484–1498, 2013.

13. Salim, S. and Moh, S., "On-demand routing protocols for cognitive radio ad hoc networks," *EURASIP Journal on Wireless Communications and Networking*, Vol. 2013, pp. 102, 2013. doi:10.1186/1687-1499-2013-102.

14. Watkins, C. J. C. H., "Learning from Delayed Rewards," PhD thesis, Cambridge University, 1989.

15. Cesanaa, M., Cuomob F., and Ekicic E., "Routing in cognitive radio networks: Challenges and solutions," *Ad Hoc Networks*, Vol. 9, No. 3, pp. 228–248, 2011.

16. Pyo, C. W. and Hasegawa, M., "Minimum weight routing based on a common link control radio for cognitive wireless ad hoc networks," *Proceedings of the 2007 International Conference on Wireless Communications and Mobile Computing*, Shanghai, People's Republic of China, pp. 399–404, September 21–25, 2007.

17. Hu, T. and Fei, Y., "QELAR: A machine-learning-based adaptive routing protocol for energy efficient and lifetime extended underwater sensor networks," *IEEE Transactions on Mobile Computing*, Vol. 9, No. 6, pp. 796–809, 2010.

18. Kumar, S. and Miikkulainen, R., "Dual reinforcement Q-routing: An on-line adaptive routing algorithm," *Proceedings of the Artificial Neural Networks in Engineering Conference*, St. Louis, MO, November 9–12, 1997.

19. Mellouk, A., Hoceini, S., and Zeadally, S., "A bio-inspired quality of service (QoS) routing algorithm," *IEEE Communications Letters*, Vol. 15, No. 9, pp. 1016–1018, 2011.

20. Dowling, J., Curran, E., Cunningham, R., and Cahill, V., "Using feedback in collaborative reinforcement learning to adaptively optimize MANET routing," *IEEE Transactions on Systems, Man and Cybernetics, Part A: Systems and Humans*, Vol. 35, No. 3, pp. 360–372, 2005.

21. Barve, S. S. and Kulkarni, P., "Dynamic channel selection and routing through reinforcement learning in cognitive radio networks," *Proceedings of the IEEE International Conference on Computational Intelligence and Computing Research*, India, pp. 1–7, December 18–20, 2012.

22. Al-Rawi, H. A. A. and Yau, K.-L. A., "Route selection for minimizing interference to primary users in cognitive radio networks: A reinforcement learning approach," *Proceedings of the IEEE Symposium on Computational*

Intelligence for Communication Systems and Networks, Spain, pp. 24–30, June 5–7, 2013.

23. Zheng K., Li, H., Qiu, R. C., and Gong, S., "Multi-objective reinforcement learning based routing in cognitive radio networks: Walking in a random maze," *Proceedings of the International Conference on Computing, Networking and Communications*, Hawaii, pp. 359–363, January 30–February 2, 2012.
24. Jiang, S. M., He, D. J., and Rao, J. Q., "A prediction-based link availability estimation for mobile ad-hoc networks," *Proceedings of the 20th Annual Joint Conference of the IEEE Computer and Communications Societies*, Vol. 3, pp. 1745–1752, April 2001.
25. Cano, J.-C. and Manzoni, P., "A performance comparison of energy consumption for mobile ad hoc network routing protocols," *Proceedings of the International Symposium on Modeling, Analysis and Simulation of Computer and Telecommunication Systems*, San Francisco, CA, pp. 57–63, August 29–September 1, 2000.

Biographical Sketches

Ling Hou received her bachelor's degree in communication engineering from Beijing University of Posts and Telecommunications, Beijing, People's Republic of China, in 2008. Since September 2009, she has been a PhD student of the department of electronic engineering at City University of Hong Kong, Kowloon, Hong Kong. The main research interests of her PhD study include routing protocol design, performance modeling, and security and reinforcement learning application in cognitive radio networks. She has also been working as a research assistant at City University of Hong Kong since September 2013. Her research work is mainly on traffic prediction, architecture identification, and performance modeling for future photonic metro networks.

K. H. Yeung received his PhD in information engineering at the Chinese University of Hong Kong, Hong Kong in 1995. He has spent 30 years in teaching, managing, designing, and research on different areas of computer networks. He is an associate professor at City University of Hong Kong, Kowloon, Hong Kong. Being an active researcher, he has research interests on network infrastructure security, wireless networks, and Internet systems. Yeung also frequently provides consultancy services to the networking industry. One notable consultancy project was to develop a 900 MHz global system for mobile communications (GSM) mobile handset for a listed company in Hong Kong. He has professional qualifications including Chartered IT Professional (CITP), Cisco Certified Network Professional (CCNP), Cisco Certified Academy Instructor (CCAI), Certified PenTest Laboratory Tester (CPLT), EC-Council Certified Security Analyst (ECSA), and Certified Ethical Hacker (CEH). He is the coauthor of the book *Network Infrastructure Security* (published by Springer) and about 100 papers in referee journals and conference proceedings.

Angus K. Y. Wong received his BSc and PhD degrees in information technology from City University of Hong Kong, Hong Kong. Since he began his academic career in Macao Polytechnic Institute, Macao, People's Republic of China, he has constantly carried out his active research work and provided consultancy services. He was commissioned by the Bureau of Telecommunications Regulation of Macao as the principle investigator to write a consultant report on the development of IPv6 for Macao in 2011. One of the recommendations in the report has led to the Bureau to establish an IPv6 research laboratory in Macao. After serving the institute for 10 years, he returned to Hong Kong and joined the School of Science and Technology at the Open University of Hong Kong, Hong Kong, in 2013 as an associate professor and the leader of the Engineering Sciences Team. His research interests include Internet systems, network infrastructure security, cognitive radio networks, and network science.

PART V
Dynamic Radio Spectrum Access

12

A Novel Opportunistic Spectrum Access Protocol Based on Transmission–Throughput Trade-Off for Cognitive Radio Networks

SHAOJIE ZHANG, HAITAO ZHAO, SHENGCHUN HUANG, AND SHAN WANG

Contents

12.1 Introduction

Recently, cognitive radio networks (CRNs) have attracted an increasing amount of interest as an effective method of alleviating the spectrum scarcity problem in wireless communications by allowing access of secondary users (SUs) to frequency channels that are allocated to primary users (PUs), in a way that does not affect the quality of service

403

(QoS) of the primary networks [1,2]. The research on CRN has been encouraged by the Federal Communications Commission (FCC), which has revealed that there is a significant amount of licensed spectrum that is largely underutilized in vast temporal and spatial dimensions [3]. In order to reuse the available spectrum that is not being used under the current fixed spectrum allocation policy, the FCC has recently allowed the access of SUs to the broadcast television spectrum at locations where that spectrum is not being used by primary services, which is known as IEEE 802.22 wireless regional area network (WRAN) standard [4] that aims to provide broadband wireless Internet access to rural areas.

The greatest challenge of CRN is the coexistence of a higher priority primary network with a lower priority secondary network. Until now two main approaches have been proposed to allow SUs to access the licensed channels termed: (1) opportunistic spectrum access (OSA) [5,6], according to which SUs are allowed to access a frequency channel only if it is sensed to be idle, and (2) spectrum sharing [7,8], according to which SUs can coexist with PUs as long as the interference caused by SUs does not exceed an interference threshold specified by PUs. In this chapter, we will focus on the former approach in single-channel scenarios.

The most popular frame structure of OSA cognitive systems studied consists of a channel sensing slot and a data transmission/idle slot, while the overall duration of the frame is fixed, as depicted in Figure 12.1. Within this frame structure, each SU that attempts to launch a packet transmission at the beginning of each frame senses the state of the channel for a duration τ, whereas it uses the remaining frame duration $T-\tau$ for data transmission or keeps silent depending on the sensing result. According to the classical detection theory [9], an increase in the sensing time results in higher detection probability and lower false-alarm probability, which leads to improved utilization of the unused channel. However, the increase of the sensing time results in a decrease of the data transmission time, and hence the achievable throughput of the cognitive radio system reduces. Therefore, an essential trade-off between the channel sensing and the data transmission is required in this frame structure. Liang et al. [10,11] proved the existence of the optimal sensing durations to minimize the probability of false alarm under the constraint of a target detection probability. In [12–14], the impacts of transmission duration, PUs' traffic, and

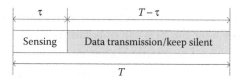

Figure 12.1 A typical frame structure for OSA cognitive systems.

access contention on the probabilities of detection and false alarm were investigated, respectively. This sensing–throughput trade-off was addressed in [15–17], where the authors studied the problem of finding the optimal sensing time that maximizes the average achievable throughput of an OSA network under a single high target detection probability constraint for the protection of the QoS of the PUs. Jafarian and Hamdi [18] analyzed the SUs' achievable throughput with a joint optimal number of sensing rounds and sensing threshold in a double-threshold sensing scheme and proposed an iterative algorithm to obtain the optimal values of the sensing time and second detection threshold. Liu et al. [19] proposed a group-based cooperative medium access control protocol called GC-MAC, which addresses the trade-off between sensing accuracy and efficiency.

However, in these studies, an ideal assumption on the PUs' behaviors was made according to which the PU has a constant occupation state throughout the entire frame duration. In other words, once the channel is sensed to be idle, it is assumed that the PUs will keep silent for the remaining frame duration and vice versa. Actually, the realistic case is that the PUs depart from the channel (or reoccupy at the channel) randomly, despite that the channel is sensed to be busy (or idle). Once considering the random departure or reoccupation of PUs during the SUs' frame duration, another type of interference to PUs will be caused by SUs' transmissions although the channel is sensed to be idle. It is obvious that the longer the SUs access the channel for transmissions, the higher the probability that PUs reoccupy the channel during the transmission, resulting in mutual interference and throughput degradation. However, from the SUs' perspective, an increase of transmission duration is expected due to an improved throughput of cognitive network. Therefore, transmission–throughput trade-off is also required in the frame structure, similar to the above sensing–throughput trade-off. In this chapter, the effect of PUs' traffic on the transmission–throughput trade-off is discussed. Moreover, an inherent problem that how long the SUs can opportunistically access a channel if the channel is sensed to be idle is constructively solved. To the best of our knowledge, this issue has not been considered in previous works.

Another remaining problem is how to coordinate multiple SUs to access the available idle slots of the channel. Some studies [12,20] investigated a single SU coexisting with PUs and utilizing all idle slots for simplification. Without considering the contention among multiple SUs and the relevant overhead, maximal achievable throughput was presented in these researches. As to the scenario with multiple SUs, Kumar et al. [21] proposed a medium access control (MAC) protocol based on OSA. Similar to distributed coordination function (DCF) introduced by IEEE 802.11 [22], a basic unit for backoff termed as physical slot as well as interframe space (IFS) was adopted to relieve the potential collisions. It is worth noting that in this protocol, besides data transmissions, the durations of idle slots are also consumed by backoff and IFSs. In order to maintain the interference threshold specified by PUs, SUs have to decrease the data transmission duration, resulting in a reduced throughput of cognitive networks. Therefore, it is of great importance to design a cognitive protocol according to which multiple SUs can utilize the idle slots effectively.

In this chapter, we propose a novel cognitive MAC protocol that overcomes the transmission–throughput trade-off and underutilization of idle slots in current protocols for CRNs. This is achieved by performing the optimization of data transmission durations and a large-scale backoff mechanism. Our main contributions in this chapter can be summarized as follows:

- We study the trade-off between the data transmission duration and the achievable throughput of CRNs, considering the random departure and reoccupation of PUs during the entire frame. Depending on traffic load distribution of the aggregated PUs and the interference threshold specified by PUs, a closed-form expression for the optimal transmission duration for SUs is presented.
- We design a cognitive MAC protocol that enables multiple SUs opportunistically contend to access the available idle slots in the channel. By performing the large-scale backoff mechanism,

which regards an entire idle slot as a basic backoff unit, the utilization efficiency for idle slots is significantly improved.

- We propose an analytical model to accurately estimate the throughput performance of our designed MAC protocol, in terms of the channel sensing and data transmission durations as well as the density of SUs.

The rest of this chapter is organized as follows: Section 12.2 discusses the transmission–throughput trade-off in the typical OSA frame structure and the optimal transmission duration is obtained to control the cognitive interference (CI) to PUs and to improve the utilization efficiency of SUs simultaneously. Section 12.3 describes the proposed cognitive MAC protocol and explains how the multiple SUs sense the channel and access the idle channel periods coordinately. Section 12.4 introduces the proposed analytical model to estimate the throughput performance of secondary network considering collisions among multiple SUs and CI between SUs and PUs. Section 12.5 presents the simulation results and discussions to validate the analytical analysis. Finally, Section 12.6 concludes the chapter.

12.2 Optimal Transmission Duration

In this chapter, we consider multiple SUs coexisting with aggregated PUs in a single-channel CRN, which means the SUs should contend to access the channel opportunistically. A typical frame structure of the SU with fixed length T is shown in Figure 12.1. In order to protect the PUs from harmful interference resulting from SUs' transmissions, the data transmission of the SUs is activated subject to the spectrum sensing results. When the sensing result of the channel state indicates idle, the SUs attempt to launch a packet transmission. Or, if the channel state is sensed to be busy, all SUs should keep silent to avoid the interference with PUs.

In [10–19], the sensing–throughput trade-off has been well investigated by analyzing the effect of sensing error on the throughput; however, the relationship between the transmission duration and the secondary throughput is still remaining challenging and will be studied in this section. Therefore, to highlight this focus, the channel

Figure 12.2   Two-slot process of channel, considering the aggregated PU.

sensing in this chapter is assumed to be ideal without considering false alarms and missed detections, and the time required is indicated by t_s.

For the sake of simplicity, we model one virtual PU in the channel by combining the channel access activities of all PUs of the respective network as shown in Figure 12.2. The channel access activities of a PU are usually modeled with a two-slot process, ON–OFF ("busy" or ON and "idle" or OFF). The state of the virtual PU is ON if at least one of the PUs of the respective network is accessing the channel. Otherwise (when all PUs are in OFF slot), the state of the virtual PU is assumed OFF.

As described on the frame structure previously, once the channel state is sensed to be idle, SUs contend to access the channel and launch a packet transmission for a duration t_{su}. Since SUs are unable to transmit a packet and sense the channel simultaneously, the transmissions launched by PUs and SUs will collide if the channel is reoccupied by PUs. The conditional collision probability that a particular packet transmission by SUs collides with the PUs' transmission is defined as *cognitive interference*, denoted by *CI*. It is obvious that an increase of SUs' transmission duration results in higher CI but an improved achievable throughput for SUs instead. Therefore, a transmission–throughput trade-off for the above frame structure is also required, and the optimal transmission duration should be carefully designed, which will be discussed in this section.

Although the OSA approach allows SUs opportunistically access all idle slots, not all OFF slots are available for the SUs to launch a packet transmission. The reason is that fractional OFF slots are likely not long enough to be sensed idle by SUs. Assuming that the time required by an ideal spectrum sensing is indicated by t_s, without considering the cases of false alarms and missed detections. Then, if the duration of one OFF slot is less than t_s, the SUs will not perceive such idle slots in the channel. Thus, the CI with the PUs resulting from SUs' transmissions is subjected to the durations of OFF slots based on the following two hypotheses:

$$H_0 : T_{off} \geq t_s \tag{12.1}$$

$$H_1 : T_{off} < t_s + t_{su} \tag{12.2}$$

By utilizing these two hypotheses, CI equals to the conditional probability that H_1 occurs under the condition that H_0 occurs during the current OFF slot. Then CI could be obtained as

$$CI = p(H_1|H_0) \tag{12.3}$$

Let $f_{T_{off}}(t)$ denote the probability density function (pdf) for the durations T_{off} of OFF slots. Since the probabilities of the two hypotheses H_0 and H_1 can be expressed as

$$p(H_0) = \int_{t_s}^{\infty} f_{T_{off}}(t)\,\mathrm{d}t \tag{12.4}$$

$$p(H_1) = \int_{0}^{t_s+t_{su}} f_{T_{off}}(t)\mathrm{d}t \tag{12.5}$$

The CI can be further deduced as

$$CI = \frac{p(H_0, H_1)}{p(H_0)} = \frac{p(t_s \leq T_{off} < t_s + t_{su})}{p(t_s \leq T_{off})} = \frac{\int_{t_s}^{t_s+t_{su}} f_{T_{off}}(t)\mathrm{d}t}{\int_{t_s}^{\infty} f_{T_{off}}(t)\mathrm{d}t} \tag{12.6}$$

To avoid the intolerable interference to the PUs' transmission, the CI is always required to be restricted below a specified interference threshold ξ. It is obvious that a transmission by SUs lasts longer; the CI trends to be higher because the PUs have more chances to reoccupy the channel. An efficient method to restrict the interference is to decrease the duration of the SUs' occupancy as low as possible. Nevertheless, from the SUs' throughput perspective, the duration of the SUs' transmission is expected to be as long as possible, to increase the ratio of transmission duration in the frame. Similar to this transmission–throughput trade-off, a compromise between the CI to PUs and the utilization efficiency of SUs is required.

In order to achieve the maximum throughput of the SUs and control the interference level to the PUs, the decision on optimal transmission duration t_{su}^* of the SU can be modeled as follows:

$$t_{su}^* = \arg \max_{t_{su}>0} \{t_{su}\} \tag{12.7}$$

$$\text{subject to} \quad CI \leq \xi$$

Without any loss of generality, we assumed that the aggregated traffic loads of the PUs are exponentially distributed with the mean of the occupied and idle slots denoted by $1/\lambda_0$ and $1/\lambda_1$, respectively. Then the pdf of T_{off} can be expressed as

$$f_{T_{off}}(t) = \begin{cases} \lambda_1 e^{-\lambda_1 t}, & t \geq 0 \\ 0, & \text{others} \end{cases} \qquad (12.8)$$

By substituting Equation 12.8 in 12.6, the CI can be expressed as a function of t_{su}:

$$CI = \frac{e^{-\lambda_1 t_s} - e^{-\lambda_1(t_s + t_{su})}}{e^{-\lambda_1 t_s}} = 1 - e^{-\lambda_1 t_{su}} \qquad (12.9)$$

Since the first derivative on t_{su} of CI is always greater than 0, CI increases with t_{su} monotonically. Thus, when CI equals the interference threshold ξ, we obtain the optimal transmission duration t_{su}^* as

$$t_{su}^* = \frac{\ln(1 - \xi)}{-\lambda_1} \qquad (12.10)$$

It is important to note that, given a specific CI threshold, the optimal transmission duration only depends on the pdf of durations of OFF slots and is not influenced by the time required by channel sensing. In Figure 12.3, optimal transmission durations for SUs at different mean durations of OFF slots are presented with several specific interference thresholds, in which t_s is set to 50 μs.

In CRNs based on OSA, an essential problem of SUs is how to choose the optimal transmission duration if the channel is sensed to be idle. Clearly, SUs' transmission duration impacts the extent of interference between the PUs and the activated SUs, and mainly dominates the throughput of the SUs. Therefore, in this section, a transmission–throughput trade-off and hence the optimization of transmission duration are investigated, considering the random departure and reoccupation of PUs. Another problem is how to coordinate multiple SUs to opportunistically access the idle slots based on the above analysis, which will be discussed in Section 12.3.

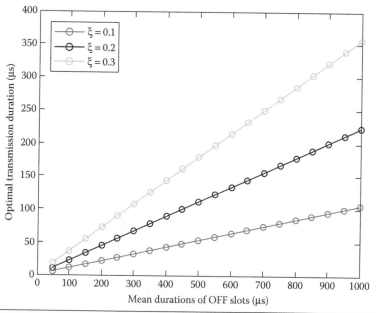

Figure 12.3 Optimal transmission duration versus mean durations of idle slots.

12.3 Cognitive MAC Protocol Design

Depending on the conclusion of the optimal transmission duration in Section 12.2, a cognitive MAC protocol is designed to coordinate multiple SUs to effectively utilize the idle slots considering the CI restraint. We ignore the influence of sensing errors by the SUs for simplicity and due to lack of space; however, the framework we provide can be extended to handle sensing errors. Here we assume that all SUs in the single-hop area expose the same channel states, and thus obtain the same channel sensing results; if there is any overlap between the transmissions of PUs and SUs, both directional transmissions will not be successfully received due to collisions.

Figure 12.4 indicates the timeline of our proposed cognitive MAC protocol, in which SUs coexist with PUs by utilizing the idle slots of the channel. In our proposed protocol, when SUs have packets to transmit, they will sense the channel continuously. Only when the channel is sensed to be idle, namely, the idle slot duration exceeds the sensing time t_s, SUs can opportunistically access the channel for t_{su}. It is to note that t_{su} comprises the transmission time of data and ACK packets, propagation delay as well as necessary IFs.

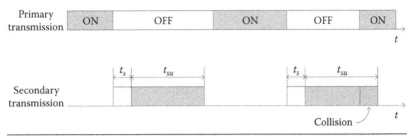

Figure 12.4   Timeline of the primary and secondary networks.

Let t_{sifs} be the length of a short interframe space (SIFS) and t_δ be the propagation delay. Let γ_h, γ_d, and γ_a be the size of header, data packet, and ACK packet, respectively. Denoting the data rate as R supported by the channel, the duration of data transmission of SUs can be estimated as

$$t_{su} = \frac{\gamma_h + \gamma_d}{R} + t_{sifs} + t_\delta + \frac{\gamma_a}{R} + t_{sifs} + t_\delta \qquad (12.11)$$

To relieve the congestion among SUs, binary exponential backoff (BEB) introduced by IEEE 802.11 is employed. Upon launching a packet transmission, the SUs need to select a backoff counter randomly from $[0, W_0]$, where W_0 denotes the minimum contention window. If the channel is sensed to be idle, the backoff counters will be decreased by 1. If the channel is sensed to be busy, the backoff counters will be frozen until the next idle slot. If the backoff counter reaches 0, the SU transmits a packet immediately after the sensing slot. The destination will respond with an ACK packet to acknowledge the successful receipt. Otherwise, once the packet transmission is failed because of collisions with either other SUs or PUs, the contention window doubles as $2^i W_0$ until reaching the maximum retransmission limit m, in which i denotes the retransmission number.

Similar to the backoff scale, the basic backoff unit in the proposed protocol is one idle slot with variable durations longer than the required sensing time, compared with a physical slot with uniform length in IEEE 802.11 protocols. Therefore, the backoff process does not waste the precious idle slots' durations. In other words, the main advantage of such design is to avoid the backoff consumption of the idle slot and further improve the SUs' utilization of the idle slots by increasing the data transmission duration.

12.4 Performance Analysis

12.4.1 Virtual Slot Model

As described in the context, the states of the channel utilized by all PUs are assumed to be alternative ON and OFF slots. SUs opportunistically contend to access the channel based on the channel sensing results. Under the proposed MAC protocol, all the OFF slots unused by PUs can be classified into four different types of virtual slots as shown in Figure 12.5, depending on the number of SUs activated during the OFF slot. The black and white regions in the figure indicate the presence and absence of PUs respectively, whereas the grey regions indicate the transmission of SUs.

1. *Unavailable slot*: In this type of virtual slots, the durations of OFF slots are smaller than the time required for a reliable channel sensing. Therefore, although there exists an OFF slot in the channel, the sensing results will always show the busy state of the channel. In other words, SUs cannot perceive this type of virtual slots to access.

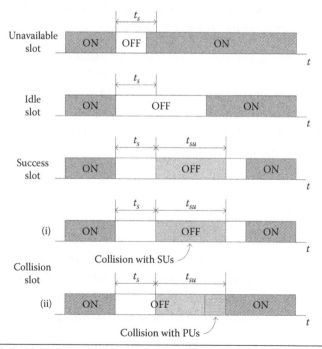

Figure 12.5 Virtual slot model to represent various OFF slots.

2. *Idle slot*: Although the durations of this type of virtual slots are longer than the channel sensing time and the sensing results show the absence of PUs, the OFF slot will still remain idle. It is achievable when no backoff counters of SUs reach 0 and all of SUs stay in the backoff procedure.

3. *Success slot*: In this slot, after the channel is sensed to be idle, only one SU decreases its backoff counter to 0 and launches a packet transmission for t_{su}. Meanwhile, the channel maintains idle until the transmission finishes. Because only one SU transmits during this OFF slot, the destination will successfully receive the packet and then acknowledge the successful receipt.

4. *Collision slot*: In this slot, after the channel is sensed to be idle, a collision occurs in two cases: (1) More than two SUs decrease their backoff counters to 0 and launch packet transmissions simultaneously, and (2) PUs return and reoccupy the channel, while the SUs' data transmission is performed. Any of these two cases can result in unsuccessful receipt. Moreover, colliding SUs cannot differentiate the essential cause of the collision and has to retransmit or drop the packet depending on the retransmission limit.

Note that, among the above four types of virtual slots, only the success slot experiences successful packet transmission, which contributes to the throughput performance of secondary network. In order to analyze the throughput of secondary network, the key problem is to obtain the stable probability distribution of each type of virtual slot, which will be discussed in Section 12.4.2.

12.4.2 Markov Chain Model

In this section, we use a Markov chain model to analyze the throughput of secondary network in the saturated case, where all SUs always have a packet to transmit. Based on previous researches [23,24], it is easy to extend this analytical model to adjust to the unsaturated situations, and hence, it is not discussed in this chapter.

Since unavailable slots make no impact on the behavior of SUs, only the latter three types of virtual slots are considered in the analytical model and collectively called available OFF slots. Clearly, the stable probability of unavailable slots equals the probability that the duration of a randomly chosen OFF slot is less than channel sensing time t_s, namely,

$$p_u = \int_0^{t_s} f_{T_{off}}(t)dt = 1 - e^{-\lambda_1 t_s} \tag{12.12}$$

Moreover, an idle slot is available for SUs with the probability $\alpha_i = 1 - p_u$. Let p be the constant and independent conditional collision probability for each packet transmitted by SUs colliding with one or more packets transmitted by other SUs. Since p and CI are statistically independent, the effective conditional collision probability for any SU in a cognitive network can be calculated as

$$p_e = p + CI - p \times CI \tag{12.13}$$

Bianchi [25] assumed that the probability τ that a node will attempt transmission in a time slot is constant and estimates it in terms of conditional collision probability p, minimum contention window W_0, and number of contention nodes n. In our proposed cognitive network, the effective collision probability is p_e, and hence, we can amend the Markov chain model by substituting p with p_e as shown in Figure 12.6.

Depending on the expression obtained by Bianchi, the probability τ_s that each SU attempts to transmit in each idle slot can be expressed as

$$\tau_s = \frac{2(1-2p_e)}{(1-2p_e)(W_0+1) + p_e W_0(1-(2p_e)^m)} \tag{12.14}$$

The collision among SUs occurs, only if more than two SUs attempt to transmit packets in the same idle slot simultaneously. Therefore, we obtain the conditional collision probability p among multiple SUs as

$$p = 1 - (1 - \tau_s)^{n-1} \tag{12.15}$$

Equations 12.13 through 12.15 represent a nonlinear system in the two unknowns τ_s and p, which can be solved using numerical techniques.

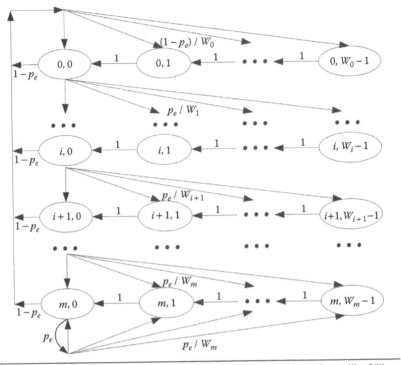

Figure 12.6   Amended Markov model for SUs in a saturated network. In the figure, $W_i = 2^i W_0$.

Let P_{tr} be the probability that there is at least one SU transmitting in a randomly chosen available OFF slot. Since each SU transmits with a probability τ_s, we obtain the expression of P_{tr} as

$$P_{tr} = 1 - (1 - \tau_s)^n \qquad (12.16)$$

However, the probability P_i that no SUs launch packet transmissions in a randomly chosen available OFF slot is

$$P_i = 1 - P_{tr} \qquad (12.17)$$

The probability p_s that a transmission launched by one of the SUs is successful is given by the probability that exactly one SU is transmitting at that OFF slot, while the PU remains idle throughout the transmission period, conditioned on the fact that at least one SU transmits:

$$P_s = \frac{n\tau_s (1 - \tau_s)^{n-1}(1 - CI)}{P_{tr}} \qquad (12.18)$$

On the contrary, the probability P_c that a transmission launched by one of the SUs is failed due to a collision with either other SUs or the PU can be recursively expressed as

$$P_c = 1 - P_s \qquad (12.19)$$

Since the latter three types of virtual slots in Figure 12.5 occur only if the corresponding OFF slots are sensed to be idle, the stable probabilities that a randomly chosen OFF slot belongs to idle, success, and collision slot, respectively, can be finally written as

$$
\left\{
\begin{aligned}
p_i &= \alpha_i P_i = e^{-\lambda_1 t_s}(1 - \tau_s)^n \\
p_s &= \alpha_i P_{tr} P_s = e^{-\lambda_1 t_s} n\tau_s (1 - \tau_s)^{n-1}(1 - CI) \\
p_c &= \alpha_i P_{tr} P_c = e^{-\lambda_1 t_s}(1 - (1 - \tau_s)^n - n\tau_s(1 - \tau_s)^{n-1}(1 - CI))
\end{aligned}
\right. \qquad (12.20)
$$

Additionally, data packets with length γ_d are successfully transmitted during each success virtual slot, whereas no data packets are successfully transmitted during the remaining three types of virtual slots. Finally, we obtain the aggregated throughput of the cognitive network as

$$S = \frac{p_s \gamma_d}{E(T_{off})} \qquad (12.21)$$

where:

$E(T_{off})=1/\lambda_1$, which indicates the mean duration of OFF slots in a long term

12.5 Simulation Results

In this section, we validate the cognitive MAC protocol and theoretical analysis in Section 12.4 using experimental results obtained from the network simulator NS2. We also present the main results of the impact of the transmission and sensing durations as well as the density of SUs on the secondary throughput. In this simulation, a background traffic load is set up to simulate the behaviors of the aggregated PU, which is exponentially distributed with the mean of the occupied and idle slots denoted by $1/\lambda_0$ and $1/\lambda_1$, respectively. We do not consider any sensing errors in the simulations. That is, if

Table 12.1 Simulation Parameters

PARAMETERS	VALUES
W_0	32
m	5
n	30
r	6 Mbps
$(1/\lambda_0)/(1/\lambda_1)$	1000 μs/1000 μs
t_{sifs}/t_δ	10 μs/1 μs
γ_h/γ_a	50 bytes/30 bytes

there is an overlap between the channel sensing duration and PU's transmission, the channel will be sensed to be busy. Moreover, if the channel is sensed to be idle, the SUs contend to access the channel for the transmission duration t_{su}.

At the start of simulations, n SUs are placed randomly in an area of size 100 m × 100 m. The simulation is run for about 1,000,000 alternative ON–OFF slots. Typical parameters from IEEE 802.11 networks are employed in the simulations as listed in Table 12.1.

12.5.1 Transmission–Throughput Trade-Off

The effect of transmission durations of SUs on the CI and secondary throughputs is illustrated in Figure 12.7, in which the labels around the curve indicate the CI computed by the analytical model. As the transmission duration is increased, the SUs trend to transmit more data during one success virtual slot. Hence, the aggregated throughput of secondary network increases significantly. Otherwise, with the increased transmission duration, there are more chances for SUs to collide with the returned PU, resulting in an increase of CI. From this figure, we can see that to protect the PUs from harmful interference caused by SUs' transmissions, the transmission–throughput trade-off is of great importance and the optimization of transmission duration for SUs is necessary in order to design a high-efficiency MAC protocol in CRNs.

Figure 12.8 indicates the relationship between the aggregated throughputs of secondary network and the CI threshold specified by the PU. In the simulations, optimal transmission duration computed

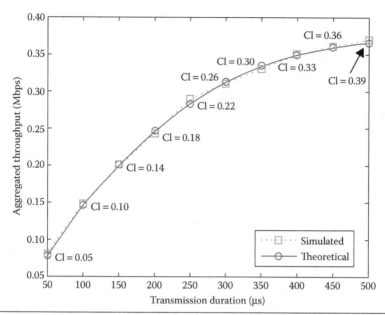

Figure 12.7 Aggregated throughputs and Cl versus transmission durations.

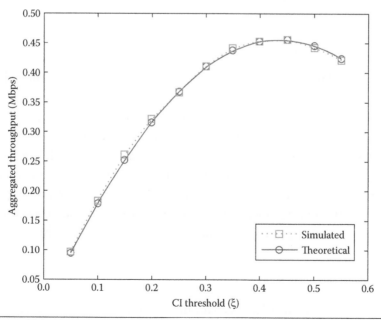

Figure 12.8 Aggregated throughputs versus Cl threshold ξ.

in Section 12.2 is employed to guarantee the CI below the specified threshold. From the figure, we can see that when the CI is below a certain value, the aggregated throughput of secondary networks increases if more CI is allowed; however, once exceeding the value, the throughput decreases gradually with the increase of the interference threshold. This is because that if more CI with PU is allowed, it becomes likely for SUs to transmit more data during one success slot, thus resulting in an improved throughput of the secondary network. The potential risk is the increase of the efficient collision probability p_e due to the increased CI defined in Equation 12.13. When the throughput sacrifice caused by collisions exceeds the benefit of long transmission duration, the throughput performance will fall to some extent. In other words, higher interference threshold and longer transmission duration allowed do not always result in higher secondary throughput due to the effect of efficient collision probability. Therefore, in order to improve the achievable secondary throughput, the optimal transmission duration should be chosen according to both the interference threshold and the effective collision probability, which will be one of our future researches.

12.5.2 Impact of Sensing Duration

Figure 12.9 indicates the relationship between the aggregated throughputs of secondary network and the duration required by an ideal channel sensing. It is to note that the throughput of secondary network trends to decrease if the time required for channel sensing increases. This is because the proportion of available idle slots out of the overall idle slots decreases due to an increased sensing duration, with which the proportion of success slots linearly decreases as Equation 12.20. Therefore, aiming at reducing the sensing time, investigation of the high-efficiency algorithms on channel sensing is helpful to improve the secondary throughput.

12.5.3 Impact of the Secondary Network Density

Since we employ the BEB mechanism similar to 802.11, the aggregated secondary throughput will be influenced by the density of SUs by making an impact on the collision probability among multiple SUs. Figure 12.10 illustrates the impact of the secondary network density on the aggregated

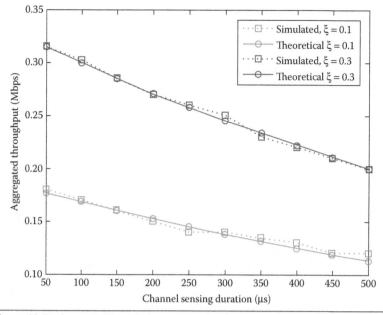

Figure 12.9 Aggregated throughputs versus sensing duration.

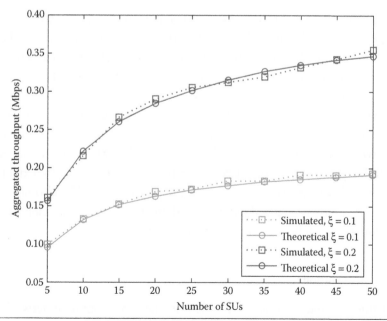

Figure 12.10 Aggregated throughputs versus number of SUs in a single-hop area.

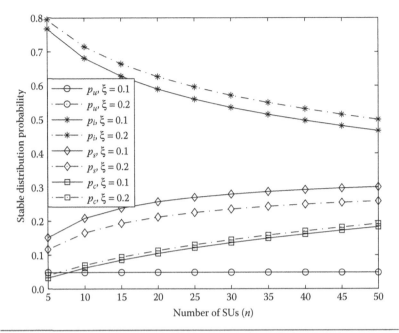

Figure 12.11   Stable distribution probabilities versus number of SUs in a single-hop area.

secondary throughput. The stable distribution probabilities of the four types of virtual slots with the SUs' density are illustrated in Figure 12.11. From Figure 12.10, we see that the secondary throughput increases with the increased number of SUs in the single-hop area. This is because as more SUs contend to access the idle slots, there are more chances for SUs to decrease their backoff counters to 0 at the beginning of the idle slots and then transmit packets, resulting in a higher success probability as shown in Figure 12.11. Although more contending SUs bring in higher collision probability p_c, the success probability p_s increases to some extent, compared with the significant fall of idle probability p_i. We also notice that the interference threshold makes different impacts on the stable distribution probabilities of different types of virtual slots.

12.6 Conclusions

In this chapter, an inherent problem in the OSA cognitive networks is that how long the SUs can opportunistically access a channel if the channel is sensed to be idle is constructively solved. Moreover, considering the random departure and reoccupation of PUs during the frame, we concluded that the proper transmission duration should be adaptively chosen to achieve higher throughput of the secondary networks at the same time to control the interference with the legal PUs at a tolerable level. Then, based on the above analysis, a high-efficient MAC protocol is designed to enable multiple SUs to opportunistically access the idle slots without harmful interference with PUs. By performing large-scale backoff, the wastage of idle slots caused by backoff in the idle slots is avoided, resulting in a throughput improvement of secondary networks. Later on, we propose an analytical model to accurately estimate the secondary throughput in terms of the transmission and sensing durations and the SUs' density of secondary networks. Finally, we validate the cognitive MAC protocol and theoretical analysis using the network simulator NS2. It has been illustrated that the analytical and the simulation results are in perfect agreement under all considered scenarios. However, all these researches are obtained in the single-channel scenarios without considering the negotiations on channel selection in multichannel cases, which offer interesting avenues for future work in designing and optimizing multichannel CRNs.

Acknowledgment

This research was supported in part by the National Natural Science Foundation of China (Grant No. 61002032).

References

1. J. Mitola, III and G. Q. Maguire, Jr., "Cognitive radios: Making software radio more personal," *IEEE Pers. Commun.*, vol. 6, no. 4, pp. 13–18, 1999.
2. S. Haykin, "Cognitive radio: Brain-empowered wireless communications," *IEEE J. Sel. Areas Commun.*, vol. 23, no. 2, pp. 201–220, 2005.
3. Federal Communication Commission, "Spectrum policy task force report, FCC 02-155," November 2002. Available: https://apps.fcc.gov/edocs_public/attachmatch/DOC-228542A1.pdf.
4. IEEE 802.22 Wireless RAN, "Functional requirements for the 802.22 WRAN Standard, IEEE 802.22-05/0007r46," September 2005. Available: https://mentor.ieee.org/802.22/dcn/05/22-05-0007-46-0000-draft-wran-rqmts-doc.doc.
5. Q. Zhao and A. Swami, "A decision-theoretic framework for opportunistic spectrum access," *IEEE Wirel. Commun. Mag.*, vol. 14, no. 4, pp. 14–20, 2007.
6. I. F. Akyildiz, W.-Y. Lee, M. C. Vuran, and S. Mohanty, "Next generation/dynamic spectrum access/cognitive radio wireless networks: A Survey," *Comput. Netw.*, vol. 50, no. 13, pp. 2127–2159, 2006.

7. A. Ghasemi and E. S. Sousa, "Fundamental limits of spectrum-sharing in fading environments," *IEEE Trans. Wirel. Commun.*, vol. 6, no. 2, pp. 649–658, 2007.

8. L. Musavian and S. Aissa, "Ergodic and outage capacities of spectrum sharing systems in fading channels," in *Proc. IEEE Global Commun. Conf.*, pp. 3327–3331, IEEE, Washington, DC, November 2007.

9. S. M. Kay, *Fundamentals of Statistical Signal Processing: Detection Theory*, vol. 2. Prentice Hall, Upper Saddle River, NJ, 1998.

10. Y. Liang, Y. Zeng, E. Peh, and A. Hoang, "Sensing-throughput tradeoff for cognitive radio networks," in *Proc. IEEE Int. Conf. on Commun.*, pp. 5330–5335, Glasgow, June 2007.

11. Y. Liang, Y. Zeng, E. Peh, and A. Hoang, "Sensing-throughput tradeoff for cognitive radio networks," *IEEE Trans. Wireless Commun.*, vol. 7, no. 4, pp. 1326–1337, April 2008.

12. W. Tang, M. Shakir, M. Imran, R. Tafazolli, and M. Alouini, "Throughput analysis for cognitive radio networks with multiple primary users and imperfect spectrum sensing," *IET Commun.*, vol. 6, no. 17, pp. 2787–2795, 2012.

13. L. Tang, Y. Chen, E. Hines, and M. Alouini, "Effect of primary user traffic on sensing-throughput tradeoff for cognitive radios," *IEEE Trans. Wirel. Commun.*, vol. 10, no. 4, pp. 1063–1068, 2011.

14. S. Zhang, H. Zhao, S. Wang, and J. Wei, "A cross-layer rethink on the sensing-throughput tradeoff for cognitive radio networks," *IEEE Commun. Let.*, vol. 18, no. 7, pp. 1226–1229, 2014.

15. Y. Gao and Y. Jiang, "Performance analysis of a cognitive radio network with imperfect spectrum sensing," in *Proc. IEEE Int. Conf. on Computer Commun. Workshops*, pp. 1–6, San Diego, CA, March 2010.

16. S. M. Almalfouh and G. L. Stuber, "Interference-aware power allocation in cognitive radio networks with imperfect spectrum sensing," in *Proc. IEEE Int. Conf. on Commun.*, pp. 1–6, May 2010.

17. S. Stotas and A. Nallanathan, "On the throughput and spectrum sensing enhancement of opportunistic spectrum access cognitive radio networks," *IEEE Trans. Wirel. Commun.*, vol. 11, no. 1, pp. 97–107, 2012.

18. J. Jafarian and K. Hamdi, "Non-cooperative double-threshold sensing scheme: A sensing-throughput tradeoff," in *Proc. IEEE Wireless Commun. and Net. Conf.*, pp. 3376–3381, Shanghai, People's Republic of China, April 2013.

19. Y. Liu, S. Xie, R. Yu, Y. Zhang, and C. Yuen, "An efficient MAC protocol with selective grouping and cooperative sensing in cognitive radio networks," *IEEE Trans. Veh. Tech.*, vol. 62, no. 8, pp. 3928–3831, 2013.

20. Y. Pei, A. T. Hoang, and Y. Liang, "Sensing-throughput tradeoff in cognitive radio networks: How frequently should spectrum sensing be carried out?" *IEEE 18th Int. Symp. on Personal, Indoor and Mobile Radio Communications*, pp. 1–5, Athens, Greece, September 2007.

21. S. Kumar, N. Shende, C. R. Murthy, and A. Ayyagari, "Throughput analysis of primary and secondary networks in a shared IEEE 802.11 system," *IEEE Trans. Wireless Commun.*, vol. 12, no. 3, pp. 1006–1017, 2013.

22. "IEEE Standards Association." Available: http://standards.ieee.org/about/get/802/802.11.html.
23. D. Malone, K. Duffy, and D. Leith, "Modeling the 802.11 distributed coordination function in nonsaturated heterogeneous conditions," *IEEE/ACM Trans. Networks*, vol. 15, no. 1, pp. 159–172, 2007.
24. K. Duffy, D. Malone, and D. J. Leith, "Modeling the 802.11 distributed coordination function in non-saturated conditions," *IEEE Commun. Let.*, vol. 9, no. 8, pp. 715–717, 2005.
25. G. Bianchi, "Performance analysis of the IEEE 802.11 distributed coordination function," *IEEE J. Sel. Areas Commun.*, vol. 18, no. 3, pp. 535–547, 2000.

Biographical Sketches

Shaojie Zhang (shaojiezhang@nudt.edu.cn) is currently pursuing his PhD degree in information and communication engineering from College of Electronic Science and Engineering, National University of Defense Technology (NUDT), Changsha, People's Republic of China. His research interests include multichannel access control and cognitive networks.

Haitao Zhao (haitaozhao@nudt.edu.cn) received his BS and PhD degrees in information and communication engineering from National University of Defense Technology (NUDT), Changsha, People's Republic of China, in 2004 and 2009, respectively. He was a visiting PhD student in the Queen's University of Belfast, Belfast, from June 2008 to September 2009. He is an associate professor of College of Electronic Science and Engineering, NUDT, People's Republic of China. His research interests include available bandwidth estimation in wireless networks, cognitive networks, and cooperative communication.

Shengchun Huang (huangsc@nudt.edu.cn) received his MS and PhD degrees from National University of Defense Technology (NUDT), Changsha, People's Republic of China, in 2008 and 2012, respectively. He was a visiting PhD student in the University of British Columbia, Vancouver, British Columbia, Canada, from October 2009 to October 2011. He is a lecture at College of Electronic Science and Engineering, NUDT, People's Republic of China. His research interests include broadband wireless access network and radio resource management.

Shan Wang (chinafir@nudt.edu.cn) received his BS and PhD degrees from National University of Defense Technology (NUDT), Changsha, People's Republic of China, in 2000 and 2006, respectively. He worked as a postdoctoral researcher in the University of Montreal, Montreal, Quebec, Canada, from September 2010 to September 2011 and as associate professor with College of Electronic Science and Engineering, NUDT, People's Republic of China. His research interests include protocol analysis, modeling and design, ad hoc networks, and network simulations and optimization.

13

A COGNITIVE ALGORITHM BASED ON THE WEATHER CONDITIONS TO ENHANCE WIRELESS OUTDOOR COMMUNICATIONS

DIANA BRI, MIGUEL GARCIA, JAIME LLORET MAURI, AND FRANCISCO RAMOS

Contents

13.1 Introduction

Cognitive networks (CNs) are relatively a recent concept in the scientific world. First researches on these networks date back to the beginning of the 2000s, and from this date until now, they have been evolving without a break. Their goal is to reach self-aware networks that are able to dynamically adapt their operational parameters in response to user needs or changing environmental conditions. Another important aspect is that they learn from these adaptations and use this knowledge to make future decisions [1]. Therefore, it is said that these networks have the ability to think, learn, and remember in order to reduce as much as possible the human intervention and maximize their self-configuration and maintenance. In order to meet this new paradigm, a cognition loop is the key and it features the

427

following states: observe, orient, plan, decide, act, and learn [2]. It can be defined as a cognitive process that can sense current reality, plan for the future, make a decision, and act accordingly.

Typical technologies limit the network ability to adapt their working using only local and reactive approaches as the network state is not shared for different network elements. By contrast, CNs seek to fulfill certain end-to-end goals of a data flow and to carry on proactive actions since they are based on a learning process to make future decisions [3]. Therefore, CNs encompass the entire network stack such that all network nodes can take actions based on the end-to-end requirements of the network. CNs goal is to enhance the next generation networks allowing them to be more complex, heterogeneous, and dynamic helping for their self-organization. Moreover, they make easier to meet user and application objectives jointly.

CNs consider that they can be accomplished with the use of a knowledge plane (KP) that transcends layers and domains to make cognitive decisions about the network. The KP adds intelligence and weight to the edges of the network and context sensitivity to its core. A KP allows the CN to learn about its own behavior over the time, by analyzing the problems, tuning its operation, and generally increasing its reliability and robustness [4].

Some authors [5] suggested to introducing a learning machine to include concepts and methods is an easy way for future developments in networks. The discipline draws on ideas from many other fields, including statistics, cognitive psychology, information theory, logic, complexity theory, and operations research but is always with the goal of understanding the computational character of learning.

Figure 13.1 shows the four phases that are required in a learning machine in order to be used as a method in a CN. The first phase is environment, which is needed to take into account because it will be responsible for the changes in network performance. This element is directly connected to performance phase, which is related to how our network is working in any of monitored elements. Knowledge phase gives us enough information to activate the learning process. This learning process will be the most intelligent box in our system, because it has to perform appropriate actions to improve the network performance through the information of knowledge and environment

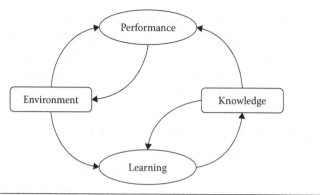

Figure 13.1 Phases of machine learning for CNs.

blocks. The general goal of learning is to improve the performance on whatever task the combined system is designed to carry out.

The goal of this work is to provide a general review of CNs in order to propose an algorithm based on a cognitive process to enhance outdoor IEEE 802.11b/g wireless communications.

The rest of the chapter is organized as follows: Section 13.2 collects some of the most important studies about CNs and cognitive radio (CR). Section 13.3 explains the main differences among these approaches and cross-layer designs. Section 13.4 details our proposal of cognitive algorithm to overcome the weather interferences by outdoor wireless local area networks (WLANs). Finally, Section 13.5 summarizes and concludes the chapter.

13.2 Related Works

In recent years, most of the studies about cognitive concepts have focused on CR and they deal with different applications: dynamic management of spectrum resources [6], cooperative medium access and cooperative communications [7], opportunistic switching among available wireless networks (cellular, WLAN, mesh, etc.) [8], and adaptive selection of available radio resources [9,10] among others. However, at the same time, a small number of research works have investigated the architecture of CNs as a whole in order to apply the cognition concept to networks far beyond only the radio aspects.

Probably, the first time that CNs were introduced as such was in 2002–2003 when Clark et al. [4,11] described KP paradigm for

CN and Dietterich and Langley [5] examined several aspects of machine learning that touch on cognitive approaches to networking. However, Mitola [2] discussed about them earlier in his dissertation on CR in 2000 as CNs are considered as an extension of software radio, but they are not analyzed in detail at all. Next, the Defense Advanced Research Projects Agency (DARPA)'s vision on the CN was presented by Ramming [12] in DARPATech 2004. From these first works, the European Union reported the latest advances on communications, which included CN as a future researching field [13] in 2005. Finally, the concept of CNs was completely defined by Thomas et al. [3] in several papers. They explained the CN concept thoroughly, clarified the utility and need for CNs [14], and discussed some possible applications for this technology [15]. Moreover, they also detailed a possible framework to deploy them and presented a practical discussion about their implementation [16].

Mihailovic et al. [17] provided a study of situation awareness mechanisms in modern telecommunication networks on the essential basis of the actual Self-NET research project effort [18]. They showed that the term *situation awareness* mainly refers to the collection of sufficient information about the environment and operational state of a system, which is a *vital* prerequisite for performing an appropriate decision-making process and executing the corresponding actions. Moreover, they presented the essential framework for structuring the appropriate situation awareness mechanisms in order to achieve an active network management system based on the network elements capable of autonomous action by using local knowledge and operator-defined policies. This framework may be helpful for junior researchers in creating CN architectures.

Another framework in [19] presents a reconfigurable platform based on an architecture specifically designed for nodes within a CN. This architecture is based on an adaptable node called reconfigurable node. This node is able to make the network and node level observations required by the cognitive engine and to respond to instructions and reconfigure all aspects of the node operations as necessary. Finally, this architecture defines three types of actions to do when a reconfigurable node needs to respond to an external condition: parametric reconfiguration, structural reconfiguration, and application reconfiguration.

Then, some specific proposals to implement CNs from different aspects have been presented in different papers. One of them is [20] that shows a quality-of-service (QoS) multicast routing protocol oriented to CN. It is called CogMRT, and it is focused on the development of cognitive protocols in cognitive wired networks. It is a distributed protocol where each node maintains local information and the routing search is in a hop-by-hop way. This protocol applies the competitive coevolutionary algorithm for the multicast tree construction. Wanga et al. [20] evaluated CogMRT and showed that it has important advantages as it exploits the cognitive help.

Zhang et al. [21] proposed the stable routing protocol (SRP) for CNs. SRP predicts the time that link will be available and chooses the link with longer available time as it is considered that the more stable the link is. The key idea is that the senders broadcast the data packets in the medium, but the forwarders with longer available time, better link quality, and light load possess the priority to relay the packets. Each sender in SRP tries to push the packet closer to the destination. The evaluation indicates that SRP performs better in terms of throughput and end-to-end delay, with a link availability prediction mechanism at the same time concerning the link quality and the traffic load for the nodes. This work concludes that SRP performs better in terms of throughput and end-to-end delay.

Another proposal makes use of CNs to enhance the behavior of multicast capacity [22]. The CN model consists of a primary hybrid network (PhN) and a secondary ad hoc network (SaN). Wang et al. devise the dynamic protection area (PA) for each primary node according to the strategy adopted in PhN. Based on the PAs, they design the multicast strategies for SaN under which the highway system acts as the multicast backbone. Under the precondition that SaN should have no negative impact on the order of the throughput for PhN, the strategy has the following merits: (1) the optimal throughput for SaN can be achieved for some cases, (2) secondary nodes can access opportunistically into the spectrum from both time and space domains, and (3) all secondary users can be served, except for some cases.

A lot of research works related to wireless networks are focusing their research on using cognitive techniques to improve their behavior. For example, Fortuna and Mohorcic [23] proposed a wireless access architecture based on the cognition concept in order to help the network operators solve the service assignment problem. Authors considered a typical network segment but introduced the concept of access point master that allows the network operators to define end-to-end goals for the system, which allows the user to select desired services, value these services, and receive meaningful notifications in case the system fails. Fortuna and Mohorcic addressed the problem and identified a suitable technology for representing the QoS knowledge for translation of application requirements to network requirements. Then, they presented an initial evaluation of the assignment engine, a component of SmartA that is responsible for optimally assigning services to be delivered by radio access interfaces. The engine uses multiobjective optimization techniques for minimizing the monetary cost and maximizing the user satisfaction.

Wang and Fu used CNs to find a better solution to improve a call admission control (CAC) to ensure QoS in wireless networks [24]. They presented a classification of different methods of CAC policies in the CN context. They showed the conceptual models for joint CAC and cross-layer optimization. Also, the benefit of cognition can only be realized fully if application requirements and traffic flow contexts are determined or inferred in order to know what modes of operation and spectrum bands to use at each point in time. The process model of cognition that involves the per-flow-based CAC is presented. Because there may be a number of parameters on different levels affecting a CAC decision, and the conditions for accepting or rejecting a call must be computed quickly and frequently, simplicity and practicability are particularly important for designing a feasible CAC algorithm. In a word, a more thorough understanding of CAC in heterogeneous wireless CNs may help one to design better CAC algorithms.

Continuing with wireless networks, a multipath routing architecture designed within the CN framework is presented, which may improve QoS for mission critical multihop wireless networks [25]. The cognitive multipath routing (CMR) implementation utilizes the multiple path management protocol by leveraging the existing functions, interfaces, and data structures for an adaptable network and user interface, modifies the protocol to conform to the cognition loop architecture, and incorporates the learning process. The learning process mainly consists of a network model and simulates the past and future network states to compare against thresholds and identify trends.

This implementation has been simulated and their analysis resulted in a message complete rate improvement ranging from 10% to 35% at a cost of throughput increasing by a factor of 2–4 times the original offered load.

Finally, Tamma et al. [26] applied the CN paradigm to the problem of development of autonomous cognitive access points (CogAP) for small-scale wireless networks. First, they presented an architecture of our autonomous CogAP. Then they introduced their algorithmic solution, in which a neural network-based traffic predictor makes use of historical traffic traces to learn network traffic conditions and predicts traffic loads on each of 802.11 b/g channels. The cognitive decision engine makes use of traffic forecasts to dynamically decide which channel is best for CogAP to operate on for serving its clients. They built a prototype CogAP device and carried out the performance evaluation of the proposed CogAP system. The obtained results showed that the proposed CogAP is effective in achieving performance enhancements with respect to the state-of-the-art channel selection strategies.

13.3 CNs versus Cross-Layer Designs and CRs

Although CNs, CRs, and even cross-layer designs have the goal of improving the network performance in some aspects (resource management, QoS, security, access control, etc.), each one of them tries it in a different way and with a different scope.

CNs go beyond CRs and cross-layer designs. Thus, while CNs consider the whole network stack to fulfill the end-to-end network goals, CRs are only focused on radio goals, and as cross-layer designs, they only try to reach local and single goals. Therefore, CNs cover all network elements that are involved in the transmission of a data flow (subnets, routers, switches, mediums, interfaces, etc.), and thus, they are more cooperative in nature than the other cognitive approaches. CNs only present limitations in terms of applicability according to the adaptability of the network elements and the flexibility of the cognitive process. Moreover, obviously CRs are only used for wireless networks, whereas CNs can include wired and wireless networks, and thus, they let integrate heterogeneous networks [27].

However, CNs and CRs have some aspects in common. First of all, they share the concept of cognitive process, the key of which is to learn from past observations and taken decisions and to use them to influence on future behavior. Moreover, both approaches are goal driven and require a software tunable platform that is controlled by the cognitive process to translate the goals into a form understandable for network and to be able to provide output in the form of a set of actions that can be implemented in the modifiable elements of the network. In the case of CNs, it is called software adaptable network (SAN) and it is in charge of this translation by both an external interface accessible to the CN and network status sensors. These devices are used to provide control and feedback from/to the network. By contrast, software-defined radio (SDR) becomes the platform of choice for the CR [28]. An SDR is a radio in which the properties of carrier frequency, signal bandwidth, modulation, any necessary cryptography, and source coding are defined by software.

Comparing cross-layer designs and CNs, we can highlight that both of them infringe the traditional layered approach since they allow nonadjacent layers communicate directly or share internal information between them. CNs include a cross-layer approach in the point of network adaptations that can be performed in different layers to provide observations of current conditions. These observations are delivered to the cognitive process and it then determines what is optimal for the network and changes the configurations of network elements' protocol stacks.

Despite similarities, CNs broaden the scope of cross-layer designs. The main difference is that while CNs balance multiple goals in order to perform multiobjective optimization, cross-layer designs typically consider single-objective optimizations. In the case of cross-layer designs, they can only perform independent objective optimizations, so they do not deal with network-wide performance goals. Therefore, we can say that cross-layer designs are less optimal than CNs that try to achieve all goals jointly in the optimization process, and thus, they include more network elements. By contrast, cross-layer designs can suppose conflicts between adaptations in a node when they come from different objective optimizations. Another different point between them is the learning concept.

As explained earlier, CNs have the ability to learn from taken decisions in the past and to apply this learning to future decisions. However, the learning process is not supported by cross-layer designs, so past adaptations' performance is not taken into account for future decisions. This is an important drawback of the cross-layer concept as network intelligence is limited. It always faces a set of inputs in the same way, although its performance was bad or poor in the past.

Last differences, such as CR, are referred to the scope of the goals and observations. Observations used in the cognitive process include all nodes of the network as optimization is carried out considering the goals for all nodes; by contrast, cross-layer design performs optimization and observations over one only node. Therefore, goals are centered in each node independently. CNs are aware of the whole network and this global information allows the cognitive process to adapt network function much better than when network visibility is limited to one only node, and it is unaware of the rest of network nodes' conditions.

Finally, it should be highlighted that including cognition brings additional charges for overheads, architecture, and operation, so it should always be lower than the level of performance improvement; otherwise, it does not make sense.

Table 13.1 makes a comparison among the three schemes proposed until now to add intelligence to the future networks in order

Table 13.1 Comparison among CNs, CRs, and Cross-Layer Designs

FEATURE	CN	CR	CROSS-LAYER DESIGNS
Scope	End to end	Local	Local
Optimization	Multiobjective optimization	Single-objective optimizations	Single-objective optimizations
Nature of network	Wired and wireless	Wireless	Wireless
Knowledge	Self-aware (capability to learn from past decisions and use this learning to influence future behavior)	Self-aware (capability to learn from past decisions and use this learning to influence future behavior)	Memoryless
Protocol stack	Communication between nonadjacent layers	Traditional layered approach	Communication between nonadjacent layers
Action	Proactive	Proactive	Reactive

to fulfill their extended requirements due to their increased complexity. As we can see, these three approaches share some features and they are different from others. However, CNs have the capability of including CR and cross-layer concepts for their approaches.

13.4 Overcome Weather Interferences by CNs

In this section, we show a proposal to CN using a KP. The KP is implemented using machine learning as we can see in the following text.

Our proposal is based on previous researches [29,30] about how the weather affects the network performance. As we have shown in papers [29,30], the weather affects significantly the rate of successes/failures of control frame (CF) transmissions over the media access layer (MAC) layer. Therefore, we consider the approach of CNs as a good option to overcome this negative influence. First of all, meteorological conditions must be measured together with network performance. In our study, the rate of success/failure of control frame transmissions is considered for the network performance. These rates can be calculated from counters gathered at access point' MIB (management information base) that can be consulted by simple network management protocol (SNMP). We have shown in our previous papers that the influence of the weather on other kind of parameters such as management frames [31] or transport layer's parameters (jitter, delay, etc.) [32] are not so clear, so we focused on CFs from that moment.

The most common formulation focuses on learning knowledge for the performance task of classification or regression. Classification involves assigning a test case to one of a finite set of classes, whereas regression predicts the case's value on some continuous variables or attributes. Our machine learning is based on classification. This classification has been done previously in our previous works, where we found high correlations between weather conditions and wireless network performance. According to our work [29], we can see that some weather conditions have higher influence than others on CFs in IEEE 802.11b/g WLAN. These influences have been calculated using Spearman correlation coefficient as CF counters do not follow a lineal distribution, so a nonparametric method is required to calculate the correlation levels. Table 13.2 shows the level of influence of each weather variable on WLAN performance.

Table 13.2 Relationship between Weather Variables and WLAN Performance

WEATHER VARIABLE	WLAN PERFORMANCE
Temperature	High
Solar radiation	Medium/High
Wind speed	Medium
Humidity	Low

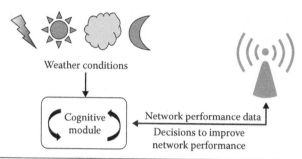

Figure 13.2 KP's agents involved in our cognitive wireless network.

Taking into account this information, we design a KP for developing a cognitive WLAN that considers the weather as an input to improve its performance. Our proposed KP consists of the elements showed in Figure 13.2. Some elements take weather data from their environment and others collect network performance data from time to time. Next, a cognitive module according certain defined rules processes this information in order to decide if some change should be applied to network function. Finally, if it is required, this same module send orders to the involved APs in order to perform some changes and actions to improve the wireless communication. Figure 13.2 represents the elements that make up our KP.

Modules that form our cognitive module and their connections are shown in Figure 13.3. The explanation about this cognitive module is performed sequentially. First, our cognitive module is collecting information from weather [weather variables such as temperature (T), solar radiation (S), wind speed (W), and humidity (H)] and network performance [CF of IEEE 802.11MIB] in a synchronous way. In order to explain their functionality, we are focusing on only one weather variable, for example, temperature. The other weather conditions will perform following the same scheme.

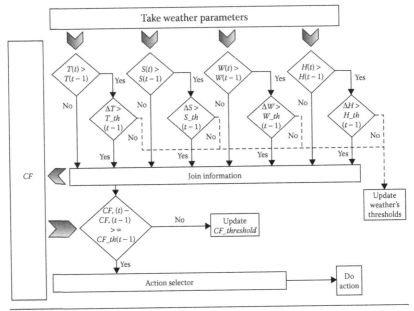

Figure 13.3 Cognitive module in detail.

Our algorithm runs every t instant and it has memory, for example, it is capable of storing the information read in $t - 1$. In an instant t, our cognitive module reads the information of weather conditions. If $T(t) > T(t - 1)$, then the temperature will rise. Our algorithm will go to the next module where ΔT is greater than a threshold. In our case, this threshold is called $T\_th(t - 1)$. If this comparison is false, our algorithm goes to the block called update weather's threshold. This block updates every threshold if it is necessary according to the evolution of each weather variable. It will allow having an adaptive algorithm and evolves correctly according to our changing environment. In case ΔT was higher in terms of absolute value than the threshold, ΔT_input would be activated in our block called join information, which will be explained as follows: Each weather variable will be able to activate its input directly in the join information block. This process is done by each weather variable and it permits to activate several inputs in our join information block.

According to the activated inputs, our join information block will be able to determine which CFs are most affected by some changes in weather conditions. This information is known by our classification process, which has been done in our previous work [29].

This classification process allows selecting the adequate CF counters. These selected CF counters will be used in our next comparative block. Following the example of temperature, we are going to think that our actual environment situation activates only *T_input* in the join information block. *RTS_failure* is the most affected CF by the changes of temperature (see Figure 13.4). Therefore, this CF will be selected and evaluated by our algorithm. Figure 13.5 shows how other

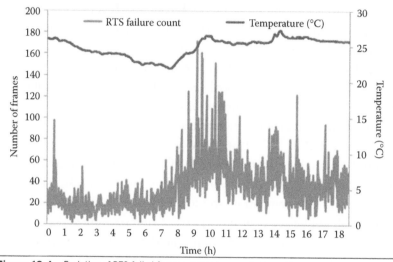

Figure 13.4   Evolution of RTS failed frames versus temperature over time.

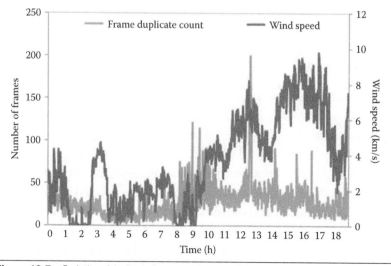

Figure 13.5   Evolution of duplicate frame count versus wind speed over time.

weather variable such wind speed impacts on the number of duplicate frames over time.

Our next block will carry out a subtraction between $CF\_RTS\_failure(t)$ and $CF\_RTS\_failure(t - 1)$. If the result of this operation is greater than or equal to a specific threshold $CF\_th(t - 1)$, it will be activated by the block called action selector. Otherwise, the block called update $CF\_x\_threshold$ will determine if it is necessary to do an update. This block updates every $CF\_x\_threshold$ if it is necessary according to the evolution of each CF counter. This block will allow having an adaptive algorithm and it evolves correctly according to our changing CF counters.

Finally, our block called action selector will have a set of actions to do in order to improve the global network behavior. These actions will improve the wireless communications without changing any parameter of IEEE 802.11b/g/n standard, so it means that any device that implements IEEE 802.11 standard could use this solution. Some actions that can increase the strength of the wireless communications are (1) changing the modulation, (2) changing the packet size, and (3) changing the contention window of IEEE 802.11b/g/n. These actions will not be taken all at once. The action performed depends on the counters that exceed their set threshold, this phase corresponds to the block called do action.

In this way, we will achieve our goal of improving the network performance by considering the weather interference using the CN approach.

13.5 Conclusion

Several published papers have shown that the weather affects the performance of WLANs. This information was taken from the errors in CFs. Based on this, we proposed an algorithm that uses a cognitive process in order to adapt the WLAN and enable intelligence to the network. It compares the weather conditions with the performance of the CF transmissions, and according to this, it is decided if it is necessary to change some transmission features. We proposed to modify the packet size, the modulation scheme, or the window size when it is required. These actions are considered when the weather data and the network performance as inputs indicate that it is necessary to perform some change in the transmission features. The main advantage of our proposed system is that it can be implemented in any wireless node of any WLAN as it does not modify the WiFi standard. The possible actions defined in our system are collected from the standard definition, so it is very easy to include it in any WLAN. The next step of our research in this field is to find the features changes that must be performed in each case and determine the level of improvement that supposes incorporate our proposal in wireless nodes.

Acknowledgments

This work has been partially supported by the Ministerio de Ciencia e Innovación by the Plan Nacional de I+D+i 2008–2011 in the "Subprograma de Proyectos de Investigación Fundamental" project TEC2011-27516. It has also been partially supported by the Vicerrectorado de Investigación, Innovación y Transferencia of the Universitat Politècnica de València (UPV), by the Programs VLC/CAMPUS and FPI-UPV. We thank the Information and Communications Systems Office (ASIC) of the UPV for allowing us to perform this research on campus WLAN, Borja Opticos Enterprise for providing meteorological data, and Azimut Electronics Company for proving its head office as a point link.

References

1. Q. Mahmoud, *Cognitive Networks: Towards Self-Aware Networks*, Wiley, New York, 2007.
2. J. Mitola, Cognitive Radio: An Integrated Agent Architecture for Software Defined Radio, PhD Thesis, Royal Institute of Technology (KTH), Stockholm, Sweden, 2000.
3. R. W. Thomas, L. A. DaSilva and A. B. MacKenzie, "Cognitive networks," in *1st IEEE International Symposium on New Frontiers in Dynamic Spectrum Access Networks*, Baltimore, MD, 2005. doi:10.1109/DYSPAN .2005.1542652.
4. D. D. Clark, C. Partridge, J. C. Ramming and J. T. Wroclawski, "A knowledge plane for the internet," in *SIGCOMM*, Karlsruhe, Germany, 2003. doi:10.1145/863955.863957.
5. T. Dietterich and P. Langley, "Machine learning for cognitive networks: Technology assessment and research challenges," Department of Computer Science, Oregon State University, Corvallis, OR, 2003. Available: http://hdl.handle.net/1957/35437.

6. S. Haykin, "Cognitive radio: Brain-empowered wireless communications," *IEEE Journal on Selected Areas in Communications*, vol. 23, no. 2, pp. 201–220, 2005.

7. Z. Zhang and H. Zhang, "A variable-population evolutionary game model for resource allocation in cooperative cognitive relay networks," *IEEE Communications Letters*, vol. 17, no. 2, pp. 361–364, 2013.

8. P. Mähönen, M. Petrova, J. Riihijärvi and M. Wellens, "Cognitive wireless networks: Your network just became a teenager," in *IEEE INFOCOM*, Barcelona, Spain, 2006. doi:10.1.1.114.9203.

9. K. Hamdi, W. Zhang and K. B. Letaief, "Opportunistic spectrum sharing in cognitive MIMO wireless networks," *IEEE Transactions on Wireless Communications*, vol. 8, no. 8, pp. 4098–4109, 2009.

10. W. Bin, W. Jinkuan, S. Xin and L. Fulai, "Q-learning-based adaptive waveform selection in cognitive radar," *Int. J. Communications, Network and System Sciences*, vol. 7, pp. 669–674, 2007.

11. D. Clark, "A new vision for network architecture," 2002. Available: http://www.isi.edu/know-plane/DOCS/DDC_knowledgePlane_3.pdf. Accessed on March 4, 2014.

12. C. Ramming, "Cognitive networks," in *DARPATech*, Anaheim, CA, 2004. Available: http://archive.darpa.mil/DARPATech2004/proceedings.html.

13. P. Kavassalis, "Key Technologies for Europe: Communications," European Commission DG Research: Key Technologies Experts Group, 2005. Available: ftp://ftp.cordis.europa.eu/pub/technology-platforms/docs/kte_communications.pdf.

14. R. W. Thomas, D. H. Friend, L. A. DaSilva and A. B. MacKenzie, "Cognitive networks: Adaptation and learning to achieve end-to-end performance objectives," *IEEE Communications Magazine*, vol. 44, no. 12, pp. 51–57, 2006.

15. R. W. Thomas, D. H. Friend, L. A. DaSilva and A. B. MacKenzie, "Cognitive networks," in *Cognitive Radio, Software Defined Radio, and Adaptive Wireless Systems*, H. Arslan (Ed.), Springer-Verlag, Secaucus, NJ, 2007, pp. 17–41.

16. R. W. Thomas and L. A. DaSilva, "Cognitive networking," in *Cognitive Radio Technology*, Elsevier, 2009, pp. 723–741.

17. A. Mihailovic, I. P. Chochliouros, E. Georgiadou, A. S. Spiliopoulou, E. Sfakianakis and M. Belesioti, "Situation awareness mechanisms," in *International Conference on Ultra Modern Telecommunications & Workshops*, St. Petersburg, Russia, 2009. doi:10.1109/ICUMT.2009.5345485.

18. S.-N. E. Project, "Deliverable D2.1 first report on mechanisms for situation awareness of cognitive network elements and decision making mechanisms for goal-oriented task planning," INFSO-ICT-224344, 2009. Available: https://www.ict-selfnet.eu. Accessed on April 2014.

19. P. Sutton, L. E. Doyle and K. E. Nolan, "A reconfigurable platform for cognitive networks," in *CROWNCOM*, Mykonos Island, Greece, 2006. doi:10.1109/CROWNCOM.2006.363467.

20. X. Wanga, H. Chengb and M. Huanga, "QoS multicast routing protocol oriented to cognitive network using competitive coevolutionary algorithm," *Expert Systems with Applications*, vol. 41, no. 11, p. 4513–4528, 2014.

21. Y. Zhang, J. Guan, C. Xu and H. Zhang, "The stable routing protocol for the cognitive network," in *IEEE Globecom Workshops*, Anaheim, CA, 2012. doi:10.1109/GLOCOMW.2012.6477730.

22. C. Wang, X.-Y. Li, S. Tang and C. Jiang, "Multicast capacity scaling for cognitive networks: general extended primary network," in *IEEE 7th International Conference on Mobile Ad-Hoc and Sensor Systems*, San Francisco, CA, 2010. doi:10.1109/MASS.2010.5663912.

23. C. Fortuna, M. Mohorcic, "Advanced access architecture for efficient service delivery in heterogeneous wireless networks," in *3rd International Conference on Communications and Networking, ChinaCom*, People's Republic of China, 2008. doi:10.1109/CHINACOM.2008.4685235.

24. J. Wang and X. Fu, "Challenges of CAC in heterogeneous wireless cognitive networks," *Physics Procedia*, vol. 25, pp. 2218–2224, 2012.

25. R. E. Tuggle, "Cognitive multipath routing for mission critical multi-hop wireless networks," in *42nd Southeastern Symposium on System Theory*, Tyler, TX, 2010.

26. B. R. Tamma, B. S. Manoj and R. Rao, "An autonomous cognitive access point for Wi-Fi hotspots," in *IEEE Global Telecommunications Conference*, Honolulu, HI, 2009.

27. C. Fortuna and M. Mohorcic, "Trends in the development of communication networks: Cognitive networks," *Computer Networks*, vol. 53, no. 9, pp. 1354–1376, 2009.

28. B. A. Fette, *Cognitive Radio Technology*, Academic Press, Burlington, MA, 2009.

29. D. Bri, M. Fernandez-Diego, M. Garcia, F. Ramos and J. Lloret, "How the weather impacts on the performance of an outdoor WLAN," *IEEE Communications Letters*, vol. 16, no. 8, pp. 1184–1187, 2012.

30. D. Bri, F. Ramos, J. Lloret and M. Garcia, "The influence of meteorological variables on the performance of outdoor wireless local area networks," in *IEEE International Conference on Communications*, Ottawa, Ontario, Canada, 2012. doi:10.1109/ICC.2012.6364448.

31. D. Bri, S. Sendra, H. Coll and J. Lloret, "How the atmospheric variables affect to the WLAN datalink layer parameters," in *6th Advanced International Conference on Telecommunications*, Barcelona, Spain, 2010. doi:10.1109/AICT.2010.15.

32. D. Bri, S. Sendra, M. Garcia and J. Lloret, "Do sensed atmospheric variables affect to the network QoS parameters in WLANs?" in *4th International Conference on Sensor Technologies and Applications*, Venice, Italy, 2010. doi:10.1109/SENSORCOMM.2010.70.

Biographical Sketches

Diana Bri was born in Gandia, Spain. She received her BSc and MEng in telecommunications engineering from Universitat Politècnica de València, València, Spain, in 2007 and 2010, respectively. She obtained her PhD in telecommunications engineering about cognitive wireless networks at the department of communications of the Universitat Politècnica de València. She works as a researcher at the Research Institute for Integrated Management of Coastal Areas in Gandia, València, Spain. Bri has coauthored more than 30 scientific papers published in international conferences and some of them in international journals. She has also been a technical committee member in several conferences and journals and she has been in the organization committee of several conferences. She is an associate editor of the international journal *Network Protocols and Algorithms*. Her research interests include wireless local area networks, outdoor network performance, wireless interferences, and cognitive networks.

Miguel Garcia was born in Alicante, Spain. He received his MSc in telecommunications engineering in 2007, his master's degree in communications technologies, systems and networks in 2008, and his PhD degree about wireless sensor networks in 2013, all of them from the Universitat Politècnica de València, València, Spain. Currently, he works as an assistant professor in the department of computer science at the Universitat de València, Spain. His research work is focused on wireless sensor networks, multimedia networks, computer networks, and cloud computing. Dr. Garcia has coauthored more than 45 scientific papers published in international conferences and more than 30 papers published in international journals (most of them included in Journal Citation Report). He has been a technical committee member in several conferences and journals, and he has also been in the organization committee of several conferences. He is an associate editor of several journals.

Jaime Lloret Mauri received his MSc in physics in 1997 and his MSc in electronic engineering in 2003, both in the Universitat de València, València, Spain, and his PhD in telecommunications engineering (Dr. Ing) in 2006 at Universitat Politècnica de València, València, Spain. He is currently an associate professor in the Universitat Politècnica de València. He is the head of the "Communications and remote sensing" research group of the Integrated Management Coastal Research Institute and the "Active and collaborative techniques and use of technologic resources in the education (EITACURTE)"

Innovation Group, both belonging to the Universitat Politècnica de València. He is the director of the University Expert Certificate "Redes y Comunicaciones de Ordenadores," the University Expert Certificate "Tecnologías Web y Comercio Electrónico," and the University Master "Digital Post Production." He is currently the chair of the Internet Technical Committee (IEEE Communications Society and Internet society) and the Working Group of the Standard IEEE 1907.1. He has authored 12 books and has published more than 250 papers in national and international conferences, and international journals (more than 90 Science Citation Index [SCI] Indexed). He has been general chair (or cochair) of 23 international workshops and conferences. He is the IEEE senior and International Academy, Research, and Industry Association (IARIA) fellow. He is the editor in chief of the international journal Networks Protocols and Algorithms and IARIA Journals Board Chair (eight journals).

Francisco Ramos was born in València, Spain. He received his MSc and PhD degrees in telecommunications engineering from the Universitat Politècnica de València, València, Spain, in 1997 and 2000, respectively. Since 1998, he has been with the department of communications at the same university, where he is now a professor. He has participated in several national and European research projects on areas such as optical access networks, broadband wireless systems, and optical networking. Professor Ramos has coauthored more than 100 papers in international journals and conferences, and he has acted as a reviewer for the Institution of Electrical Engineers (IEE), Institute of Electrical and Electronics Engineers (IEEE), Optical Society of America (OSA), Elsevier, and Taylor & Francis publishers. He is also the recipient of the Prize of the Telecommunication Engineering Association of Spain for his thesis dissertation on the application of optical nonlinear effects in microwave photonics. His current research interests include complex systems, chaos theory, and cybernetics.

PART VI

Vehicular Cognitive Networks and Applications

14

COGNITIVE HANDOVER FOR VEHICULAR HETEROGENEOUS NETWORKS

Issues and Challenges

ALI SAFA SADIQ, KAMALRULNIZAM
ABU BAKAR, KAYHAN ZRAR
GHAFOOR, JAIME LLORET MAURI,
AND NORSHELIA BINTI FISAL

Contents

14.1 Introduction

Recently, the wireless networks are managed through utilizing a fixed spectrum policy, whereby licenses were assigned officially to provide the network services for large geographical areas. In wireless access in vehicular environments (WAVEs), the communication methods deployed based on two standards: vehicle-to-vehicle communication (V2V) and vehicle-to-roadside (V2R) base station (BS) or access point (AP) communication. For instance, in the United States and Europe, the bands at 5.85–5.925 GHz of the wireless spectrum were assigned for both V2V and V2R. However, the transmitting process was operated based on the standards of IEEE 802.11p and IEEE 1609.4 [1,2]. In the meanwhile, in realistic urban environments, there are many applications functioning by vehicles, such as safety applications in additional to traffic monitoring application systems.

One of the challenges is the need for stable bandwidth and low delay in order to maintain these applications within heterogeneous vehicular environments. This type of challenge is quite sensible particularly in the heterogeneous wireless network environments, whereby different quality of services (QoSs) are received from multiple access links. Basically, any limitation within wireless spectrum for IEEE 802.11p-based vehicular networks will result in degrading the real-time applications (e.g., voice-over-Internet protocol [VoIP] and video streaming).

In cognitive radio network (CRN) technology [3], an opportunistic technology for spectrum utilization is introduced. This technology is known as dynamic spectrum access that directly contributes in several forms of vehicular communication. The functionalities of CRN are implemented in a way to manage the available spectrum in the vicinity. In other words, the CRN-based vehicular environment detects the available spectrum over digital television (TV) frequency bands in the range of ultrahigh frequency. Moreover, it decides the channel that can be used as a next optimal candidate based on the desired QoS of ongoing applications. Besides, it is always expected that during the vehicle roaming, there will be no any harmful interference to the licensed owners of the spectrum in CRNs.

However, in CRNs there are many different characteristics that include additional concerns than only placing a cognitive radio (CR) within a vehicle. For instance, in a vehicular environment, the availability of wireless spectrum for CR systems perceived by each available roadside BS and AP is changing dynamically by the time. This dynamic changing is not only due to the activities of the licensed or primary users of CRN but also based on the relative movement of vehicles. Subsequently, the spectrum measurements of available CRNs need to be undertaken over the total movement track of the vehicles. Moreover, the spectrum sensing or detecting is another key challenge facing CRNs within the vehicular environment. In other words, it is quite needed that the vehicles are aware about the free spectrum and sharing it without noticeable interference with other vehicles. However, the spectrum management is highly required by monitoring the best available spectrum from the surrounding roadside access links to meet the communication requirements.

The remainder of this chapter is organized as follows: Section 14.2 provides a summary of the related works to cognitive handover-based

CRNs. Section 14.3 categorizes the CRNs depending on their architecture, characteristics, and features. Finally, Section 14.4 concludes the chapter by highlighting the main issues and challenges that are facing the cognitive handover in vehicular heterogeneous network-based CRNs.

14.2 Related Work

In order to enable multichannel operations in the WAVE, the IEEE 1609.4 standard was proposed. Using this standard, the wireless channels are periodically synchronized into control and service intervals [4]. Within the WAVE system that is functioning based on the spectrum of 75 MHz, seven channels are operating. One is control channel and the other six operate as service channels. When the vehicular users need to transmit the information in the 5.9 GHz band, they have to compete for the channel access in order to use it. Yet, as a way to comprehend the possible vehicular communications, vehicles must be able to communicate each other utilizing V2V or V2R communication. This can be achieved by utilizing a wide range of spectra and networks such as WiFi networks, cellular networks, WiMAX networks, ad hoc networks, TV bands, and satellite networks. The availability of multispectrum choices depends on the accessible wireless networks in the surrounding area in addition to the location of vehicles.

Therefore, several recent works were proposed as advances in CRNs such as [3–9]. In CRNs the users are able to sense and handoff from one network to another based on the needs and the available environment with the support of CRs. It is worth mentioning that in a vehicular environment, the high density of vehicles especially in urban areas can cause irregular network status. Moreover, the high speed of vehicles in addition to the unpredictable behavior of human inside them leads to challenging the network selection in heterogeneous networks.

Qiben Yan et al. [10] employed the channel selection technique within a network. A theoretical study of the throughput called mobile content distribution (MCD) in vehicular ad hoc networks (VANETs) was introduced. The MCD method was proposed in spite of using of packet-level network coding (PLNC) and symbol-level network coding (SLNC) due to the existing lacks in the basic comprehension of the limits of MCD protocols by the use of network coding in

VANETs. Therefore, a theoretical model was developed [10] in order to calculate the viable throughput of cooperative MCD in VANETs. This was achieved by using SLNC and considering a road topology of one dimension with deployed APs as the content source. The derivative of predicted achievable throughput for a vehicle with respect to distance from an AP was calculated and employed for both PLNC and SLNC. Figure 14.1 shows the graphs that were modeled in [10]. The model design of PLNC is presented in Figure 14.1a, whereby the packet transmission in the proposed method stores the packet flow that is received by each node until the transmission opportunity is obtained using the scheduling scheme. However, Figure 14.1b illustrates the virtual of multihypergraph, where by the process of one packet injected to one hyperarc. The proposed theoretical method is not applicable in heterogeneous vehicular networks since different characteristics can be experienced by vehicles during their roaming.

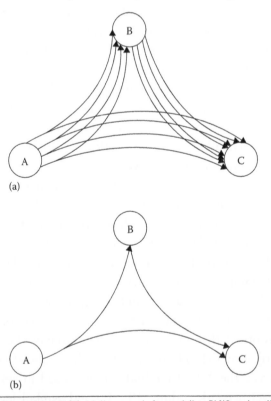

Figure 14.1 The modeled PLNC (a) Hypergraph for modeling PLNC and multihypergraph for modeling SLNC (b). (Data from Yan, Q, et al., *IEEE Journal on Selected Areas in Communications,* 30(2), 484–492, 2012.)

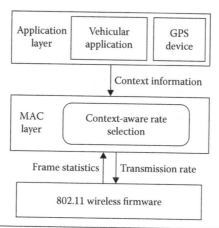

Figure 14.2 The system architecture of the proposed CARS. GPS, global positioning system; MAC, media access control layer. (Data from Shankar, P. et al., CARS: Context-aware rate selection for vehicular networks, *IEEE International Conference on Network Protocols*, pp. 1–12, Orlando, FL, October 19–22, 2008.)

However, in order to address the related issues to the rapid variations in the link quality due to fading and mobility at vehicular high-speed environment, in addition to the transmission rate adaptation. A novel context-aware rate selection (CARS) algorithm was implemented and evaluated by Pravin Shankar et al. [11]. In the proposed CARS algorithm, the context information that is represented by the vehicle speed and distance from the neighbor was utilized to address the aforementioned issues, while increasing the link throughput. Figure 14.2 presents the architecture design of the proposed CARS algorithm [11]. However, this type of solution is not really applicable in heterogeneous vehicular environment due to the network characteristics that are different from one to another.

14.3 Architecture, Characteristics, and Features of CRNs

The CRN networks are categorized into three classes based on architectures' deployment for CRNs as presented in Figure 14.3. For instance, Figure 14.3a shows that the wireless network is formed between vehicles only that rely on the information sharing among them using the beaconing system. Figure 14.3b illustrates the second type of CRN communication that is based on the heterogeneous roadside interactive system. In this kind of communication, the APs and BSs act as

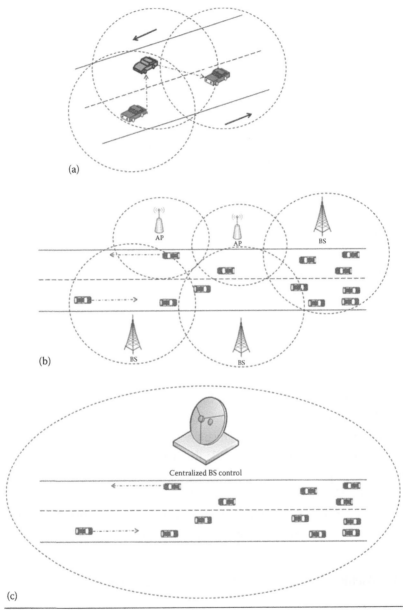

Figure 14.3 Three main scenarios of architectures' deployment for CRNs. (a) Vehicle-to-vehicle communications; (b) multi local heterogeneous wireless communications; (c) centralized wireless communications.

data sources that supply and coordinate the vehicles in the deployed scenario. Eventually, a centralized wireless communication is demonstrated in Figure 14.3c, in which the central BS separately performs the handover decisions to be processed with selected wireless channel. In other words, the distributed information by vehicles is not elaborated in the obtained handover decisions in this kind of architectures.

In this section, the characteristics and features of the CRN communication architectures that are presented in Figure 14.3 are discussed. The mobility impact is one of the important factors that the cognitive handover relies on during handover making decision-based CRNs. For instance, in CRNs based on the Federal Communications Commission in the United States, local sensors are deployed in which the vehicles collect various sensing samples at different locations. Thus, the obtained sensing samples could be demonstrated at different levels of relationship that normally depends on the characteristics of the existence of buildings or obstacles in addition to the velocity of the vehicles. Therefore, one of the main challenging points is to integrate these different data samples in the correct form [12]. To this end, in order to achieve an efficient cognitive handover in heterogeneous networks, the traffic and vehicles' mobility speed should be carefully considered.

However, coordinating the process spectrum usage is one of the important key issues that are normally facing the cognitive handover-based vehicular heterogeneous network environment [13]. Therefore, there are some efforts done in such a way to establish the two modes of integrated spectrum databases. The concept of these databases is to keep the vehicles update their spectrum vacancies' list so that they can easily maintain the cognitive handover to the most available spectrum in the vicinity. One of the favored options is the interconnection that was made between the geolocation information and its database access, which is used in identifying the available channels, power and direction restriction information to be continuously updated into database form. Thus, the vehicles should obtain a dedicated out-of-band radio interface to request the available databases. As another solution, a multiple roadside connection link should be deployed, which can supply the saved information to the onboard radio channels. Therefore, the key challenge of spectrum sensing-based opportunistic spectrum access should be optimally addressed to achieve more efficient cognitive handover in heterogeneous networks.

As another aspect, the security issue in CRNs is one of the concerns that could be faced in the vehicular environment. In order to improve the sensing accuracy of the available spectrum, the interaction among the vehicles is extremely needed to keep the spectrum information continuously updated. Thus, security issues can be raised up during this particular process [14]. Throughout the process of CRNs that allows the users exchange their sensing reports such as the current visibility on each channel, this process could issue another challenge to make it more protected. The proposed security methods for CRNs are tried to address this issue by assigning weight values to each node [14]. Though, in vehicular heterogeneous networks, the situation is more complicated. The reason is that the surrounded vehicles are rapidly changing their link of connection by the time of traveling. Hence, a fast detection method should be developed to recognize in a fast way the malicious vehicles that could send fake sensing reports.

Furthermore, the existing collaboration techniques can cause privacy issues to the users. Though spying the transmitted spectrum reports that are normally exchanging among vehicles, the attackers could reveal the identity and movement pathway of drivers. Therefore, it is really challenging to develop a system that is able to identify the existence of malicious vehicles that send false sensing reports during their traveling. Besides, providing more privacy for driver identity and movement pathway while vehicles are cooperating is a big issue that needs to be addressed.

Finally, in order to summarize the main characteristics and features of cognitive handover, Table 14.1 lists them based on the utilized architecture of vehicular heterogeneous networks.

Table 14.1 Main Characteristics and Features of Cognitive Handover for Vehicular Heterogeneous Networks

ARCHITECTURE	CHARACTERISTICS AND FEATURES
Vehicle-to-vehicle communications	Low implementation complexity; no network support required; roles of cooperation are required
Multilocal heterogeneous wireless communications	High receiver protection; higher accuracy; fixed infrastructure is needed
Centralized wireless communications	BS separately decides the channels to be used by the vehicle, not relying on information from the vehicles; overhead issues

14.4 Conclusion

In this chapter, the current state-of-the-art research on CRNs is discussed. In CRNs, the main aim is to achieve high throughput under multiple constraints and maintain the availability of licensed spectrum, transmission mode selection, and link scheduling. Considering the special features of cooperative communications, this technology can achieve high-bandwidth multimedia applications. Though the research area of cognitive handover in vehicular heterogeneous networks is still at an initial phase, more concern need to be addressed. In other words, the previously discussed related works in this chapter, which are concerning in maintaining a better spectrum management functionality based CRNs, these methods are still need to be reconsidered. This can be done by taking into account the different aspects and features of vehicular heterogeneous networks behaviors, for instance, the impact of mobility on spectrum management during the handover processes. Moreover, the probabilities of cooperation among the vehicles share the spectrum information while they are traveling in different speeds. Furthermore, more effort need to be devoted to evaluate the cognitive handover processes based CRNs in terms of the impact of buildings by considering different vehicular environments. This effort is demanded as a way to achieve more realistic CRNs that totally support the vehicular heterogeneous networks.

References

1. IEEE, Amendment 6: Wireless access in vehicular environments (WAVE), part 11, IEEE standard 802.11p, 2010. Available: http://ieeexplore.ieee .org/xpl/mostRecentIssue.jsp?punumber=5514473. Accessed: May 2, 2014.
2. IEEE, IEEE standard for wireless access in vehicular environments (WAVE) multi-channel operation, std 1609.4, 2010. Available: http:// ieeexplore.ieee.org/xpl/mostRecentIssue.jsp?punumber=5712767. Accessed: May 2, 2014.
3. Ian F. Akyildiz, Won-Yeol Lee, Mehmet C. Vuran, and Shantidev Mohanty. Next generation/dynamic spectrum access/cognitive radio wireless networks: A survey. *Computer Networks*, 50(13): 2127–2159, 2006.
4. Danda B. Rawat, Yanxiao Zhao, Gongjun Yan, and Min Song. CRAVE: Cognitive radio enabled vehicular communications in heterogeneous networks. In *IEEE Radio and Wireless Symposium,* pp. 190–192, Austin, TX, January 20–23, 2013.

5. Gustavo Marfia, Marco Roccetti, Alessandro Amoroso, Mario Gerla, Giovanni Pau, and J.-H. Lim. Cognitive cars: Constructing a cognitive playground for VANET research testbeds. In *Proceedings of the 4th International Conference on Cognitive Radio and Advanced Spectrum Management*, Barcelona, Spain, October 26–29, 2011.

6. Xiao Yu Wang and Pin-Han Ho. A novel sensing coordination framework for CR-VANETs. *IEEE Transactions on Vehicular Technology*, 59(4): 1936–1948, 2010.

7. Yanxiao Zhao, Min Song, Chunsheng Xin, and Manish Wadhwa. Spectrum sensing based on three-state model to accomplish all-level fairness for co-existing multiple cognitive radio networks. In *IEEE Proceedings of the INFOCOM*, pp. 1782–1790, Orlando, FL, March 25–30, 2012.

8. Yingzhe Li, Xinbing Wang, Xiaohua Tian, and Xue Liu. Scaling laws for cognitive radio network with heterogeneous mobile secondary users. In *IEEE Proceedings of the INFOCOM*, pp. 46–54, Orlando, FL, March 25–30, 2012.

9. Danda B. Rawat, Bhed B. Bista, and Gongjun Yan. CoR-VANETs: Game theoretic approach for channel and rate selection in cognitive radio VANETs. In *Proceedings of the 7th International Conference on Broadband, Wireless Computing, Communication and Applications*, pp. 94–99, Victoria, BC, November 12–14, 2012.

10. Qiben Yan, Ming Li, Zhenyu Yang, Wenjing Lou, and Hongqiang Zhai. Throughput analysis of cooperative mobile content distribution in vehicular network using symbol level network coding. *IEEE Journal on Selected Areas in Communications*, 30(2): 484–492, 2012.

11. Pravin Shankar, Tamer Nadeem, Justinian Rosca, and Liviu Iftode. CARS: Context-aware rate selection for vehicular networks. In *IEEE International Conference on Network Protocols*, pp. 1–12, Orlando, FL, October 19–22, 2008.

12. Alexander W. Min and Kang G. Shin. Impact of mobility on spectrum sensing in cognitive radio networks. In *Proceedings of the ACM workshop on Cognitive Radio networks*, Beijing, People's Republic of China, pp. 13–18, 2009.

13. Haesik Kim and Aarne Mämmelä. Cognitive radio and networks for heterogeneous networking. *Cognitive Communications: Distributed Artificial Intelligence, Regulatory Policy & Economics, Implementation*, pp. 17–52, 2012.

14. Ian F. Akyildiz, Brandon F. Lo, and Ravikumar Balakrishnan. Cooperative spectrum sensing in cognitive radio networks: A survey. *Physical Communication*, 4(1): 40–62, 2011.

Biographical Sketches

Ali Safa Sadiq received his BSc degree in computer science from Al-Mustansiriya University Baghdad, Iraq (2005); MSc degree in computer science, specialized in wireless communications from

Universiti Teknologi Malaysia, Johor, Malaysia (2011); and PhD degree in wireless heterogeneous networks from Universiti Teknologi Malaysia, Johor, Malaysia (2014). He has published several scientific/research papers in international journals and conferences. His current research interests include intelligent handover decision making in heterogeneous wireless networks, vehicular and mobile ad hoc networks, artificial intelligent applications, routing protocols, wireless sensor networks, media access control (MAC) layer design, mobility handover in mobile IPV6 and video communications, and cognitive vehicular networks. Dr. Sadiq served as a reviewer for several prestigious international journals and conferences. He also served as a technical program committee chair of Smart Sensor Protocols and Algorithms (SSPA 2013) Conference in Dalian, China. He is currently serving as a PC member of Multimedia Wireless ad hoc Networks 2014 Conference in Benidorm, Spain, and International Conference on Emerging Wireless Communications and Networking (October 06–08, 2014) in Erbil, Iraq. Sadiq was awarded Pro-Chancellor Academic Award in conjunction with the 47th Convocation 2011 as a best master student in his batch. He was also awarded the UTM International Doctoral Fellowship (IDF). Dr. Sadiq is currently serving as a senior lecturer in computer system and networking department at Faculty of Computer systems & Software Engineering, Universiti Malaysia Pahang.

Kamalrulnizam Abu Bakar obtained his PhD degree in computer science, specialized in wireless communications from Aston University, Birmingham, in 2004. Currently, he is an associate professor of computer science at Universiti Teknologi Malaysia, Johor, Malaysia, and a member of the Pervasive Computing Research Group. He involves in several research projects and is the referee for many scientific journals and conferences. His specialization includes mobile and wireless computing, information security, and grid computing.

Kayhan Zrar Ghafoor received his BSc degree in electrical engineering from Salahaddin University, Erbil, Iraq (2003); his MSc degree in remote weather monitoring from Koya University, Koya, Iraq (2006); and his PhD degree in wireless networks from Universiti

Teknologi Malaysia (UTM), Johor, Malaysia (2011). He is working as a senior lecturer in the department of software engineering at Koya University, Iraq. He has published over 30 scientific/research papers in prestigious international journals and conferences. Dr. Ghafoor served as a guest editor of the "Special Issue on Network Protocols and Algorithms for Vehicular Ad Hoc Networks" in *Mobile Networks and Applications* (*MONET*) journal. He is currently working as a general chair of a workshop named "Smart Sensor Protocols and Algorithms" in conjunction with "The 9th International Conference on Mobile Ad-Hoc and Sensor Networks (MSN 2014)" that will be held in Hungary. He also served as an associate editor, an editorial board member, and a reviewer for numerous prestigious international journals; appeared as a workshop general chair for international workshops and conferences; and worked as a technical program committee member for more than 30 international conferences. He is the recipient of the UTM Chancellor Award in the 48th UTM convocation in 2012. He was also awarded the UTM International Doctoral Fellowship (IDF) and Kurdistan Regional Government (KRG) scholarship (Ahmad Ismail Foundation). His current research interests include routing over VANETs and tactical wireless networks, cognitive vehicular networks as well as artificial intelligence and network coding applications. He is a member of Institute of Electrical and Electronics Engineers (IEEE), Vehicular Technology Society, IEEE Communications Society, Internet Technical Committee (ITC), and International Association of Engineers (IAENG).

Jaime Lloret Mauri received his MSc in physics in 1997 in electricity, electronics and computer Science and his MSc in electronic engineering in 2003, both in the Universitat de València, València, Spain, and his PhD in telecommunications engineering (Dr. Ing) in 2006 at Universitat Politècnica de València, València, Spain. He is currently an associate professor in the Universitat Politècnica de València. He is the head of the "Communications and remote sensing" research group of the Integrated Management Coastal Research Institute and the "Active and collaborative techniques and use of technologic resources in the education (EITACURTE)" Innovation Group, both belonging to the Universitat Politècnica de València. He is the director of the University Expert Certificate "Redes y Comunicaciones de Ordenadores," the University Expert Certificate "Tecnologías Web y Comercio Electrónico," and the University Master "Digital Post Production." He is currently the chair of the Internet Technical Committee (IEEE Communications Society and Internet society) and the Working Group of the Standard IEEE 1907.1. He has authored 12 books and has published more than 250 papers in national and international conferences, and international journals (more than 90 Science Citation Index [SCI] Indexed). He has been general chair (or cochair) of 23 international workshops and conferences. He is the IEEE senior and International Academy, Research, and Industry Association (IARIA) fellow. He is the editor in chief of the international journal *Networks Protocols and Algorithms* and IARIA Journals Board Chair (eight journals).

Norsheila Fisal received her BSc in electronic communication from the University of Salford, Manchester, UK, in 1984; MSc degree in telecommunication technology in 1986; and PhD degree in data communication from the University of Aston, Birmingham, UK, in 1993. Currently, she is a professor with the Faculty of Electrical Engineering, Universiti Technologi Malaysia, leading as a Director of UTM MIMOS CoE in Telecommunication Technology and head of Telematic Research Group (TRG).

She is actively involved in research focusing on work related to broadband networking in wired and wireless network, multimedia communication, and teletraffic engineering. Her current research interests are in wireless sensor networks, wireless mesh and relay networks, cognitive radio networks, LTE advanced network and WiMaX network.

Her is a member of Institute Electrical and Electronic Engineers (IEEE), Malaysian National Confederation of Computers (MNCC), Institute For Information Processing (IFIP), and Board of Engineer Malaysia.

15

Cognitive Radio-Enabled Vehicular Networking for Transportation Cyber-Physical Systems

DANDA B. RAWAT, CHANDRA BAJRACHARYA, JAIME LLORET MAURI, AND KAYHAN ZRAR GHAFOOR

Contents

15.1 Introduction

Wireless communication for vehicular networking is regarded as the backbone for intelligent transportation cyber physical systems (CPSs) that allow the forwarding of upcoming traffic information to the drivers in a timely manner. Vehicular ad hoc network (VANET) is one of the successful applications of mobile ad hoc networks (MANETs).

Vehicular communication has attracted the attention of academia and industry all over the world (Car-to-Car Communication Consortium 2007; Festag et al. 2008; Fuentes et al. 2011; Hartenstein and Laberteaux 2010). Road traffic crashes are one of the largest problems being faced not only in the United States but also all over the world. The studies of Wang and Thompson (1997) have shown that "about 60% roadway collisions could be avoided if the operator of the vehicle was provided warning at least one-half second prior to a collision." The number of death and injuries, and the excessive cost of traffic collisions can be significantly lowered if the drivers are provided with upcoming traffic information in a timely manner (Car-to-Car Communication Consortium 2007; Festag et al. 2008; Ghafoor et al. 2013a, 2013b; Hartenstein and Laberteaux 2010; Lloret et al. 2013; Moskvitch 2011; Thiagarajan et al. 2009; US NHTSA 2012; Watfa 2010) Wireless communication is proposed as a reliable information dissemination mechanism for vehicular networks. Moving vehicles act as nodes in a network and communicate with each other via vehicle-to-vehicle (V2V) communications through VANETs, as well as with roadside base stations via vehicle-to-roadside (V2R), along with possible roadside-to-roadside (R2R) communications. For instance, when a road accident occurs, emergency responders could provide information about road closure and the estimated time of reopenings using vehicular communications. In a vehicular network, messages have to be routed from the information source to one or several destinations as quickly as possible and efficiently (Fuentes et al. 2011; Rawat et al. 2011a, 2011b; Watfa 2010). To accomplish vehicular communications, the standard, IEEE 802.11p, also known as wireless access in vehicular environments (WAVE), operates at 5.9 GHz. The IEEE 802.11p has seven 10 MHz channels: six of them are data channels and one of them is a control channel. There are many challenges in vehicular communications including security, privacy, trust, reliable communications, robust connectivity, and routing among fast-moving vehicles. Recent research in this area is bridging the gap between theory and practice (Car-to-Car Communication Consortium 2007; Rawat et al. (in press); US DOT 2005; Watfa 2010). The US DOT (2005) plans to develop an architecture for vehicle infrastructure integration based largely on roadside equipment for collecting data from passing vehicles and for disseminating it to other interested vehicles or concerned authorities. The cellular telephone system was studied at the University of Virginia, Charlottesville, Virginia, for vehicular traffic monitoring in limited deployments (Smith et al. 2003). Santa et al. (2008) have successfully tested cellular network based on universal mobile telecommunications system (UMTS) for vehicular communications. Since the information exchange between participating vehicles happens through V2R and R2V with possible R2R communications, one can expect higher delay (few milliseconds to a few seconds). Because of the high delay/latency introduced by cellular systems, it is not suitable for forwarding emergency messages in vehicular networks. In addition, cellular networks use licensed bands and users need to pay a usage fee. Unlicensed communication such as WiFi (i.e., wireless local area network) and Bluetooth would provide a better alternative for communication as users do not have to pay licensing fees. However, because of the installation cost and limited availability of WiFi networks, it is not an attractive choice (Little and Agarwal 2005; Trivedi et al., 2011; Watfa 2010). In this context, V2V-based communications through VANETs, where devices and associated technologies use unlicensed bands, are a suitable option.

Technology based on IEEE 802.11p standard works well for dedicated short-range communications (up to 1000 m) in vehicular networks. As mentioned earlier, six data channels and one control channel that are used for vehicular communications could be easily overloaded when there are many vehicles in situations such as city and traffic light stops. Similarly, when the vehicle density is sparse, short-range-based communication could lead to frequent network breakage among fast-moving vehicles. This limits the scalability of vehicular networks. Thus, vehicular users should be able to sense the idle spectrum bands in a wide-spectrum regime and identify spectrum opportunities to use them dynamically without creating any harmful interference to licensed primary users that have the legal permission to use those bands exclusively.

This chapter presents the cognitive radio (CR)-enabled vehicular communications to sense the channels to identify spectrum opportunities to forward upcoming traffic information (Rawat et al. 2012; Rawat and Popescu 2012).

The rest of the chapter is organized as follows: Section 15.2 presents the general framework for transportation CPS including communication models. Section 15.3 discusses the CR-enabled vehicular communication for transportation CPS. Section 15.4 presents the collaborative decision making for transportation CPS. Section 15.5 concludes the chapter.

15.2 Transportation CPS Framework

CR-enabled transportation CPS and applications have the capability to interact with and expand the capabilities of the physical world through computation, communication, and control as shown in Figure 15.1. Transportation CPS for information dissemination is expected to use a variety of wireless technologies such as WiMAX, WiFi, Bluetooth, ZigBee, WAVE, cellular, and satellite. Each vehicle senses the radio spectrum to identify the spectrum opportunities, adapt its transmit parameters suitable to the spectrum opportunities, and use them opportunistically without creating harmful interference to licensed owners.

In transportation CPS, vehicles that participate in communications are considered to be equipped with computing and communicating devices. Communications among vehicles in vehicular networks could be accomplished using peer-to-peer networks or through the roadside infrastructures as discussed in Sections 15.2.1 and 15.2.2.

15.2.1 Vehicle-to-Roadside-to-Vehicle Model

In vehicular networks, vehicles can exchange information using V2R and roadside-to-vehicle (R2V) communications as shown in Figure 15.2.

In this model, vehicles communicate with each other using roadside infrastructures as relay units to forward information to vehicles. However, V2R- and R2V-based wireless communications in vehicular networks introduce the latency/delay (Santa et al. 2008), as message has

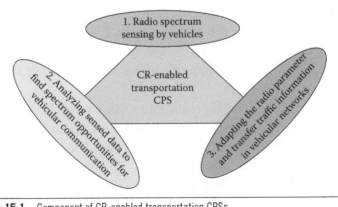

Figure 15.1   Component of CR-enabled transportation CPSs.

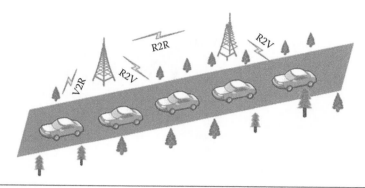

Figure 15.2 V2R, R2R, and R2V communications for transportation CPS.

to be transmitted from a source vehicle to a roadside unit and from the roadside unit to a destination vehicle with possible intermediate R2R communications. Vehicles could use existing cellular infrastructure to forward the information. However, when cellular infrastructures are used, vehicular users will have to pay usage fee as cellular bands are licensed to service providers. As a result, they may not be able to remain online all the time to transmit information. Thus, this method introduces high delay, as vehicular users have to set up a call before transmitting the actual information every time. Alternatively, IEEE 820.11 wireless network infrastructure could be used since there is no usage cost for industry, scientific and medical (ISM) bands. However, there are no such IEEE 820.11 wireless network infrastructures installed throughout the highways or roads to cover all areas for vehicular communications. Installation or usage cost and delay introduced by the roadside units and message forwarding using V2R and R2V communications are also not a good choice for vehicular communications. Thus, the vehicles using CRs would be able to sense the radio spectrum, identify the idle bands, and use them opportunistically using V2V communications.

15.2.2 V2V Model

When roadside infrastructures are not available for vehicular communications, vehicles directly communicate with each other through single hop or multiple hops by forming an ad hoc network as shown in Figure 15.3. In this model, as vehicles do not use any roadside units, there is no cost for the installation of roadside infrastructures. As message forwarding is done using direct communications, vehicles

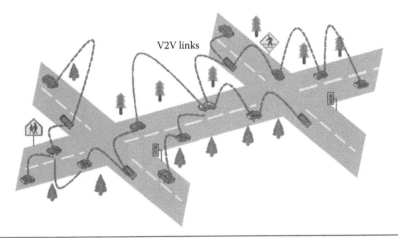

V2V links

Figure 15.3 V2V communications using single hop and multiple hops to forward messages in the transportation CPS.

could transmit the information more quickly than the vehicle-to-roadside-to-vehicle (V2R2V)-based communications. V2V-based communication is the best way of delivering the emergency messages using single-hop communications. This V2V-based communication is also applicable for the situations such as evacuations when all other road infrastructures are overloaded. Based on the needs, vehicles could sense the channel and use the least jammed/overloaded channels for their communications. Vehicles could use ISM bands for V2V communications without paying any licensing fees associated with radio-frequency (RF) spectrum with the technologies such as Bluetooth, ZigBee, WiFi technology in ad hoc mode, and ultra wide band (UWB) (Arslan et al. 2006; Bluetooth Technology 2012; IEEE 2003; ZigBee Alliance 2009; ZigBee Alliance 2012).

However, in the V2V model, without implementing robust security and privacy-aware systems, vehicular users could pretend to be someone else or could insert malicious messages in the vehicular network to mislead the communications (Rawat et al. 2011a). As vehicles could join the network and leave it at any time depending on the drivers' will. It would be almost impossible to track down malicious vehicles since there is no central unit to keep records of all participating vehicles. Thus, we need to provide anonymity to participating vehicles or users and implement robust security and privacy-aware systems (Rawat et al. 2011a).

15.3 CR-Enabled Vehicular Communications

The selection of wireless technology used in vehicular communication depends on the application that the vehicular network is envisioned to support. There are several factors that influence networks of fast-moving vehicles. Vehicular network has the following peculiar features (Li et al. 2009):

- Network topology changes constantly because of the fast-moving vehicles.
- Human/driver behavior affects the network topology of vehicular network since drivers can join or take exit at any time when it is possible to do so.
- The density of vehicles depends on the location and penetration ratio. For instance, in an urban area, more vehicles would be present, whereas in a rural area, vehicle density would be very low.
- Most of the existing wireless access technologies are not designed for high-speed vehicles.
- Within the vehicle, there is virtually unlimited power, unlimited storage, and unlimited computing capabilities, making vehicular network different from other wireless networks.
- Vehicular communication requires low latency for safety applications to forward emergency messages in a timely manner.
- Information and entertainment (infotainment) multimedia messages are bandwidth hungry; thus, the multimedia applications could easily suffer in low-bandwidth vehicular communication networks.

It is important to note that the choice of the wireless technology depends on the application that the vehicular network is proposed to support. When vehicles have choices of wireless technologies, they should be able to choose the best one using some best network selection methods. The vehicles with virtually unlimited power, storage, and computing capabilities could implement CR technology and identify to use the suitable wireless frequencies. This allows vehicular users access the wireless networks that meet the requirement of vehicular networks. Depending on the availability of the RF based on the sensed information by CR embedded on the vehicle, suitable

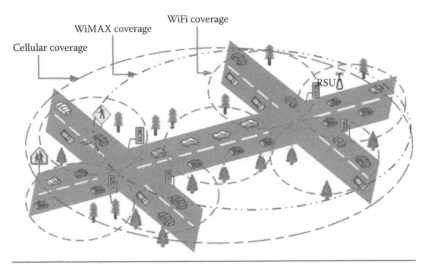

Figure 15.4 Wireless coverage of different networks for CR-enabled transportation CPS.

bands or wireless networks will be chosen to provide reliable wireless communications as shown in Figure 15.4.

15.3.1 Applications of CR-Enabled Transportation CPS

One of the main goals of transportation CPS is to provide safety and comfort for passengers and drivers by forwarding upcoming traffic information in a timely manner. Using CR, individual vehicles would be able to identify a suitable frequency and adapt their transmit parameters to transmit safety messages and infotainment information in a timely manner and with high reliability. In a broad sense, vehicular communication has three main applications: safety, comfort, and distributed cloud computing

15.3.1.1 Emergency and Safety-Related Applications Transportation CPS is intended to forward the upcoming traffic information in a timely manner to the vehicles to inform drivers so that they could make a wise decision to avoid accidents and/or delays (Rawat et al. 2011b). The emergency- and safety-related applications include collision alert, emergency vehicle approaching, deceleration alert, road condition warnings, merge assistance, and so on. Note that the emergency- and safety-related applications require messages to be propagated from the point of occurrence to the destination vehicles with very low latency.

15.3.1.2 Infotainment Applications Infotainment applications in transportation CPS aim to improve passenger comfort and traffic efficiency and include information about roadside facilities (such as shopping malls, fast foods, hotels, parking spots, and gas station/price), electronic payments, weather condition, and interactive multimedia communications. These applications are bandwidth hungry but are delay torrent.

15.3.1.3 Distributed Cloud Computing Applications As vehicles are equipped with virtually unlimited power, storage, and computing capabilities, they could form a cloud for distributed computing on the fly and provide computing services for vehicular networks as proposed by Olariu et al. (2011).

15.4 Radio Spectrum Sensing for Vehicular Communications

Based on the received signal, each vehicle uses two hypotheses for the received signal $r(t)$ to test whether the given channel is idle. The received signal is just the noise (hypothesis H_0) when channel is idle; otherwise, it will have both noise and signal (hypothesis H_1), that is,

$$r(j) = \begin{cases} n(j) & H_0 \\ x(j) + n(j) & H_1 \end{cases} \tag{15.1}$$

To decide whether there is a signal vehicle can use energy detection against a given threshold λ as

$$L(r(j)) = \frac{1}{W} \sum_{j=0}^{W-1} |r(j)|^2 \quad H_1 > \lambda < H_0 \tag{15.2}$$

If there is a signal, the hypothesis H_1 is true, which implies that $L(r(j)) = 1$, otherwise $L(r(j)) = 0$. Using this approach, each vehicle can identify whether the channel is idle. We formulate the available probability of the jth link in the nth time interval $[t_n, t_{n+1})$ following a Bernoulli process as

$$\pi_j^n = \frac{P_{j10}^n}{P_{j01}^n + P_{j10}^n} \tag{15.3}$$

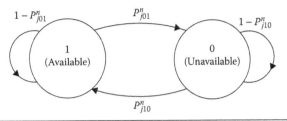

Figure 15.5 State transition diagram of opportunistic link.

The state transition diagram with transition probabilities is shown in Figure 15.5 with two different statuses of the channel (available or unavailable).

There are chances of getting miss-detection of the signal or false alarm. This can be avoided using cooperative sensing. Note that dedicated short range communication (DSRC) mandates that each vehicle should send its periodic broadcast message (which includes its speed, acceleration, geolocation, direction, etc.) to other neighboring vehicles. With that periodic message, each vehicle could incorporate the spectrum occupancy information of the channel and make a decision collaboratively. Thus, a collaborative decision could be made using logical AND operation or OR operation.

$$L_{FC} = \text{AND}_{\forall i} L(r(j))_i \qquad (15.4)$$

where:

AND is the AND operator

When all vehicles report $L(r(j))_i = 1$, then the collaborative decision L_{FC} will be 1 (hypothesis H_1); Otherwise L_{FC} will be 0 (hypothesis H_0). Alternatively, we could use logical OR operator as

$$L_{FC} = \text{OR}_{\forall i} L(r(j))_i \qquad (15.5)$$

When any of the participating vehicles report true, that is, $L(r(j)) = 1$, then the collaborative decision L_{FC} will be 1 even if all other vehicles report $L(r(j)) = 0$. When all report idle, then the L_{FC} will be 0. The OR operator could be little less conservative since when one vehicle reports $L(r(j)) = 1$, the collaborative decision L_{FC} becomes 1. To avoid this problem, vehicles could use majority voting instead of OR operation as

If $L_{MFC} > 0.5$, then $L_{FC} = 1$ (hypothesis H_1),

otherwise $L_{FC} = 0$ (hypothesis H_0)

where:

L_{MFC} is the average value of at fusion center and is computed as

$$L_{MFC} = \frac{1}{N} \sum_{n=1}^{N} L(r(j))_n \qquad (15.6)$$

When a vehicle finds the idle spectrum using the above-mentioned methods, it could set up a connection in a identified idle channel and tune its transmit parameters with the help of CR to transmit the traffic information.

Two different scenarios exist in vehicular communications when the V2V-based model is used:

- *Scenario 1 (one-way traffic where vehicles move in the same direction)*: In this scenario, vehicles move in the same direction and with almost constant speeds, and the relative speed between the vehicles is very small. For instance, the relative speed of two vehicles moving with 60 miles/hr in the same direction is 0, and thus, one vehicle could be seen as stationary to a moving vehicle. The smaller the relative speed, the longer the time after spectrum sensing for connection setup and data exchange between the vehicles will be as shown in Figure 15.6.

- *Scenario 2 (two-way traffic, i.e., vehicles move in both directions)*: The relative speed of two vehicles moving in opposite directions will be the sum of the speeds of two vehicles. For instance, two vehicles traveling in opposite directions on a highway at a speed of 60 miles/hr will be moving at a relative speed of 120 miles/hr, and these vehicles will have a very short time for connection setup and data exchange using wireless devices for a given transmission range. Furthermore, spectrum sensing time also plays a significant role to have successful communications.

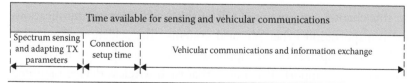

Figure 15.6 Spectrum sensing and opportunistic vehicular communications. TX, Transmitter.

15.5 Summary and Conclusions

According to Cops (2006), program manager of Vehicle Infrastructure Integration Consortium, 50% of vehicles on the road will have communication devices on board. By 2030, almost all vehicles will have the communication equipment, and thus we expect fully operational vehicular networks. Until then, there are several challenges that still need to be addressed for fully realizing the capabilities of vehicular communication for intelligent transportation systems, which include the following:

1. *Security and privacy*: Note that the identity of the vehicle is linked with the owner or renter, and thus privacy and security in vehicular network is important. Vehicular network users would expect the same level of privacy and security as they have with the legacy road driving.
2. *Low latency and reliable connectivity*: Emergency messages should be forwarded in a timely manner to educate drivers who are approaching the accident areas. Connectivity among fast-moving vehicles is another challenge to be addressed to realize the full potential of vehicular networks for reliable and fast delivery of traffic messages.
3. *Authentication*: Authentication process for vehicles as well as messages is one of the major challenges in vehicular networks.
4. *Vehicular data ownership*: Messages containing the identity of drivers should not be in the public domain so that no one can trace individual vehicles or drivers.
5. *Human behaviors*: Drivers have significant role in vehicular network since the network topology changes according to the driving habit of the drivers and their destinations. Thus, the driver behavior results in variation of vehicle density on the road and network topology.

In this chapter, we presented CR-enabled vehicular communications for transportation cyber physical systems. The selection of wireless technology depends on the application that the vehicular network is envisioned to support. It is noted that each wireless device mounted on a vehicle should be able to use radio spectrum opportunities and adapt the transmit parameters according to its operating wireless environment. Delivering real-time traffic information to drivers through wireless communication in vehicular networks can assist in avoiding traffic accidents and congestions.

References

Arslan, H., Chen, Z. N., and Di Benedetto, M. G. (Eds.). (2006). *Ultra Wideband Wireless Communication.* John Wiley & Sons, Chicago, USA.

Bluetooth Technology. (2012). Bluetooth Wireless Technology: Simple, Secure, Everywhere. Retrieved on February 2, 2014, from https://www.bluetooth.org/Building/overview.htm.

Car-to-Car Communication Consortium. (2007). CAR 2 CAR Communication Consortium Manifesto: Overview of the C2C-CC System. Retrieved on February 2, 2014, from http://www.car-to-car.org/index.php?id=11&L=oksjfr.

Cops, M. (2006). VII Strategy for Safety and Mobility Program. Retrieved on February 2, 2014, from http://www.sigmobile.org/workshops/vanet2006/slides/Cops_VANET06.pdf.

Festag, A., et al. (2008). "NoW—Network on Wheels": Project Objectives, Technology and Achievements. *Proceedings of the 5th International Workshop on Intelligent Transportation,* Hamburg, Germany, pp. 211–216.

Fuentes, J. M., González-Tablas, A. I., and Ribagorda, A. (Eds.). (2011). *Handbook of Research on Mobility and Computing: Evolving Technologies and Ubiquitous Impacts,* Chapter 56, Overview of Security Issues in Vehicular Ad-Hoc Networks, IGI Global, pp. 894–911.

Ghafoor, K. Z., Abu Bakar, K., Lloret, J., Ke, C.-H., and Lee, K. C. (2013a). Intelligent Beaconless Geographical Forwarding for Urban Vehicular Environments. *Wireless Networks,* 19(3), pp. 345–362.

Ghafoor, K. Z., Lloret, J., Abu Bakar, K., Sadiq, A. S., and Ben Mussa, S. A. (2013b). Beaconing Approaches in Vehicular Ad Hoc Networks: A Survey. *Wireless Personal Communications,* 73(3), pp. 885–912.

Hartenstein, H. and Laberteaux, K. (Eds.). (2010). *VANET: Vehicular Applications and Inter-Networking Technologies.* Wiley, Hoboken, NJ.

IEEE. (2003). IEEE 802.15.4 ZigBee Standard. Retrieved on February 2, 2014, from http://standards.ieee.org/getieee802/download/802.15.4-2003.pdf.

Li, Z., Wang, Z., and Chigan, C. (2009). "Security of Vehicular Ad Hoc Networks," in *Wireless Technologies for Intelligent Transportation Systems,* Chapter 6, pp. 133–174, Nova Science Publishers, New York.

Little, T. D. and Agarwal, A. (2005). An Information Propagation Scheme for VANETs. *Proceedings of the IEEE Intelligent Transportation Systems,* pp. 55–160, 13–15 September, Vienna, Austria.

Lloret, J., Ghafoor, K. Z., Rawat, D., and Xia, F. (2013). Advances on Network Protocols and Algorithms for Vehicular Ad Hoc Networks. *Mobile Networks and Applications,* 18(6), pp. 749–754.

Moskvitch, K. (2011). "Talking" Cars Could Reduce Motorway Pile-Ups. Retrieved on June 14, 2012, from http://www.bbc.com/news/technology-14125245.

Olariu, S., Khalil, I., and Abuelela, M. (2011). Taking VANET to the Clouds. *International Journal of Pervasive Computing and Communications*, 7(1), pp. 7–21.

Rawat, D. B., Bista, B. B., and Yan, G. (2012). CoR-VANETs: Game Theoretic Approach for Channel and Rate Selection in Cognitive Radio VANETs. *Proceedings of the 7th International Conference on Broadband and Wireless Computing, Communication and Applications*, Victoria, Canada, November 12–14, 2012.

Rawat, D. B., Bista, B. B., Yan, G., and Weigle, M. C. (2011a). Securing Vehicular Ad-Hoc Networks against Malicious Drivers: A Probabilistic Approach. *Proceedings of the International Conference on Complex, Intelligent, and Software Intensive Systems*, Seoul, Korea, June 30–July 2.

Rawat, D. B. and Popescu, D. (2012). Precoder Adaptation and Power Control for Cognitive Radios in Dynamic Spectrum Access Environments. *IET Communications*, 6(8), pp. 836–844.

Rawat, D. B., Popescu, D. C., Yan, G., and Olariu, S. (2011b). Enhancing VANET Performance by Joint Adaptation of Transmission Power and Contention Window Size. *IEEE Transactions on Parallel and Distributed Systems*, 22(9), pp. 1528–1535.

Rawat, D. B., Yan, G., Bista, B. B., and Weigle, M. C. (in press). Trust on the Security of Wireless Vehicular Ad-Hoc Networking. *Ad Hoc & Sensor Wireless Networks Journal*.

Santa, J., Moragon, A., and Gomez-Skarmeta, A. F. (2008). Experimental Evaluation of a Novel Vehicular Communication Paradigm Based on Cellular Networks. *Proceedings of the IEEE Intelligent Vehicles Symposium*. June 4–6, Eindhoven, the Netherlands

Smith, B. L., Zhang, H., Fontaine, M., and Green, M. (2003). Cell Phone Probes as an ATMS Tool. Center for Transportation Studies at the University of Virginia. Retrieved on February 2, 2014, from http://www.cts.virginia.edu/docs/UVACTS-15-5-79.pdf.

Thiagarajan, A., Ravindranath, L., LaCurts, K., Toledo, S., Eriksson, J., Madden, S., and Balakrishnan, H. (2009). VTrack: Accurate, Energy-Aware Road Traffic Delay Estimation Using Mobile Phones. *Proceedings of the 7th ACM Conference on Embedded Networked Sensor Systems*, November 4–6, Berkeley, CA.

Trivedi, H., Veeraraghavan, P., Loke, S. W., Desai, A., and Singh, J. (2011). Routing Mechanisms and Cross-Layer Design for Vehicular Ad Hoc Networks: A Survey. *Proceedings of the IEEE Symposium on Computers & Informatics*, March 20–23, Kuala Lumpur, Malaysia, pp. 243–248.

US Department of Transportation (DOT). (2005). Vehicle Infrastructure Integration. Retrieved on June 14, 2012, from http://www.vehicle-infrastructure.org/documents/VII%20Architecture%20version%201%201%202005_07_20.pdf.

US National Highway Traffic Safety Administration (NHTSA). (2012). Budget Overview of FY 2012 Congressional Submission. Retrieved on December 2, 2013, from http://www.nhtsa.gov/staticfiles/administration/pdf/Budgets/FY2012_Budget_Overview-v3.pdf.

Wang, C. D. and Thompson, J. P. (1997). Apparatus and Method for Motion Detection and Tracking of Objects in a Region for Collision Avoidance Utilizing a Real-Time Adaptive Probabilistic Neural Network. US Patent No. 5613039.

Watfa, M. (Ed.). (2010). *Advances in Vehicular Ad-Hoc Networks: Developments and Challenges.* IGI Publishers, Piscataway, NJ.

ZigBee Alliance. (2009). ZigBee Overview. Retrieved on February 2, 2014, from https://docs.zigbee.org/zigbee-docs/dcn/07-5482.pdf.

ZigBee Alliance. (2012). ZigBee Specification: Features At-a-Glance. Retrieved on February 2, 2014, from https://docs.zigbee.org/zigbee-docs/dcn/07-5299.pdf.

Biographical Sketches

Danda B. Rawat received his PhD in electrical and computer engineering from Old Dominion University, Norfolk, Virginia. He is currently an assistant professor in the department of electrical engineering at Georgia Southern University, Statesboro, Georgia. He is the founding director of Cybersecurity, Wireless Systems and Networking Innovations (CWiNs) Lab at Georgia Southern University. His research focuses on wireless communication systems and networks. His current research interests include design, analysis, and evaluation of cognitive radio networks, cyber physical systems, a vehicular/wireless ad hoc networks, OpenFlow-based networks, software-defined networks, wireless sensor networks, and wireless mesh networks. His research interests also include information theory, mobile computing, and network security. He has published over 70 scientific/technical papers on these topics. He has authored or edited 4 books and over 10 peer-reviewed book chapters related to his research areas. Dr. Rawat has been serving as an editor for over six international journals. He also served as a lead guest editor for a "Special Issue on Recent Advances in Vehicular Communications and Networking" (Elsevier) published in *Ad Hoc Networks* (2012), a guest editor for a "Special Issue on Network Protocols and Algorithms for Vehicular Ad Hoc Networks" (ACM/Springer) published in *Mobile Networks and Applications* (*MONET*) (2013), and a guest editor for a

"Special Issue on Recent Advances in Mobile Ad Hoc and Wireless Sensor Networks" (Inderscience) published in *International Journal of Wireless and Mobile Computing* (2012). He served as a program chair, a conference chair, and a session chair for numerous international conferences and workshops, and served as a technical program committee member for several international conferences including IEEE GLOBECOM, CCNC, GreenCom, AINA, WCNC, and VTC conferences. He has previously held an academic position at Eastern Kentucky University, Richmond, Kentucky; Old Dominion University; and Tribhuvan University, Kathmandu, Nepal. He is the recipient of the Best Paper Award at the International Conference on Broadband and Wireless Computing, Communication and Applications 2010 (BWCCA 2010) and the Outstanding Ph.D. Researcher Award 2009 in Electrical and Computer Engineering at Old Dominion University among others. Dr. Rawat is a senior member of IEEE and a member of ACM and ASEE. He has been serving as a vice chair of the Executive Committee of the IEEE Savannah Section since 2013.

Chandra Bajracharya received her PhD in electrical and computer engineering from Old Dominion University, Norfolk, Virginia. During her doctoral dissertation, she was associated with the Frank Reidy Research Center for Bioelectrics, Norfolk, VA, USA and worked on several projects funded by the US Air Force and National Institutes of Health (NIH). Her PhD dissertation was on the characterization of near field focusing on impulse reflector antenna and its applications in target detection, imaging, and electromagnetic waves into biological targets. She also has an extensive teaching experience of over 5 years, as a lecturer at the department of electrical engineering, Tribhuvan University, Nepal, India. Her research interests include medical cyber physical systems, transportation cyber physical systems, numerical electromagnetics, biological effects of electromagnetic fields, UWB antennas, antenna design, power electronics, alternative energy, communication systems, signal processing, and smart grid. She has published several scientific/technical papers on these topics. She is a member of IEEE and has served as a technical program committee member and reviewer of several conferences.

Jaime Lloret Mauri received his MSc in physics in 1997 and his MSc in electronic engineering in 2003, both in the Universitat de València, València, Spain, and his PhD in telecommunications engineering (Dr. Ing) in 2006 at Universitat Politècnica de València, València, Spain. He is currently an associate professor in the Universitat Politècnica de València. He is the head of the "Communications and remote sensing" research group of the Integrated Management Coastal Research Institute and the "Active and collaborative techniques and use of technologic resources in the education (EITACURTE)" Innovation Group, both belonging to the Universitat Politècnica de València. He is the director of the University Expert Certificate "Redes y Comunicaciones de Ordenadores," the University Expert Certificate "Tecnologías Web y Comercio Electrónico," and the University Master "Digital Post Production." He is currently the chair of the Internet Technical Committee (IEEE Communications Society and Internet Society) and the Working Group of the Standard IEEE 1907.1. He has authored 12 books and has published (more than 250 papers in national and international conferences, and international journals (more than 90 Science Citation Index [SCI] Indexed). He has been general chair (or cochair) of 23 international workshops and conferences. He is the IEEE senior and International Academy, Research, and Industry Association (IARIA) fellow. He is the editor in chief of the international journal *Networks Protocols and Algorithms* and IARIA Journals Board Chair (eight journals).

Kayhan Zrar Ghafoor received his BSc degree in electrical engineering from Salahaddin University, Erbil, Iraq (2003); his MSc degree in remote weather monitoring from Koya University, Koya, Iraq (2006); and his PhD degree in wireless networks from Universiti Teknologi Malaysia (UTM), Johor, Malaysia (2011). He is working as a senior lecturer in the department of software engineering at Koya University, Iraq. He has published over 30 scientific/research papers in prestigious international journals and conferences. Dr. Ghafoor served as a guest editor of the "Special Issue on Network Protocols and Algorithms for Vehicular Ad Hoc Networks" in *Mobile Networks and Applications* (*MONET*) journal. He is currently working as a general chair of a workshop named "Smart Sensor Protocols and Algorithms" in conjunction with "The 9th International Conference on Mobile Ad-hoc and Sensor Networks (MSN 2014)," which will be held in Hungary. He also served as an associate editor, editorial board member, and reviewer for numerous prestigious international journals; appeared as a workshop general chair for international workshops and conferences; and worked as a technical program committee member for more than 30 international conferences. He is the recipient of the UTM Chancellor Award in the 48th UTM convocation in 2012. He was also awarded the UTM International Doctoral Fellowship (IDF) and Kurdistan Regional Government (KRG) scholarship (Ahmad Ismail Foundation). His current research interests include routing over vehicular ad hoc networks and tactical wireless networks, cognitive vehicular networks as well as artificial intelligence and network coding applications. He is a member of Institute of Electrical and Electronics Engineers (IEEE) Vehicular Technology Society, IEEE Communications Society, Internet Technical Committee (ITC), and International Association of Engineers (IAENG).

Index

Printed and bound by CPI Group (UK) Ltd, Croydon, CR0 4YY

18/10/2024

01776256-0017